ALGEBRA

An Introductory Course

One-Volume Edition

Morris Bramson

Former Chairperson, Department of Mathematics
Martin Van Buren High School, New York City

AMSCO SCHOOL PUBLICATIONS, INC.
315 Hudson Street / New York, N.Y. 10013

IMPORTANT FEATURES OF THIS BOOK

Gradual presentation aids students who have difficulty with mathematics.

Simple, concise language clearly explains ideas and procedures.

Visual aids and real-life situations promote understanding.

Illustrative problems offer step-by-step guidance.

Ample exercises provide drill and reinforcement of learning.

"Check your skills" sections encourage maintenance and power growth.

Text-referenced cumulative reviews monitor progress.

Topics emphasized on competency examinations are included.

Glossary of terms helps students master mathematical vocabulary.

Dedicated to my grandchildren Danny, Amy, and Lisa, who, I hope, will develop an appreciation and enjoyment of mathematics in the years ahead.

Please visit our Web site at:

www.amscopub.com

When ordering this book, please specify *either:*
R 415 P *or* ALGEBRA: AN INTRODUCTORY COURSE, ONE-VOLUME EDITION, PAPERBACK
   or
R 415 H *or* ALGEBRA: AN INTRODUCTORY COURSE, ONE-VOLUME EDITION, HARDBOUND

ISBN 0-87720-259-1 (Paperback edition)
ISBN 0-87720-261-3 (Hardbound edition)

Copyright © 1987 by Amsco School Publications, Inc.
No part of this book may be reproduced in any form without written permission from the publisher.

Printed in the United States of America

Preface

Algebra: An Introductory Course provides students with a gradual, progressive development of algebraic concepts, skills, and applications. Although this book should be particularly helpful to students who have experienced some difficulty in learning mathematics, it places strong emphasis on basic algebra.

The first three chapters introduce students to formulas, simple equations, and the use of equations in the solution of simple verbal problems. The only prerequisite for these topics is some skill in the arithmetic of positive whole numbers.

For those students who lack the necessary basic skills, however, a complete review of the arithmetic of whole numbers is provided in Appendix I and a review of arithmetic problem solving is provided in Appendix II.

Further reviews of arithmetic skills and concepts are interspersed throughout this book. Chapter 4 provides a review of arithmetic fractions followed by an introduction to algebraic fractions and the solution of simple fractional equations. Chapter 5 follows a similar sequence of arithmetic review and algebraic development for decimal fractions and decimal equations. Chapter 6 develops the concepts of ratio and proportion and makes use of algebraic skills in the solution of problems of these types. Chapter 7 reviews the basic skills and concepts of percentage and illustrates the use of simple equations in solving percentage problems.

In the treatment of topics throughout *Algebra: An Introductory Course*, excessive abstraction and rigorous theory are avoided. Mathematical ideas and processes are explained in concrete terms and grow out of a need for them in everyday situations. For those students interested in a glimpse of a more theoretical, rigorous approach to algebra, an introduction to set theory is provided in Appendix III.

In addition to algebra and related arithmetic, basic geometric concepts are illustrated and explained. (Appendix V contains a review of the basic concepts and terminology of geometry.) Chapter 8 reviews linear measure and focuses on the metric system. The concepts of area and volume are carefully developed here, and they are applied to a variety of situations. Exponents are introduced in connection with the area of a square and the volume of a cube.

The algebra of signed numbers, followed by monomial and polynomial expressions, is developed in Chapters 9 and 10. The solution of equations is then expanded in Chapter 11 to use signed numbers and the new skills. Then equations are used to solve more difficult verbal problems of assorted types.

Chapter 12 introduces students to the techniques of graphing linear equations and the graphic solution of two linear equations. This skill is then applied in Chapter 13 to the graphic solution of inequalities.

In Chapter 14, the algebraic solutions of two linear equations are presented in detail. In Chapter 15, these techniques are applied to the solution of a variety of verbal problems related to science, business, social science, industry, and other areas.

In Chapter 16, the algebraic operations explained earlier are extended to include polynomial expressions. These operations are then applied in Chapter 17 to algebraic functions and in Chapter 18 to the solution of fractional equations. In Chapter 19, roots and radicals are considered, and the techniques developed here are then applied to the solution of quadratic equations in Chapter 20. In Chapter 21, numerical trigonometry is developed and applied to indirect measurement related to the right triangle.

As each topic is developed, it is followed by a variety of illustrative problems. An ample set of graded exercises is then provided. These exercises vary sufficiently in difficulty so as to provide for individual differences in student ability. Sections called "Check your skills" and "Cumulative review" offer material for continuous review throughout the book.

Chapter 22 introduces the basic concepts of statistics and probability, topics that are appearing in the mathematical syllabi of several states. Problems related to these topics are now included in competency examinations given in many states. Appendix IV sets forth a review of statistical graphs.

The author expresses thanks and appreciation to each of the following persons, who carefully read the manuscript and made many helpful criticisms and valuable suggestions: Harold Baron, Assistant Principal, Mathematics Department, August Martin High School, New York City; Sidney Cabin, former Chairperson of the Mathematics Department, Martin Van Buren High School, New York City; Marilyn Occhiogrosso, former Assistant Principal, Mathematics Department, Erasmus Hall High School, New York City.

MORRIS BRAMSON

Contents

4 Common Fractions and Simple Fractional Equations

5 Decimal Fractions and Simple Decimal Equations

6 Ratio and Proportion

11 Equations and Problem Solving

12 Graphing Linear Equations

13 Inequalities

1

The Formula

1. MEANING OF A FORMULA

In everyday living, we frequently use abbreviations and shorthand expressions to represent groups of words. Most people know that UN represents United Nations and that NFL means National Football League. If your history teacher refers to 600 BC, what does BC represent? What is FBI an abbreviation of?

We also use shorthand expressions for ideas in mathematics, science, social science, business, and industry. When these ideas are mathematical, they usually involve mathematical symbols as well as letters. For example, if it *costs* a man $40 to buy a radio and he wishes to make a *gain* of $10, he must *sell* the radio for $50. The selling price is found by adding the cost to the gain. Using words, we can write the rule as

$$\text{Selling price} = \text{Cost} + \text{Gain}$$

Using letters, we can write this rule in a shorthand manner as

$$S = C + G$$

where S represents *selling price*, C represents *cost*, and G represents *gain*. The terms are related by the mathematical symbols $=$ and $+$. This total expression is called a **formula**. In a formula, the letters are not only abbreviations for words, but also represent quantities.

In the work that follows in Chapters 1, 2, and 3, it is assumed that you have some basic arithmetic skills in handling whole numbers and fractions. If you find that you are having difficulty, review the arithmetic of whole numbers in Appendix I on page 419. If you are having trouble with fractions, review these in Chapters 4 and 5.

2. USING FORMULAS

Let us use the formula $S = C + G$ to solve a simple problem.

ILLUSTRATIVE PROBLEM 1:   An electric toaster costs a dealer $72.50 and he wishes to make a gain of $12.25. For how much must he sell it?

Solution: $S = C + G$
$S = \$72.50 + \$12.25 = \$84.75$ (*answer*)

In this solution, we perform the following steps:

Step 1: Write the formula.
Step 2: Substitute the given values in the formula.
Step 3: Perform the mathematical operations indicated by the symbols in the formula.

The use of letters to represent numbers makes up one of the basic features of **algebra**. Later, we are going to use algebra to solve more difficult problems of this type, and we shall find that using the preceding steps is most helpful in getting our solutions.

Many of you have already learned some of the properties of a **triangle** (a 3-sided figure as shown in the diagram). The **perimeter** of the figure is the distance around it, or the sum of the lengths of the sides. If we represent the lengths of the sides by r, s, and t, we may find the perimeter, p, by the formula:

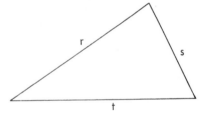

$$p = r + s + t$$

If you are having difficulty recalling simple geometric facts, review Appendix V on page 461.

ILLUSTRATIVE PROBLEM 2:   If the lengths of the sides of a triangle are 12 inches, 15 inches, and 21 inches, find its perimeter.

Solution: $p = r + s + t$
$p = 12 + 15 + 21$
$p = 48$ inches (*answer*)

If the three sides of a triangle are equal, it is called an **equilateral triangle**, as in the figure. If we represent each side by s, then the perimeter formula tells us that

$$p = s + s + s$$

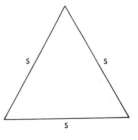

We may write this as $p = 3 \times s$ or $p = 3 \cdot s$ or $p = 3s$.

In algebra, we usually avoid the "times" sign ($\times$) because it resembles the letter x and may lead to confusion. When writing a product of numbers, we prefer the center dot ($\cdot$) to the $\times$ sign. When writing a product of letters or of numbers and letters, we usually omit the multiplication sign entirely. The preceding formula is usually written $p = 3s$.

ILLUSTRATIVE PROBLEM 3: Find the perimeter of an equilateral triangle when the length of each side is $6\frac{1}{2}$ inches.

Solution:

$$p = 3s$$

$$p = 3 \cdot 6\tfrac{1}{2}$$

$$p = 19\tfrac{1}{2} \text{ inches} \quad (answer)$$

ILLUSTRATIVE PROBLEM 4: A retailer sells ties for 8 dollars each.

a. Write a formula for the amount of money, A, in dollars that she receives for selling t ties.

b. Use the formula from part **a** to find the amount of dollars she receives for 15 ties.

Solution:

a. The amount of dollars she receives is the product of 8 and t. Thus,

$$A = 8t \quad (answer)$$

b. $A = 8t$
$A = 8 \cdot 15$
$A = 120$ dollars (*answer*)

Exercises

1. A mechanic's assistant earns \$6 an hour.
   a. If he works h hours, write a formula for his salary, S.
   b. Using this formula, find his salary for a 40-hour week.
2. An auto travels on a highway at 55 miles per hour.
   a. Write a formula for the distance, d, in miles that the auto travels in t hours.
   b. What distance does the auto travel in 4 hours?

3. A plane is cruising at 250 miles per hour.
   a. Write a formula for the distance, d, in miles flown by the plane in t hours.
   b. How many miles will the plane fly in 6 hours?

4. a. Write a formula for the distance, d, traveled by an object in t hours if its rate of travel is r miles per hour.
   b. A satellite is traveling around the earth at the rate of 10,000 miles per hour. How many miles does it travel in 15 minutes? (Hint: 15 minutes = $\frac{1}{4}$ hour.)

5. a. A merchant buys shirts for $6.50 each. Write a formula for his cost, C, of n shirts.
   b. What is his cost per dozen for these shirts?

6. a. The driver of an auto determines how much gas she uses per mile (consumption, C) by dividing the distance covered in miles (d) by the number (N) of gallons of gas she uses. Write a formula for finding C.
   b. How many miles per gallon does an auto travel if it uses 8 gallons to travel 100 miles?

7. a. Write a formula for the number of inches, I, in F feet.
   b. How many inches long is a $4\frac{1}{2}$-foot board?

8. a. A classroom has r rows of seats with s seats in each row. Write a formula for the number, N, of seats in the room.
   b. Find the number of seats in a room with 6 rows and 7 seats in each row.

---

3. COMBINING LIKE TERMS

In determining the perimeter of an equilateral triangle, we say that $p = 3s$ is the same as $p = s + s + s$ because we know that adding a number to itself 3 times is the same as multiplying it by 3. When we multiply a letter by a number, we call the number the **numerical coefficient** of the letter. Thus, in the product $3s$, 3 is the numerical coefficient of s. In the expression $5x$, what is the numerical coefficient of x? In the expression $\frac{1}{4}d$, what is the numerical coefficient of d? The d in this expression is called the **literal factor**.

When we have a product of letters or of numbers and letters, the product is called a **term**. Thus, $3s$ is a term and $5b$ is a term. The expression $3s + 5b$ is a sum of two terms. How many terms are in $2a + 3b - 4c$?

The terms $5a$ and $3a$ are called **like terms** because they have a **common literal factor**. It is possible to add or subtract like terms if we recall how we added or subtracted like quantities in arithmetic. For example,

$$5 \text{ eggs} + 2 \text{ eggs} = 7 \text{ eggs}$$

$$8 \text{ feet} - 3 \text{ feet} = 5 \text{ feet}$$

$$4 \text{ hours} + 6 \text{ hours} = 10 \text{ hours}$$

In a like manner, we say that

$$5a + 2a = 7a$$

$$8f - 3f = 5f$$

$$4h + 6h = 10h$$

We can add or subtract like terms vertically just as we do in arithmetic. Thus,

$$\text{Add:} \quad \begin{array}{r} 4b \\ 6b \\ \hline 10b \end{array} \qquad \text{Subtract:} \quad \begin{array}{r} 5r \\ 3r \\ \hline 2r \end{array}$$

RULE: To add or subtract like terms, we add or subtract their numerical coefficients and multiply the result by their common literal factor.

Note that we cannot add 5 eggs and 2 shoes. In a like manner, we cannot combine $5a + 2b$ into one term.

However, we can add 5 nickels and 1 penny to 2 nickels and 3 pennies to get a sum of 7 nickels and 4 pennies. In algebra, we can show the addition like this:

$$\begin{array}{r} 5n + 1p \\ 2n + 3p \\ \hline 7n + 4p \end{array}$$

We usually write $1p$ as just p, with the 1 understood.

EXAMPLES

1. $4m + 3m = 7m$

2. $8t - 3t = 5t$

3. Add:
$$\begin{array}{r} 7k \\ 4k \\ \hline 11k \end{array}$$

4. Add:
$$\begin{array}{r} 5x \\ 8x \\ \hline 13x \end{array}$$

5. $7h + 4h - h = 11h - h = 10h$

6. Add: $2a + 7b$
$\ \underline{4a + 2b}$
$\ 6a + 9b$

7. $6r - 6r = 0 \cdot r = 0$

8. $10a - 3b + 4a = 14a - 3b$

9. $12x - 4x + 7y = 8x + 7y$

Exercises

1. State the numerical coefficient of each of the following:

   a. $4t$ **b.** $7m$ **c.** $\frac{1}{2}p$ **d.** r

2. State whether the expressions are *like* terms or *unlike* terms and write the sum in each case:

   a. $3x$ and $5x$ **b.** $4t$ and t **c.** $4m$ and $3n$ **d.** $2y$ and $\frac{1}{2}y$

3. Combine like terms:

   a. $4r + 7r =$ **b.** $3s + 5s - 2s =$ **c.** $8x - x =$
   d. $5y - 5y =$ **e.** $6t + 2\frac{1}{2}t =$ **f.** $8p - 2p + 5p =$

4. Add: $7m$
$\ \underline{8m}$

5. Add: $4y$
$\ 5y$
$\ \underline{3y}$

6. Add:

   a. $2x + 3y$ **b.** $a + 6b$ **c.** $r + 2s + t$ **d.** $3m + 9$
   $\ \underline{5x + 4y}$ $\ \underline{3a + 5b}$ $\ \underline{4r + 3s + 5t}$ $\ \underline{2m + 11}$

7. Simplify the following formulas by combining like terms wherever possible:

   a. $p = 2a + 3a + 5b$ **b.** $T = 3m + d + 2m$
   c. $c = 4r - 2r + 3s$

4. ORDER OF OPERATIONS

In order to join a certain club, a person has to pay $5 down and then $2 every month. The formula for the total payment, T, after n months is:

$$T = 5 + 2n$$

Now suppose we wish to find the total payment after 10 months. Substituting in the formula, we obtain

$$T = 5 + 2 \times 10$$

If we add the 5 and 2 first and then multiply by 10, we obtain 70. If we multiply 2 by 10 first and then add 5, we obtain 25. Which is correct?

Apparently it is not clear here as to which operation to perform first. In order to avoid this kind of confusion, mathematicians have agreed upon the following rule for the order of operations:

RULE: Reading from left to right, first perform the operations of multiplication and division. Then perform the operations of addition and subtraction.

Using this rule in the preceding problem, we write

$$T = 5 + 2 \times 10$$
$$T = 5 + 20$$
$$T = 25$$

The dues for 10 months are $25.

EXAMPLES: **1.** $3 \cdot 5 + 8 = 15 + 8 = 23$
2. $12 \div 2 - 3 = 6 - 3 = 3$

Exercises

In 1 to 7, find the value of each expression.

1. $6 + 5 \times 7$ **2.** $9 \div 3 - 1$ **3.** $5 \times 6 + 7 \times 8$

4. $9 \times 2 - 3 \times 4$ **5.** $8 \times 5 \div 20$ **6.** $18 - 6 \div 3$

7. $6 + 6 \times 6 - 6 \div 6$

8. In the triangle shown, two sides are equal and are named by the letter r. The third side is named by the letter s.
   a. Write a formula for the perimeter, p, of this triangle.
   b. Find the perimeter of the triangle if $r = 8$ and $s = 10$.
   c. Find the perimeter of the triangle if $r = 12\frac{1}{2}$ and $s = 9\frac{1}{4}$.

9. In the four-sided figure, two sides are equal and are named by the letter c. The other two sides are named by the letters a and b.
   a. Write a formula for the perimeter, p.
   b. Find p if $a = 12$, $b = 15$, and $c = 10$.

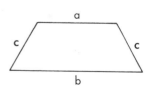

10. An auto rental service charges $15 per day to rent a car and $.09 per mile for each mile driven. The formula for the cost in dollars, C, of renting a car for d days and driving m miles is

$$C = 15d + .09m$$

   a. Find the cost of renting a car for 4 days and driving 300 miles.

   b. Find the cost of renting a car for 6 days and driving 460 miles.

11. A movie theater has an admission charge of $5 for adults and $3 for children.

   a. Write a formula for the total receipts, R, in dollars on a day when there are a adults and c children admitted.

   b. Using the formula, find the total receipts on a day when 456 adults and 237 children are admitted.

12. On a certain test, students are awarded 5 points for each correct answer but are penalized for a wrong answer by having 2 points deducted. The formula for computing the total number of points is

$$P = 5c - 2w$$

where c is the number of correct answers and w is the number of wrong answers.

   a. How many points does a student receive for 15 correct answers and 5 wrong answers?

   b. How many points does a student receive for 10 correct answers and 10 wrong answers?

---

5. USING PARENTHESES IN FORMULAS

Suppose we wish to find the perimeter of a rectangle of length ℓ and width w, as in the figure. We may write the formula for the perimeter as

$$p = \ell + w + \ell + w$$
$$\text{or} \quad p = 2\ell + 2w$$

Another way of writing this is

$$p = 2(\ell + w)$$

where the quantity *in the parentheses* is to be obtained *first*; that is, the numerical values of ℓ and w are to be added *first*, and then the sum is to be multiplied by 2.

Let us assume that the length of a rectangle is 30 inches and its width is 20 inches. The perimeter would then be

$$p = 2(\ell + w)$$

$$p = 2(30 + 20)$$

$$p = 2 \cdot 50 = 100 \text{ inches}$$

Note that the same result would be obtained by the other form of the formula:

$$p = 2\ell + 2w$$

$$p = 2 \cdot 30 + 2 \cdot 20$$

$$p = 60 + 40 = 100 \text{ inches}$$

These examples illustrate the following rule:

RULE: When several operations are indicated in a numerical expression, those grouped within parentheses are to be completed *first*. Then multiplication and division are done before addition and subtraction.

ILLUSTRATIVE PROBLEM 1: Find the value of $3 \times (9 - 7) + 2$.

Solution: Perform the operation in parentheses first, giving $3 \times 2 + 2$. Now the accepted rule of operations tells us to multiply first and then add, giving $6 + 2 = 8$. We arrange our work like this:

$$3 \times (9 - 7) + 2$$

$$3 \times 2 + 2$$

$$6 + 2$$

$$8 \quad (answer)$$

Note: We frequently write $3 \times (9 - 7)$ as $3(9 - 7)$, where it is understood that the quantity in parentheses is to be multiplied by 3.

ILLUSTRATIVE PROBLEM 2: The cost of sending a parcel by railway between two towns is $2.00 for the first 3 pounds and $.30 per pound for each additional pound.

a. Write a formula for finding the cost, C, of sending a package of p pounds.

b. Find the cost of sending a package of 8 pounds.

Solution:

a. The cost of the first 3 pounds is a flat $2.00. The remaining $(p - 3)$ pounds then cost $.30 per pound. Therefore, the total cost, C, is given by the formula

$$C = 2.00 + .30(p - 3) \quad (answer)$$

b. If $p = 8$, we have

$$C = 2.00 + .30(8 - 3)$$
$$C = 2.00 + .30 \times 5$$
$$C = 2.00 + 1.50$$
$$C = \$3.50 \quad (answer)$$

Exercises

1. Find the value of:
   a. $(4 + 3) \times 7$ **b.** $10 + 4 \times (5 - 2)$ **c.** $18 \div (4 + 2)$
   d. $(25 - 10) \div 2$ **e.** $(6 + 11 - 3) \div 7$ **f.** $(12 - 3) \times (7 + 4)$

2. In the formula $L = 5 + d(n - 1)$:
   a. Find L, when $d = 2$ and $n = 10$.
   b. Find L, when $d = 3$ and $n = 20$.
   c. Find L, when $d = 2\frac{1}{2}$ and $n = 15$.

3. The formula used to convert from Fahrenheit readings (F) on a thermometer to Celsius readings (C) is
$$C = \tfrac{5}{9}(F - 32)$$

   a. The boiling point of water is 212°F. Convert this Fahrenheit reading to a Celsius reading.
   b. Convert a Fahrenheit reading of 68°F to a Celsius reading.

4. A car service charges 90¢ for the first mile and 60¢ for each additional mile. The formula for the cost, C, of riding m miles is then
$$C = 90 + 60(m - 1)$$

   a. Find the cost of making a trip of 8 miles.
   b. Find the cost of a $4\frac{1}{2}$-mile trip.

5. The cost of sending a certain telegram is 80¢ for the first 10 words and 15¢ for each additional word. The formula for finding the cost, C, in cents, of sending a telegram of n words is
$$C = 80 + 15(n - 10)$$

   Find the cost of sending a telegram of
   a. 18 words.   **b.** 25 words.

6. **a.** Write a formula for the cost, c, of borrowing a book from a library for d days if the cost for the first 3 days is 25¢ and the cost for each additional day is 5¢.

   b. Find the cost of borrowing a book for a week.

---

6. OBTAINING FORMULAS FROM TABLES

The owner of an amusement park posts a table to be used by the employees to find the cost of different numbers of admission tickets. Here is the table:

Number of tickets (n)	1	2	3	4	5	6
Cost of tickets (c)	1.25	2.50	3.75	5.00	6.25	7.50

From this table, we can readily determine a formula for c in terms of n. Note that the cost per ticket is \$1.25 and, therefore, the formula is $c = 1.25n$.

Using the formula, we can find the cost of any number of tickets.

Thus, for $n = 20$, $c = 1.25 \times 20 = \$25.00$.

The formula, therefore, is a shorthand way of summarizing the table.

An airline lists the following table, showing the time in London (L) that corresponds to the time in New York (N):

London time (L)	6	7	8	9	10	11	12	13
New York time (N)	1	2	3	4	5	6	7	8

This table is based on the 24-hour clock used by airlines. Can you write a formula to show how L is related to N? Use either $L = N + 5$ or $N = L - 5$. As before, the formula is a shorthand summary of the table.

Exercises _____

In 1 to 5, write a formula to express the relationship shown in each table.

1.

Number of tickets (n)	1	2	3	4	5
Cost in dollars (c)	1.50	3.00	4.50	6.00	7.50

2.

Number of inches (i)	12	24	36	48	60
Number of feet (f)	1	2	3	4	5

3.

H	1	2	3	4	5
W	8	9	10	11	12

4.

K	9	11	13	15	17
L	6	8	10	12	14

5.

Husband's age (h)	27	30	33	36	40
Wife's age (w)	23	26	29	32	36

6. The following table shows the conversion from British pounds (P) to U.S. dollars (D) at one time.

British pounds (P)	1	2	3	4	5
U.S. dollars (D)	2.40	4.80	7.20	9.60	12.00

a. Write a formula relating D to P.

b. How many dollars would be received for 10 British pounds?

_____

Chapter Review Exercises

1. A pentagon is a five-sided figure. In the pentagon shown, all five sides are equal and are marked s.

a. Write a formula for the perimeter, p, of this pentagon.

b. Find the perimeter of the base of the Pentagon building in Washington, D.C., each side of which is 921 feet.

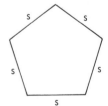

2. In a certain test consisting of Parts I and II, a student gets 3 points for each correct answer in Part I and 2 points for each correct answer in Part II.

   a. Write a formula for the total number of points, P, received for a correct answers in Part I and b correct answers in Part II.

   b. Using the formula, find the total number of points received by a student getting 15 correct answers in Part I and 16 correct answers in Part II.

3. **a.** Using the formula $C = \frac{5}{9}(F - 32)$, find the Celsius reading corresponding to 32°F, the freezing point of water.

   b. Normal body temperature is 98.6°F. What is this reading in the Celsius scale?

4. The cost of a long-distance telephone call is 95¢ for the first three minutes and 40¢ for each additional minute.

   a. Write a formula for the total cost, T, of a telephone call lasting m minutes.

   b. Using the formula, find the total cost of an 8-minute call.

5. A worker earns $8 per hour regular pay and $1\frac{1}{2}$ times this rate for overtime pay.

   a. Write a formula for the salary, S, earned by a worker who works r hours at regular pay and V hours overtime.

   b. Using the formula, find the salary of a worker who works 40 hours in a week at regular pay and 8 hours overtime.

6. **a.** Write a formula for the average (A) of three numbers: r, s, t.

   b. Using the formula, find the average of three test scores: 78, 83, 90.

7. The following table shows a worker's weekly salary, S, for working h hours per week:

h	10	15	20	25	30	35	40
S	$60	$90	$120	$150	$180	$210	$240

   a. Write a formula for S in terms of h.

   b. Using the formula, find the amount earned for 38 hours of work.

CHECK YOUR SKILLS

1. Add:

   a. 129
      958
      787
      436

   b. 9497
      6364
      4269
      9785

   c. 357
      22
      1845
      639

2. Subtract:

   a. 637
      −265

   b. 805
      −276

   c. 835
      −397

3. Multiply:

   a. 27
      ×6

   b. 43
      ×27

   c. 27
      ×40

   d. 220
      ×30

   e. 807
      ×28

4. Divide:

   a. $4\overline{)56}$

   b. $11\overline{)352}$

   c. $31\overline{)9.641}$

   d. $69\overline{)483}$

   e. $25\overline{)2300}$

5. A lumber yard had 12,280 feet of a certain plank. If 6252 feet are used, how many feet of this plank remain?

6. Find the total weight of 6875 metal blocks if each block weighs 97 pounds.

7. A job requires 480 hours. How many 8-hour days is this?

8. Find the value of $(8 + 5) \times 9$.

9. Write a formula to find the average, A, of four numbers, w, x, y, and z.

10. Add: $3x + 2y$
     $5x + 7y$

2

Solving Simple Equations

1. MEANING OF AN EQUATION

Consider the following statements:

$$4 + 3 = 7 \qquad 8 \times 3 = 12 \times 2 \qquad 9 - 3 = 6$$

Each statement shows that the two quantities are equal and, therefore, each is a statement of equality. In mathematics, a statement of equality is called an **equation**.

Which of the following are equations?

$$10 + 2 = 12 \qquad 19 - 8 \qquad 7 + 4 \qquad \tfrac{120}{6} = 20$$

The equation $5 + 4 = 9$ consists of two quantities, $5 + 4$ and 9. The quantity $5 + 4$ is on the left side of the equals sign and is called the **left member** of the equation. What do you think we call the quantity on the right side of the equals sign?

In the following equations, identify the left member and the right member:

$$15 = 5 \times 3 \qquad\qquad n = 14 + 3$$

$$9 \times 4 = 18 \times 2 \qquad\qquad \frac{p}{4} = 3$$

$$5y = 35 \qquad\qquad x + 8 = 12$$

In some of these equations, a letter has been used to represent a missing number. In $x + 8 = 12$, it is clear that 4 is the value of x that makes the equation true because $4 + 8 = 12$. In $5y = 35$, it is also clear that 7 is the value of y that makes this equation true because $5 \times 7 = 35$.

The value of the letter that makes the equation a true statement is called the **root** of the equation. In $\frac{p}{4} = 3$, when p is replaced by 12, the equation is a true statement. Thus, 12 is the root of the equation.

Exercises

By trying various numbers, find the root of each equation.

1. $6m = 18$ **2.** $x + 3 = 9$ **3.** $y - 3 = 12$ **4.** $\frac{r}{3} = 5$

5. $14 + t = 20$ **6.** $s + \frac{1}{4} = 1$ **7.** $15 = x + 9$ **8.** $21 = p - 2$

2. SOLVING EQUATIONS BY ADDITION OR SUBTRACTION

Find the roots of the following equations:

$$y + 7 = 10 \qquad y + 5 = 8 \qquad y + 4 = 7 \qquad y + 2 = 5 \qquad y = 3$$

Note that in each equation the root is 3. Equations having the same root are called **equivalent equations.** In order to find the root of a given equation, we must change the given equation to an equivalent equation where the root is obvious, as in the last equation of the preceding set, $y = 3$.

We may think of an equation as a **balance scale,** where the left and right pans of the scale are similar to the left and right members of an equation. If the scale is balanced to begin with, we can maintain the balance if we add equal weights to or remove equal weights from both pans. Likewise, if we add or subtract the same number to or from both members of an equation, we will obtain an equivalent equation.

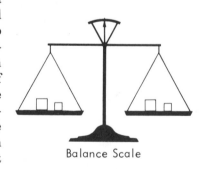

Balance Scale

This principle is very useful in finding the roots of many equations. For example, consider the equation

$$y + 5 = 12$$

Since the y and 5 are added, we can obtain y alone on the left side by subtracting 5. But the **balance principle** requires that we subtract 5 from the right member also to obtain an equivalent equation. Hence we have

$$y + 5 = 12 \quad \text{(original equation)}$$
$$\underline{-5 = -5} \quad \text{(subtract 5 from both members)}$$
$$y \quad\;\; = \quad 7 \quad \text{(root)}$$

We can check this equation by replacing y by 7 in the original equation.

Check: $y + 5 = 12$

$$7 + 5 = 12$$

$$12 = 12 \; \checkmark$$

Let us find the root of the equation $x + 3 = 10$:

$$x + 3 = \quad 10$$
$$\underline{-3 \quad\;\; -3}$$
$$x \quad\;\; = \quad 7 \quad \text{(root)}$$

Clearly, $x = 7$ checks in the original equation. We refer to the principle used here as the **subtraction principle of equality.** In like manner, we may also use the **addition principle of equality.** For example, if we start with the equation $m - 5 = 11$, we would now add 5 to both members of the equation. Thus,

$$m - 5 = \quad 11 \quad \text{(original equation)}$$
$$\underline{+5 \quad\;\; +5} \quad \text{(add 5 to both members)}$$
$$m \quad\;\; = \quad 16 \quad \text{(root)}$$

Principle 1: If we add the same number to both members of an equation, the result is an equivalent equation.

Principle 2: If we subtract the same number from both members of an equation, the result is an equivalent equation.

When we find the root of an equation by changing it to an equivalent equation whose root is obvious, we say that we are **solving** the equation or finding its **solution**.

Exercises

Solve and check each equation.

1. $y + 7 = 18$ **2.** $m - 14 = 20$ **3.** $x + 3 = 12$

4. $r + 2\frac{1}{2} = 5\frac{1}{2}$ **5.** $s - 2.7 = 5.5$ **6.** $10 + t = 20$

7. $14 = a + 11$ **8.** $k - 2 = 8\frac{1}{4}$ **9.** $k + 12 = 12$

10. $4\frac{3}{4} = n + \frac{3}{8}$ **11.** $p - .13 = 2.9$ **12.** $w - \frac{2}{3} = 1\frac{1}{2}$

---

3. SOLVING EQUATIONS BY DIVISION

Determine by inspection the root of each of the following equations:

$$6n = 24 \quad 3n = 12 \quad 2n = 8 \quad n = 4$$

Note that these equations are **equivalent equations** since 4 is a root of all four equations.

Recall that, when we write $5x$, both 5 and x are **factors** of the expression; that is, the 5 and the x are *multiplied* together to obtain the product $5x$. In such a product, we refer to the numeral 5 as the **coefficient** of x. In the expression $6a$, 6 would be the coefficient of a.

Consider the equation $6n = 24$. Since the 6 and n are multiplied, we can obtain n alone on the left side by dividing the left member by the coefficient 6. The balance principle requires that we divide the right member by 6 as well.

Hence, we have

$$6n = 24 \quad \text{(original equation)}$$

$$\frac{6n}{6} = \frac{24}{6} \quad \text{(divide both members by 6)}$$

$$n = 4 \quad \text{(root)}$$

We refer to the principle used here as the **division principle of equality**.

Principle 3: If we divide both sides of an equation by the same number, the result is an equivalent equation.

ILLUSTRATIVE PROBLEM 1: Solve and check: $8x = 32$

Solution: By what number must we divide $8x$ in order to get x? Obviously, by 8. Hence, we must also divide the right member by 8.

$8x = 32$ (original equation)   *Check:*

$\dfrac{8x}{8} = \dfrac{32}{8}$ (division principle $8x = 32$ (original equation)

of equality) $8 \cdot 4 = 32$ (substitute 4 for x)

$x = 4$ (*answer*) $32 = 32$ ✔

ILLUSTRATIVE PROBLEM 2: Solve and check: $6y = 45$

Solution: *Check*:

$6y = 45$ $6y = 45$

$\dfrac{6y}{6} = \dfrac{45}{6}$ $6 \cdot 7\frac{1}{2} = 45$

                               $45 = 45$ ✔

$y = 7\frac{1}{2}$ *(answer)*

Note that, in equations of this type, we always divide both members of the equation by the coefficient of the unknown term. In illustrative problem 1, we divided by 8, the coefficient of x. In illustrative problem 2, we divided by 6, the coefficient of y.

Exercises

Find the root of each equation and check.

1. $5x = 20$	**2.** $8y = 96$	**3.** $7d = 28$	**4.** $6m = 50$
5. $2b = 7.4$	**6.** $12z = 100$	**7.** $.2r = 68$	**8.** $.04t = .76$
9. $15 = 3k$	**10.** $11n = 0$	**11.** $15p = 225$	**12.** $0 = 5s$

4. SOLVING EQUATIONS BY MULTIPLICATION

We can use a method similar to that used in section 3 to solve equations such as $\dfrac{n}{5} = 7$. We know that $3 \times \dfrac{11}{3} = 11$ and that $5 \times \dfrac{8}{5} = 8$.

These examples indicate that, in general,

$$5 \times \frac{n}{5} = n$$

and suggest the approach to use in solving equations such as $\dfrac{n}{5} = 7$.

Just as we were able to *divide* both members of an equation by the same quantity and arrive at an equivalent equation, we can also *multiply* both members by the same quantity and arrive at an equivalent equation whose solution is obvious. Let us solve the equation $\dfrac{n}{5} = 7$:

$$\frac{n}{5} = 7 \qquad \text{(original equation)}$$

$$5 \times \frac{n}{5} = 7 \times 5 \quad \text{(multiply both members by 5)}$$

$$n = 35 \qquad \text{(root)}$$

Check:

$$\frac{n}{5} = 7 \quad \text{(original equation)}$$

$$\frac{35}{5} = 7 \quad \text{(substitute 35 for } n\text{)}$$

$$7 = 7 \ \textrm{✔}$$

This illustration suggests the following **multiplication principle of equality:**

Principle 4: If we multiply both sides of an equation by the same number, the result is an equivalent equation.

Exercises

In 1 to 6, solve and check.

1. $\dfrac{n}{4} = 8$ **2.** $\dfrac{b}{11} = 7$ **3.** $\dfrac{c}{3} = 9$

4. $14 = \dfrac{k}{5}$ **5.** $\dfrac{3r}{4} = 2\frac{1}{2}$ **6.** $\dfrac{x}{10} = 3.4$

7. What is $5 \times \frac{1}{5}n$ equal to? **8.** What is $\frac{2}{3} \times \frac{3}{2}k$ equal to?

In 9 to 14, solve and check.

9. $\frac{1}{5}x = 12$ **10.** $\frac{1}{3}y = 9$ **11.** $\frac{3}{2}t = 21$

12. $\dfrac{x}{.5} = 12.6$ **13.** $\dfrac{p}{4} = 0$ **14.** $1 = \dfrac{3r}{9}$

5. SOLVING EQUATIONS BY TWO STEPS

In solving some equations, it becomes necessary to use two principles of equality in succession before arriving at an equivalent equation whose solution is obvious.

For example, let us solve the equation $4x - 3 = 17$:

$$4x - 3 = 17$$
$$\underline{+3 \quad +3} \text{ (addition principle of equality)}$$
$$4x \quad\quad = 20$$

This equation is equivalent to the original equation, but requires further simplification. Thus,

$$4x = 20$$

$$\frac{4x}{4} = \frac{20}{4} \text{ (division principle of equality)}$$

$$x = 5 \quad \text{(root)}$$

If we used the division principle first and then the addition principle, we would obtain the same result here. At this stage, however, it is simpler to use the addition principle first. Now let us check this solution in the original equation.

Check:

$$4x - 3 = 17$$

$$4 \cdot 5 - 3 = 17 \text{ (substitute 5 for } x\text{)}$$

$$20 - 3 = 17$$

$$17 = 17 \text{ ✔}$$

ILLUSTRATIVE PROBLEM: Solve for y and check: $\dfrac{y}{4} + 3 = 15$

Solution: Let us begin by first subtracting 3 from both members.

$$\frac{y}{4} + 3 = 15$$
$$\underline{\phantom{\frac{y}{4}} -3 \quad\quad -3}$$
$$\frac{y}{4} \quad\quad = 12$$

Now multiply both members by 4.

$$4 \times \frac{y}{4} = 12 \times 4$$

$$y = 48 \quad (\textit{answer})$$

Check:

$$\frac{y}{4} + 3 = 15 \text{ (original equation)}$$

$$\frac{48}{4} + 3 = 15 \text{ (substitute 48 for } y\text{)}$$

$$12 + 3 = 15$$

$$15 = 15 \text{ ✔}$$

Exercises _____

In 1 to 18, solve and check.

1. $2p - 7 = 11$ **2.** $3y + 5 = 17$ **3.** $4r + 1 = 22$

4. $2r - 8 = 9$ **5.** $7n + 3 = 3$ **6.** $8 + 5t = 28$

7. $\dfrac{s}{4} + 5 = 15$ **8.** $10 + \dfrac{k}{3} = 18$ **9.** $2a - 3.8 = 5.0$

10. $3y + .2 = 9.2$ **11.** $3p - \frac{1}{4} = 8$ **12.** $5t + .2 = 10.7$

13. $\dfrac{m}{3} + 2 = 6$ **14.** $6r - 2 = 5.2$ **15.** $\frac{2}{3}x + 1 = 1$

16. $\frac{5}{8}z = 50$ **17.** $4a - \frac{1}{2} = 11\frac{1}{2}$ **18.** $.04x + 3 = 3.08$

6. COMBINING LIKE TERMS IN EQUATIONS

In Chapter 1, we learned how to add or subtract *like terms.* For example, $2a + 5a = 7a$ and $8m - 2m = 6m$.

We may use this principle in solving an equation that has a sum or difference of like terms in one of its members.

ILLUSTRATIVE PROBLEM 1: Solve and check: $2x + 3x = 15$

Solution: $2x + 3x = 15$

$$5x = 15 \quad \text{(combine like terms)}$$

$$\frac{5x}{5} = \frac{15}{5} \quad \text{(divide both members by 5)}$$

$$x = 3$$

Check: $2x + 3x = 15$

$$2(3) + 3(3) = 15 \quad \text{(substitute 3 for } x\text{)}$$

$$6 + 9 = 15$$

$$15 = 15 \ \checkmark$$

Answer: $x = 3$

Recall here that 2(3) is another way of writing $2 \cdot 3$ or 2×3. This notation was already used in Chapter 1.

ILLUSTRATIVE PROBLEM 2: Solve and check: $7y - 4y = 21$

Solution:

$$7y - 4y = 21$$

$$3y = 21 \quad \text{(combine like terms)}$$

$$\frac{3y}{3} = \frac{21}{3} \quad \text{(divide both members by 3)}$$

$$y = 7$$

Check:

$$7y - 4y = 21 \quad \text{(original equation)}$$

$$7(7) - 4(7) = 21 \quad \text{(substitute 7 for } y\text{)}$$

$$49 - 28 = 21$$

$$21 = 21 \ \checkmark$$

Answer: $y = 7$

In checking, it is important to substitute in the *original* equation. If we substitute in an equivalent equation, we may have already made a mistake in obtaining the equivalent equation from the original.

Exercises ——————————————————————————

In 1 to 12, solve and check.

1. $5m + 2m = 14$ **2.** $32 = 10y - 2y$ **3.** $x + 4x = 30$

4. $8 = 2.5k + 1.5k$ **5.** $p + 2p + 3p = 180$ **6.** $\frac{3}{2}t + \frac{5}{2}t = 50$

7. $14.2r - 4.2r = 30$ **8.** $390 = 4n + 2n$ **9.** $\frac{8}{5}y - \frac{7}{5}y = 6$

10. $2.40 = c + .20c$ **11.** $3t + \frac{1}{2}t = 28$ **12.** $4\frac{1}{2}x - 2\frac{3}{4}x = 14$

7. MORE DIFFICULT EQUATIONS WITH UNKNOWN TERMS ON BOTH SIDES

ILLUSTRATIVE PROBLEM: Solve and check: $5x = 21 + 2x$

Here terms involving the unknown quantity are in both members of the equation. We must, therefore, subtract $2x$ from both sides so as to bring like terms to one side.

Solution: *Check:*

$5x = 21 + 2x$ $5x = 21 + 2x$ (original equation)

$\underline{-2x = \quad -2x}$ $5(7) = 21 + 2(7)$ (substitute 7 for x)

$3x = 21$ $35 = 21 + 14$

$\dfrac{3x}{3} = \dfrac{21}{3}$ $35 = 35$ ✔

$x = 7$ *(answer)*

Exercises

In 1 to 12, solve and check.

1. $7y = 15 + 2y$ **2.** $9m = 33 - 2m$ **3.** $5t = 24 + t$

4. $3 - p = 8p$ **5.** $1.1k = 30 + .5k$ **6.** $3\frac{1}{2}x = 18 + 1\frac{1}{2}x$

7. $11r = 36 - 7r$ **8.** $3t + 6 = 55 - 4t$ **9.** $4b + 8 = 20 + b$

10. $2.3y = 40 - .2y$ **11.** $8z + 24 = 5z - 12$ **12.** $4c - 3 = 47 - c$

Chapter Review Exercises

1. In each of the following, name the numerical coefficient:
   a. $7y$ **b.** $12x$ **c.** $\frac{1}{2}a$ **d.** $\frac{2}{3}r$ **e.** $.3k$

2. In problem **1a**, by what number must we divide $7y$ to get just $1y$ or y? Name such divisors for **1b**, **1c**, **1d**, and **1e**.

3. By what number should both members of each of the following equations be multiplied or divided to solve the equation?

   a. $8x = 56$ **b.** $3n = 33$ **c.** $\dfrac{k}{4} = 12$ **d.** $\frac{1}{6}y = 10$

   e. $.5n = 40$ **f.** $5p = 32$ **g.** $7m = 0$ **h.** $9t = 7.2$

4. Solve and check each equation in exercise **3**.

5. What number must be added to or subtracted from both members of each of the following equations in order to solve the equation?
   a. $r + 7 = 15$ **b.** $t - 9 = 11$ **c.** $8 + x = 20$
   d. $y - 2\frac{1}{2} = 7\frac{1}{2}$ **e.** $n + 2.3 = 17$ **f.** $18 = k - 7$
   g. $r + 5 = 5$ **h.** $y - .07 = 4.2$

6. Solve and check each equation in exercise **5**.

7. Solve and check the following equations:

    a. $2x + 3 = 15$ **b.** $4 + 8n = 44$ **c.** $3p - 6 = 12$

    d. $2y + \frac{1}{2} = 6\frac{1}{2}$ **e.** $.5 + 2k = 7.1$ **f.** $5a + 9 = 9$

    g. $\frac{5}{8}y = 40$ **h.** $\dfrac{3z}{10} = 12$ **i.** $3n - n + 4 = 15$

    j. $28 + 4r = 54$ **k.** $8x + 4 - x = 25$ **l.** $\frac{1}{2}t + 1 = 17$

    m. $2a - 6 = 0$ **n.** $\frac{2}{3}t - 7 = 17$ **o.** $5r + 3r = 80$

    p. $.3c + .02c = 6.4$ **q.** $7p + 4 - p = 34$ **r.** $6x = 36 - 3x$

    s. $5y - 4 = 2y + 8$ **t.** $12z - 19 = 7z + 11$

CHECK YOUR SKILLS

1. Evaluate: **a.** $10 - 4 + 2 + 6$

              b. $14 + 16 - 3 + 10 - 4 - 6$

2. Add:

a. 8768	**b.** 9,568	**c.** 85,416	**d.** 63,929
2436	76	78,959	94,873
6832	800	20,048	51,020
4095	19,875		

3. Subtract:

a. 390	**b.** 709	**c.** 2807	**d.** 8700
-169	-594	-998	-5008

4. Multiply:

a. 89	**b.** 509	**c.** 847	**d.** 699
$\times 76$	$\times 87$	$\times 79$	$\times 507$

5. Divide:

    a. $21\overline{)882}$ **b.** $82\overline{)2952}$ **c.** $23\overline{)943}$ **d.** $78\overline{)3978}$

6. A tank contained 5356 gallons of oil. If 686 gallons were used, how many gallons remained?

7. A bar of iron 18 feet long weighs 1134 pounds. What is the weight per foot?

8. A building is 28 stories high. Each story is 14 feet high. What is the height in feet of the building?

9. Solve and check: $4x + 5x = 72$

10. Solve and check: $7x = 36 + 4x$

3

Solving Problems by Using Equations

1. ALGEBRAIC REPRESENTATION

The equation provides us with a very powerful tool for solving many mathematical problems. But, before we can solve these problems, we must first become familiar with the way certain words or phrases in English are **translated** into the symbols and letters of algebra. If you have difficulty solving verbal problems in arithmetic, review Appendix II on page 436.

In Chapter 1, we learned how to use letters to represent numbers. If n represents a certain number, $7n$ represents seven times that number. What is 7 more than the number n? Seven more than 4 is $4 + 7$, or 11. Hence, 7 more than n is $n + 7$.

What is 7 less than n? Again, think of what you would do with two numbers. Seven less than 10 is $10 - 7$, or 3. Thus, 7 less than n is $n - 7$.

Certain algebraic symbols may be expressed in words in several different ways. A few of these ways are listed below:

Algebraic Expression	Word Expression
$3n$	3 times n 3 multiplied by n the product of 3 and n
$x + 7$	7 more than x the sum of x and 7 x increased by 7
$t - 5$	t diminished by 5 t decreased by 5 5 less than t the difference of t and 5 5 subtracted from t

$$\frac{m}{9} \text{ or } m \div 9 \qquad \left\{ \begin{array}{l} \text{one-ninth of } m \\ m \text{ divided by } 9 \\ \text{the quotient of } m \text{ and } 9 \end{array} \right.$$

ILLUSTRATIVE PROBLEMS:

1. Bill is 3 years older than Tom. If Tom is t years old, express Bill's age algebraically.

   Answer: Bill is $(t + 3)$ years old.

2. Jim's father is 3 times as old as Jim is. If Jim is y years old, express his father's age algebraically.

   Answer: His father is $3y$ years old.

3. Maria has n dollars. Anne has 3 less than twice what Maria has. Express Anne's amount algebraically.

   Answer: Anne has $(2n - 3)$ dollars.

4. How many feet are in p inches?

Solution: How would we change 24 inches to feet? We would divide 24 by 12. Likewise, we divide p by 12 so that p inches $= \dfrac{p}{12}$ feet.   (*answer*)

Exercises _____

1. If n represents a number, write algebraically:
   a. 10 more than n
   b. 3 less than n
   c. 5 times n
   d. 7 more than twice n
   e. n decreased by 9
   f. 30 diminished by 5 times n
   g. one-half of n
   h. the product of n and 12
   i. the quotient of 48 divided by n
   j. 17 less than four times n
   k. $\frac{2}{3}$ of n increased by 12

2. Represent algebraically:
   a. 7 more than t
   b. 18 less than k
   c. 5 times r
   d. s decreased by 14
   e. p divided by 9
   f. 8 less than 6 times x

g. y increased by 3   **h.** 25 decreased by m
i. one-half of q **j.** n subtracted from 8
k. p increased by q **l.** r decreased by s
m. 5 less than 8 times t **n.** 32 more than $\frac{9}{5}$ of c

3. If x and y represent two numbers, write algebraically:
   a. the product of the two numbers
   b. the quotient of the first number divided by the second
   c. the sum of the two numbers
   d. the first number diminished by twice the second number

4. State in words a meaning of each of the following algebraic expressions:
   a. $r + 4$ **b.** $m - 7$ **c.** $6t$ **d.** $2n - 5$
   e. $\dfrac{y}{12}$ **f.** $\frac{1}{2}x + 8$ **g.** $13 - k$ **h.** $\dfrac{p}{4} - 6$
   i. $a - b$ **j.** $p + 2q$

In 5 to 22, express the result algebraically.

5. A man is now k years old.
   a. Express his age 10 years from now.
   b. Express his age 10 years ago.

6. Jane had 60 cents and spent c cents. How many cents does she have left?

7. An apple costs p cents. How much will a dozen apples cost?

8. **a.** If 10 toys cost $30, how much does each toy cost?
   b. If 10 toys cost x dollars, express the cost of each toy.

9. A man is h years old and his wife is w years old. How much older is the man than his wife?

10. The length of a rectangle is f feet and the width is 3 feet shorter. Express the width.

11. How many feet are there in
    a. 1 yard?   **b.** 5 yards?   **c.** 7 yards?   **d.** y yards?

12. How many cents are there in
    a. 1 dime?   **b.** 3 dimes?   **c.** n dimes?

13. A bus travels 40 miles per hour. How far does it go in
    a. 3 hours?   **b.** 5 hours?   **c.** t hours?

14. The side of a square is s inches in length. Express its perimeter.

15. If we represent a whole number by n, how do we represent the next whole number?

16. How many yards are there in
    a. 3 feet?   **b.** 6 feet?   **c.** 12 feet?   **d.** f feet?

17. Bob has b dollars.
    a. If Anita has 8 dollars more than Bob, represent Anita's amount.
    b. Represent the sum of Bob's amount and Anita's amount.

18. Eduardo is x years old. His father is 3 times as old as Eduardo.
    a. Represent his father's age.
    b. Represent the difference between the father's and son's ages.

19. Eggs cost c cents per dozen.
    a. How much does 1 egg cost?
    b. Represent the cost of 5 eggs.

20. The width of a rectangle is k feet. If the length is 10 more than twice the width, represent the length.

21. Represent the sum of p and q diminished by twice r.

22. A girl bought 8 items, each costing c cents. How much change should she receive from a $5.00 bill?

---

2. WRITING ALGEBRAIC EQUATIONS

We are now prepared to solve problems by means of equations. We will start with some simple problems that may easily be done by arithmetic processes alone. In order to develop greater skill in algebra, however, we will form equations and solve them.

ILLUSTRATIVE PROBLEM 1: If 12 is added to a certain number, the result is 35. Find the number. Form an equation for this problem; solve and check it.

Solution: Let n = the certain number.
        Then:

$$\begin{array}{rl} n + 12 = & 35 \quad \text{(equation)} \\ -12 = & -12 \\ \hline n \quad\quad = & 23 \quad (answer) \end{array}$$

Check: To check the solution to a verbal problem, check it in the *original* problem. Thus, if 12 is added to 23, the result is 35, and this checks.

ILLUSTRATIVE PROBLEM 2: Five times a number diminished by 18 is 42. Find the number.

Solution: Let x = the number.
Then:

$$5x - 18 = 42 \quad \text{(equation)}$$
$$\underline{+18 \quad +18}$$
$$5x \quad\quad = 60$$
$$\frac{5x}{5} = \frac{60}{5}$$
$$x = 12 \quad (answer)$$

Check: Five times 12 = 60; 60 − 18 = 42. ✔

ILLUSTRATIVE PROBLEM 3: A boy, who is 12 years old, is $\frac{1}{3}$ as old as his father. How old is his father?

Solution: Let y = the father's age in years.
Then:

$$\tfrac{1}{3} y = 12 \quad \text{(equation)}$$

Check: 12 is $\frac{1}{3}$ of 36.

Multiply both members by 3:

$$y = 36 \quad (answer)$$

SOLVING WORD PROBLEMS

The steps in solving "word problems" are as follows:

a. Read the problem carefully.

b. Determine what is given and what is to be found.

c. Represent one of the unknown quantities by a letter.

d. Represent any other unknown quantities in terms of this letter.

e. Write the equation that expresses the meaning of the problem.

f. Solve the equation.

g. Check the answer in the wording of the original problem.

Exercises

In 1 to 20, solve each problem by forming an equation. Then solve the equation and check in the *original* problem. Give the equation as part of your answer.

1. If 8 is added to a certain number, the sum is 84. Find the number.

2. Eleven times a certain number is 132. Find the number.

3. One-third of a number is 83. Find the number.

4. If 14 is subtracted from a certain number, the result is 72. Find the number.

5. Twice a certain number increased by 15 is 35. Find the number.

6. If three times a certain number is diminished by 17, the result is 28. Find the number.

7. There are 32 students in a class. The number of girls is 3 times the number of boys. How many boys are in the class?

8. Two-thirds of a number is 12. Find the number.

9. Three-sevenths of a number is 24. Find the number.

10. A football player scored one-fourth of the team's final score. If he scored 18 points, what was the team score?

11. One-third of a certain number increased by 12 is equal to 18. Find the number.

12. The larger of two numbers is 7 times the smaller. If n represents the smaller, then:

    a. Represent the larger in terms of n.

    b. If the sum of the two numbers is 120, write an equation expressing this relation.

    c. Solve this equation for n.

    d. Write both numbers and check.

13. A board that is 80 inches long is cut into two pieces so that one piece is four times as long as the other. If y is the length in inches of the smaller piece, then:

    a. Represent the larger piece in terms of y.

    b. Write an equation to find y.

    c. Solve the equation and check.

14. What number increased by $\frac{1}{4}$ of itself is equal to 30?

15. A boy has three times as much money as does his sister. Together, they have $60. How much does each have?

16. A man paid for a TV set in 12 equal payments. If the set cost $240, how much did he pay each month?

17. One number is four times another. The difference between the larger number and the smaller number is 48. Find the two numbers.

18. Together, Mary and Nancy saved $30. In dividing it between them, Mary got $4 more than Nancy. How much did each girl receive?

19. Separate 180 into two parts so that one part will be 5 times the other.

20. A girl earned $20 more than twice as much as her brother. Together, they earned $80. How much did each earn?

3. CONSECUTIVE NUMBER PROBLEMS

Two whole numbers are said to be **consecutive** if their difference is 1. Thus, 8 is consecutive to 7. If 11 is the first of three consecutive numbers, what are the other two?

ILLUSTRATIVE PROBLEM 1: What is the sum of 8 and the next higher consecutive number?

Solution: $8 + 9 = 17$ *(answer)*

ILLUSTRATIVE PROBLEM 2: The sum of two consecutive numbers is 43. Find the numbers.

Solution:

Let n = the first number.   *Check:* $21 + 22 = 43$ ✔

Then $n + 1$ = the next consecutive number.

$$n + n + 1 = 43 \quad \text{(sum of the numbers)}$$
$$2n + 1 = 43$$
$$\underline{ -1 \quad -1}$$
$$2n = 42$$
$$n = 21 \quad \text{(first number)}$$
$$n + 1 = 22 \quad \text{(next number)}$$

Answer: The numbers are 21 and 22.

Exercises_____

In the following problems, when you are asked to *find* consecutive numbers, form an equation, solve, and check.

1. The first of three consecutive numbers is 15. Write the other two.
2. The first of three consecutive numbers is *p*. Represent the other two in terms of *p*.
3. The sum of two consecutive numbers is 61. Find the numbers.
4. The sum of three consecutive numbers is 69. Find the numbers.
5. Name the three greatest consecutive 2-digit integers.
6. The sum of the first and third of three consecutive numbers is 34. Find the numbers.
7. Numbers such as 6, 8, 10 are called consecutive *even* integers.
   a. Starting with 12, what are the next two consecutive even integers?
   b. If *x* is an *even* integer, represent the next two consecutive even integers in terms of *x*.
8. The sum of two consecutive even integers is 86. Find them.
9. Numbers such as 13, 15, 17 are called consecutive *odd* integers.
   a. Starting with 23, name the next two consecutive odd integers.
   b. If *t* is an *odd* number, represent the next two consecutive odd integers in terms of *t*.
10. Let *r* be an odd integer.
   a. Is (*r* + 1) an even or an odd integer?
   b. Is (*r* + 2) an even or an odd integer?
11. The sum of two consecutive odd integers is 72. Find the numbers.
12. The sum of three consecutive odd integers is 99. Find them.

---

4. COIN PROBLEMS

In solving problems that deal with nickels, dimes, and quarters, we often find it helpful to represent the value of the coins in cents.

The value of 4 nickels in cents is 5(4) or 20 cents.
The value of *x* dimes in cents in *x*(10) or 10*x* cents.
The value of (*n* + 2) quarters in cents is (*n* + 2)25 or 25(*n* + 2) cents.

ILLUSTRATIVE PROBLEM: Bill has 5 more pennies than dimes. He has 71¢ altogether. How many of each coin does he have?

Solution: Let d = the number of dimes.
   Then $d + 5$ = the number of pennies.
      And $10d$ = the value of the dimes in cents.

$$10d + d + 5 = 71 \quad \text{(value of dimes and pennies in cents)}$$
$$11d + 5 = 71$$
$$\underline{-5 \quad -5}$$
$$11d \quad = 66$$
$$d = 6 \text{ (number of dimes)}$$
$$d + 5 = 11 \text{ (number of pennies)}$$

Check: The value of 6 dimes = 60¢
   The value of 11 pennies = $\underline{11¢}$
      Total value = 71¢ ✔

Answer: Bill has 6 dimes and 11 pennies.

Exercises _____

1. What is the value in cents of
   a. 3 dimes?   **b.** 6 dimes?   **c.** n dimes?

2. Fritz has y dimes. What is their value in cents?

3. What is the value in cents of
   a. 3 dimes and 4 pennies?   **b.** n dimes and 3 pennies?

4. Anne has d dimes.
   a. What is their value in cents?
   b. If she has 4 more pennies than dimes, represent the number of pennies she has.
   c. Express in terms of d the total value in cents of her coins.

5. A man has 5 more pennies than dimes. If he has 93¢ altogether, how many coins of each kind does he have?

6. What is the value in cents of
   a. 3 quarters?   **b.** n quarters?
   c. 3 quarters and 2 dimes?   **d.** n quarters and 4 dimes?

7. A toy bank contains twice as many pennies as quarters. If the bank contains $1.35, how many of each coin does it have?

8. What is the value in cents of
   a. 8 nickels?   **b.** n nickels?   **c.** y nickels and 7 pennies?

9. Represent the total value in cents of q quarters, d dimes, and n nickels.

10. Bill saved only quarters and nickels in a toy bank. When the numbers of quarters and nickels were equal, he had $2.70. How many of each coin did he have?

---

5. GEOMETRIC PROBLEMS

ILLUSTRATIVE PROBLEM 1: The length of a rectangular table is twice its width. If the perimeter is 84 inches, find the length and width of the table.

Solution: Let w = width in inches.

And $2w$ = length in inches.

Make a diagram, as shown.

$w + 2w + w + 2w = 84$ (perimeter)

$6w = 84$ (combine like terms)

$w = 14$ (divide both members by 6)

$2w = 28$ (length)

Check: The perimeter is: $14 + 28 + 14 + 28 = 84$ ✓

Answer: The table is 14 inches wide and 28 inches long.

ILLUSTRATIVE PROBLEM 2: In a triangle, two angles are equal and the third angle is twice as large as each of the others. Find the angles. (Recall that the sum of the angles of a triangle is 180°.)

Solution: Let x = number of degrees in each of the smaller angles.

And $2x$ = number of degrees in the larger angle.

Make a diagram, as shown.

$2x + x + x = 180$ (sum of angles is 180°)

$4x = 180$ (combine like terms)

$x = 45$ (divide both sides by 4)

$2x = 90$ (larger angle)

Check: $45° + 45° + 90° = 180°$ ✓

Answer: The angles are $45°$, $45°$, and $90°$.

Exercises _____

1. Find the side of a square whose perimeter is 84 feet.

2. The length of a rectangle is 3 times its width. If the perimeter is 96 inches, find the length and width.

3. If the three angles of a triangle are equal to one another, find the number of degrees in each angle.

4. In a triangle, one angle is twice another and the third angle is three times as large as the smallest. Find the number of degrees in each angle.

5. The perimeter of an equilateral triangle is 69 inches. Find each side.

6. The length of a rectangle is 4 inches more than the width. If the perimeter is 56, find the length and width.

7. If each of the equal angles of an isosceles triangle is 38°, how many degrees is the third angle?

8. The length of a rectangle is 3 more than twice the width. If the perimeter is 72 inches, find the length and width.

9. The width of a rectangle is one-half of its length. If the perimeter is 93 inches, find the length and width.

---

6. USING EQUATIONS WITH FORMULAS

In Chapter 1, we learned something about the use of formulas. With our new knowledge of how to solve equations, we can now extend our use of formulas.

ILLUSTRATIVE PROBLEM: The perimeter of a square is given by the formula $p = 4s$, where s is the length of a side. If p is 92 inches, find s.

Solution:

$$p = 4s \quad \text{(perimeter formula)}$$

$$92 = 4s \quad \text{(substitute 92 for } p\text{)}$$

$$23 = s \quad \text{(divide both sides by 4)}$$

$$s = 23 \text{ inches} \quad (answer)$$

Exercises_____

1. The perimeter of a square is 72 inches. Using the formula $p = 4s$, find s.

2. The formula for the perimeter of an equilateral triangle of side s is $p = 3s$. If the perimeter of an equilateral triangle is 70 inches, find the length of each side.

3. The formula for the number of feet f in y yards is $f = 3y$. Find the number of yards in 96 feet.

4. In the formula $S = C + P$, S is the selling price of an article, C its cost, and P the profit. If $S = \$32.50$ and $C = \$21.25$, find P.

5. The formula $P = 2g + f$ gives the total number of points P scored by a basketball team making g field goals and f foul shots. If $P = 68$ and $f = 26$, find the number of field goals.

6. The perimeter of an isosceles triangle is given by $p = 2a + b$. If $p = 38$ inches and $b = 15$ inches, find a.

7. The formula $F = \frac{9}{5}C + 32$ converts Celsius temperatures to the Fahrenheit scale. If $F = 68°$, find C.

8. The formula $d = rt$ gives the distance d (in miles) covered by an object moving at r miles per hour for t hours. Find the rate r for a car that travels 215 miles in 5 hours.

9. Using the formula $d = rt$, find how many hours it takes for a train to travel 288 miles at 64 miles per hour.

10. At a certain photo lab, the formula $C = 40 + 8n$ gives the cost, in cents, of developing and printing n films. If a man pays $\$12.16$ to develop and print some films, find the number of them.

_____

Chapter Review Exercises

1. Tom is t years old.
    a. Represent his age 7 years ago.
    b. Represent his age 3 years from now.

2. Bill has n nickels and Mary has 3 more nickels than Bill.
    a. Represent the number of nickels Mary has.
    b. Represent the value in cents of Mary's nickels.
    c. Represent the number of nickels they both have.
    d. Represent the value in cents of the nickels they both have.

3. **a.** If the smallest of 3 consecutive numbers is p, represent the other two in terms of p.

   b. Represent the sum of all three numbers.

4. Six times a number is increased by 4. The result is 68. Find the number.

5. The sum of two angles is $120°$. If one angle is 4 times as large as the other, how many degrees are there in each angle?

6. The sum of two consecutive numbers is 85. What are the numbers?

7. Paul has three times as many dimes as pennies. The total he has is $1.24. How many of each coin does he have?

8. There are three consecutive numbers such that the sum of the first and third is 86. Find them.

9. In a ninth-grade class of 420 students, there are twice as many girls as boys. How many of each are there?

10. A boy is paid $4.00 per hour for baby-sitting plus $1.80 for carfare. If he receives $19.80 one evening, how many hours did he baby-sit?

11. A bottle and a cork cost $2.20. If the bottle costs $2.00 more than the cork, how much does each cost?

12. The length of a rectangle is 4 times its width. If the perimeter is 75 inches, find the length and width.

13. The average of three numbers p, q, and r is given by the formula $A = \dfrac{p + q + r}{3}$. If a boy gets 81 and 84 on two tests, what must he get on a third test to average 85?

14. Frank had $32.50. He worked for 15 hours and then had $100 altogether. How much did he earn per hour?

15. In a triangle, one angle is twice another and the third angle is $20°$ more than the smallest angle. Find the three angles.

CHECK YOUR SKILLS

1. Subtract:
   - a.   639
     −378
   - b.   400
     −238
   - c.   3457
     −2498
   - d. 7005
     −487
   - e.   7928
     −5349

2. Find the value of each of the following:
   - a. 14×0
   - b. 154×100
   - c. $\frac{860}{10}$
   - d. $26{,}000 \div 1000$

3. Multiply:
   - a.   52
     $\times 18$
   - b.   37
     $\times 34$
   - c. 940
     $\times 20$
   - d. 694
     $\times 83$
   - e.   805
     $\times 306$
   - f. 6458
     $\times 82$

4. Divide:
   - a. $549 \div 9$
   - b. $8 \overline{)\,248}$
   - c. $\frac{289}{4}$
   - d. $363 \div 4$
   - e. $\frac{304}{4}$
   - f. $7 \overline{)\,503}$
   - g. $\frac{170}{33}$
   - h. $38 \overline{)\,250}$
   - i. $25 \overline{)\,2300}$

5. A worker earns $52 per day. How much does the worker earn for 25 workdays?

6. A certain job requires 1144 hours for completion. If the work is divided equally among 26 people, how many hours must each person work?

7. In a class of 28 students, 52¢ is collected from each one. What is the total amount collected?

8. The perimeter of a square is 108 inches. Using the formula $p = 4s$, find s.

9. The sum of three consecutive integers is 66. Find the numbers.

10. Represent the sum of $2x$ and y diminished by twice z.

4

Common Fractions and Simple Fractional Equations

1. MEANING OF A FRACTION

In this chapter, we shall review some of the basic ideas and skills relating to *common fractions*.

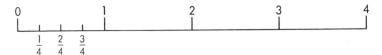

On the number line above, the space between 0 and 1 has been divided into four equal parts. Each part is then called one-fourth of the distance from 0 to 1. If this distance from 0 to 1 is 1 inch, then the first fractional part represents one-fourth of an inch; this fraction is written $\frac{1}{4}$. If we take 3 of these smaller units, the distance from 0 is then three-fourths of an inch, written $\frac{3}{4}$. The number below the fraction line tells us into how many equal parts the inch has been divided; this is the **denominator** of the fraction. The number above the fraction line tells us how many of these fractional parts to take; this is called the **numerator** of the fraction.

The numerator and denominator are called **terms** of a fraction. The fraction $\frac{5}{8}$ is read "five-eighths" and means that five of eight equal parts are being taken, as shown in the figure. When the terms of a fraction are whole numbers, we refer to the fraction as a **common fraction** or a **rational number**.

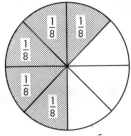

Shaded part = $\frac{5}{8}$

On the number line, we can see that the fraction $\frac{4}{4}$ brings us to 1. If we continue to divide the space from 1 to 2 into quarters, we will then have lengths of $\frac{5}{4}, \frac{6}{4}, \frac{7}{4}, \frac{8}{4}$.

Note that $\frac{8}{4}$ brings us to 2, or 8 ÷ 4. Thus, the fraction line behaves like a division sign, so that a common fraction is the indicated *division* of one whole number by another whole number.

A ***proper fraction*** is one in which the numerator is less than the denominator. Thus, $\frac{1}{7}, \frac{2}{5}$, and $\frac{5}{8}$ are proper fractions. The value of a proper fraction is always less than 1.

The following figures represent proper fractions as shown:

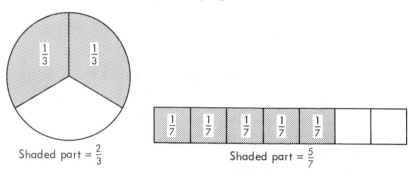

Shaded part = $\frac{2}{3}$ Shaded part = $\frac{5}{7}$

An ***improper fraction*** is one in which the numerator is greater than or equal to the denominator. Thus, $\frac{3}{2}, \frac{8}{5}$, and $\frac{7}{7}$ are improper fractions. The value of an improper fraction is always greater than or equal to 1. $\frac{5}{1}$ is an improper fraction equal to 5.

A ***mixed number*** consists of a whole number and a fraction. Thus, $4\frac{1}{2}, 7\frac{2}{3}$, and $5\frac{7}{8}$ are mixed numbers.

The mixed number $2\frac{1}{4}$ means $2 + \frac{1}{4}$, or 2 and $\frac{1}{4}$, as shown below.

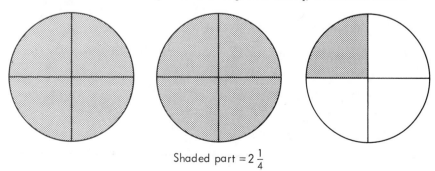

Shaded part = $2\frac{1}{4}$

Exercises

1. From the fractions $\frac{4}{7}$, $\frac{9}{13}$, $\frac{17}{12}$, $\frac{5}{3}$, $\frac{5}{8}$, $\frac{99}{100}$, $\frac{6}{3}$, $\frac{7}{1}$, $\frac{1}{7}$, choose
   a. the proper fractions.   **b.** the improper fractions.

2. **a.** Into how many equal parts is the circle divided?
   b. What fractional part of the circle is each of the equal parts?
   c. What fractional part of the circle is shaded?
   d. What fractional part of the circle is unshaded?

   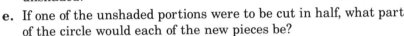

   e. If one of the unshaded portions were to be cut in half, what part of the circle would each of the new pieces be?

3. Using the digits 3, 4, 5, write
   a. a proper fraction.   **b.** an improper fraction.
   c. a mixed number.

4. **a.** A quarter is what fractional part of a dollar?
   b. A dime is what fractional part of a dollar?
   c. A nickel is what fractional part of a dollar?

5. What do we call each of the following?
   a. top number of a fraction
   b. bottom number of a fraction
   c. a whole number and a fraction
   d. a fraction less than 1 in value

6. **a.** What is the value of a fraction whose numerator and denominator are equal?

   b. What is the value of $\frac{n}{1}$ where n is a whole number?

   c. If $\frac{x}{12} = 1$, what is the value of x?

   d. If you divide a pie into 10 equal parts and take $\frac{7}{10}$ away, what fractional part of the pie remains?

7. Given the fractions $\frac{1}{3}$, $\frac{1}{4}$, $\frac{1}{5}$.
   a. Which is the smallest fraction?
   b. Which is the largest fraction?

2. REDUCING FRACTIONS TO LOWEST TERMS

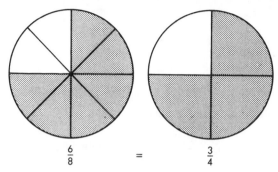

$$\frac{6}{8} \qquad = \qquad \frac{3}{4}$$

The two figures indicate that $\frac{6}{8}$ of a circle has the same value as $\frac{3}{4}$ of a circle. We refer to $\frac{6}{8}$ and $\frac{3}{4}$ as **equivalent fractions** and write $\frac{6}{8} = \frac{3}{4}$. Note that, if we divide both 6 and 8 by 2 in the first fraction, we obtain the second fraction, $\frac{3}{4}$. We say that we have **reduced** the first fraction to **lower terms**. In this example, we say that $\frac{3}{4}$ is in **lowest terms** because the 3 and 4 have no common factor other than 1, and thus the fraction cannot be reduced further.

The fraction $\frac{12}{18}$ may be reduced as follows:

$$\frac{12}{18} = \frac{12 \div 3}{18 \div 3} = \frac{4}{6} = \frac{4 \div 2}{6 \div 2} = \frac{2}{3}$$

Thus, $\frac{12}{18} = \frac{4}{6} = \frac{2}{3}$. We say $\frac{4}{6}$ is in *lower* terms and $\frac{2}{3}$ is in *lowest* terms.

RULE: The numerator and denominator of a fraction may both be *divided* by the same number without changing the value of the fraction.

Just as we can *reduce* a fraction to *lower* terms, we can also *raise* a fraction to **higher terms**. We frequently find it convenient to change a fraction to an equivalent fraction with a larger denominator. Thus, we can change $\frac{2}{3}$ to $\frac{6}{9}$ by multiplying both terms of the fraction by 3.

RULE: The numerator and denominator of a fraction may both be *multiplied* by the same number without changing the value of the fraction.

ILLUSTRATIVE PROBLEM 1: Reduce $\frac{18}{30}$ to lowest terms.

Solution: 18 and 30 are both divisible by 6.

$$\frac{18 \div 6}{30 \div 6} = \frac{3}{5} \quad (answer)$$

Note: Here, $\frac{18}{30}$ can be reduced to other fractions ($\frac{9}{15}$, $\frac{6}{10}$), but $\frac{3}{5}$ is the fraction in *lowest* terms.

ILLUSTRATIVE PROBLEM 2: Write $\frac{2}{5}$ as an equivalent fraction with 20 as a denominator.

Solution:

$$\frac{2}{5} = \frac{?}{20}$$

To change the denominator 5 to 20, we must multiply it by 4. Hence, we must also multiply the numerator 2 by 4.

$$\frac{2}{5} = \frac{2 \times 4}{5 \times 4} = \frac{8}{20} \quad (answer)$$

Exercises _____

In 1 to 15, reduce each fraction to lowest terms.

1. $\frac{15}{20}$ **2.** $\frac{24}{60}$ **3.** $\frac{18}{48}$ **4.** $\frac{35}{56}$ **5.** $\frac{18}{24}$

6. $\frac{9}{144}$ **7.** $\frac{45}{70}$ **8.** $\frac{8}{14}$ **9.** $\frac{45}{180}$ **10.** $\frac{21}{28}$

11. $\frac{80}{120}$ **12.** $\frac{23}{69}$ **13.** $\frac{45}{54}$ **14.** $\frac{150}{200}$ **15.** $\frac{104}{108}$

In 16 to 24, change each fraction to higher terms.

16. $\frac{3}{5} = \frac{?}{25}$ **17.** $\frac{3}{4} = \frac{?}{16}$ **18.** $\frac{3}{7} = \frac{?}{14}$

19. $\frac{7}{8} = \frac{?}{16}$ **20.** $\frac{4}{15} = \frac{?}{30}$ **21.** $\frac{5}{16} = \frac{?}{64}$

22. $\frac{9}{10} = \frac{?}{50}$ **23.** $\frac{7}{2} = \frac{?}{10}$ **24.** $\frac{5}{3} = \frac{?}{12}$

25. How many sixteenths of an inch are in $\frac{3}{4}$ of an inch?

26. How many sixteenths of a pound (ounces) are in one-half of a pound?

27. How many twelfths of a foot (inches) are in $\frac{2}{3}$ of a foot?

3. EQUIVALENCE OF MIXED NUMBERS AND IMPROPER FRACTIONS

It is frequently desirable to change an improper fraction to a mixed number. A recipe might call for $1\frac{1}{2}$ cups of flour but would not call for $\frac{3}{2}$ cups.

Since a fraction is an indicated division, a method is already suggested for reducing an improper fraction to a mixed number. For example, $\frac{7}{3}$ may be written as $7 \div 3$. Thus:

$$
\begin{array}{r}
2\frac{1}{3} \\
3\overline{)\,7} \\
\underline{6} \\
1
\end{array}
$$

We see that $\frac{7}{3} = 2\frac{1}{3}$.

We may *check* this by noting that, if $1 = \frac{3}{3}$, then $2 = \frac{6}{3}$ and $2\frac{1}{3} = 2 + \frac{1}{3} = \frac{6}{3} + \frac{1}{3} = \frac{7}{3}$.

RULE: To change an improper fraction to a mixed number, divide the numerator by the denominator and write the fractional part of the quotient in lowest terms.

Thus, $\frac{65}{10} = 6\frac{5}{10} = 6\frac{1}{2}$.

In computation, it is often convenient to work with a mixed number in the form of an improper fraction. This transformation, the reverse of changing an improper fraction to a mixed number, is illustrated in the problems that follow.

ILLUSTRATIVE PROBLEM 1: Change $3\frac{1}{5}$ to an improper fraction.

Solution: $3\frac{1}{5} = 3 + \frac{1}{5}$

Write 3 as an equivalent fraction with a denominator of 5:

$$
\frac{3}{1} = \frac{3 \times 5}{1 \times 5} = \frac{15}{5}
$$

Then: $3\frac{1}{5} = 3 + \frac{1}{5} = \frac{15}{5} + \frac{1}{5} = \frac{16}{5}$ (*answer*)

The process may be summarized in the following rule:

RULE: To change a mixed number to an improper fraction, multiply the whole number by the denominator of the fraction and add its numerator. The result is the numerator of the improper fraction, and the denominator remains the same.

Thus, $3\frac{1}{5} = \frac{3 \times 5 + 1}{5} = \frac{15 + 1}{5} = \frac{16}{5}$.

ILLUSTRATIVE PROBLEM 2: Change $3\frac{4}{7}$ to an improper fraction.

Solution: $3\frac{4}{7} = \frac{3 \times 7 + 4}{7} = \frac{21 + 4}{7} = \frac{25}{7}$ (*answer*)

ILLUSTRATIVE PROBLEM 3: Change 7 into an equivalent fraction having 3 as the denominator.

Solution: $7 = \dfrac{7}{1} = \dfrac{7 \times 3}{1 \times 3} = \dfrac{21}{3}$ *(answer)*

Exercises

In 1 to 12, change each improper fraction to a mixed number in simplest form or a whole number.

1. $\frac{15}{6}$ **2.** $\frac{31}{20}$ **3.** $\frac{65}{20}$ **4.** $\frac{48}{4}$ **5.** $\frac{28}{3}$ **6.** $\frac{150}{10}$

7. $\frac{35}{9}$ **8.** $\frac{34}{6}$ **9.** $\frac{18}{4}$ **10.** $\frac{27}{27}$ **11.** $\frac{7}{6}$ **12.** $\frac{88}{11}$

In 13 to 18, change the following whole numbers into improper fractions as indicated.

13. $3 = \frac{?}{8}$ **14.** $5 = \frac{?}{7}$ **15.** $12 = \frac{?}{8}$

16. $14 = \frac{?}{3}$ **17.** $20 = \frac{?}{100}$ **18.** $15 = \frac{?}{4}$

In 19 to 26, write the following mixed numbers as improper fractions.

19. $3\frac{2}{7}$ **20.** $5\frac{1}{2}$ **21.** $1\frac{1}{6}$ **22.** $7\frac{8}{15}$

23. $7\frac{3}{10}$ **24.** $8\frac{1}{6}$ **25.** $1\frac{5}{24}$ **26.** $3\frac{11}{20}$

27. How many $\frac{1}{4}$-inch parts are there in $3\frac{1}{4}$ inches?

28. How many $\frac{1}{2}$-cups of milk are there in a container holding $4\frac{1}{2}$ cups?

29. How many eighths of an inch are there in $4\frac{3}{8}$ inches?

30. A carpenter cuts $5\frac{2}{3}$ yards of molding into strips each $\frac{1}{3}$ yard long. How many strips does he obtain?

4.   ADDING LIKE FRACTIONS

Like fractions are fractions having the same denominator.

Suppose we divided a rectangle into seven equal parts (also rectangles), so that each part is $\frac{1}{7}$ of the original rectangle.

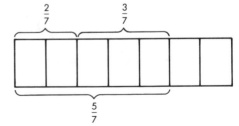

Note that $\frac{2}{7} + \frac{3}{7} = \frac{5}{7}$ and that $\frac{5}{7} + \frac{2}{7} = \frac{7}{7} = 1$.

RULE: To add like fractions, simply add the numerators and keep the same denominator.

When the result is an improper fraction, we usually reduce it to a whole number or a mixed number. It is customary to reduce all fractional results to lowest terms.

ILLUSTRATIVE PROBLEM 1: Add $\frac{3}{11}$, $\frac{5}{11}$, and $\frac{6}{11}$.

Solution:

$$\frac{3}{11} + \frac{5}{11} + \frac{6}{11} = \frac{14}{11} = 1\frac{3}{11} \quad (answer)$$

ILLUSTRATIVE PROBLEM 2: Add $\frac{4}{9}$ and $\frac{2}{9}$.

Solution:

$$\frac{4}{9} + \frac{2}{9} = \frac{6}{9} = \frac{6 \div 3}{9 \div 3} = \frac{2}{3} \quad (answer)$$

ILLUSTRATIVE PROBLEM 3: Add $3\frac{3}{8}$ and $4\frac{7}{8}$.

Solution:

$$3\frac{3}{8}$$
$$+4\frac{7}{8}$$
$$\overline{7\frac{10}{8}}$$

Change $\frac{10}{8}$ to $1\frac{2}{8} = 1\frac{1}{4}$. Thus,

$$7\frac{10}{8} = 7 + \frac{10}{8} = 7 + 1\frac{1}{4} = 8\frac{1}{4} \quad (answer)$$

Exercises —————————————————————————————

In 1 to 13, add and reduce all results to lowest terms.

1. $\frac{3}{8} + \frac{2}{8}$ **2.** $\frac{15}{19} + \frac{2}{19}$ **3.** $\frac{7}{24} + \frac{13}{24}$ **4.** $\frac{8}{11} + \frac{3}{11}$

5. $2\frac{1}{3} + 3\frac{1}{3}$ **6.** $4\frac{5}{7} + 6\frac{3}{7}$ **7.** $\frac{3}{10} + \frac{2}{10}$ **8.** $\frac{8}{15} + \frac{9}{15}$

9. $\quad 5\frac{2}{9}$ **10.** $\quad 3\frac{2}{11}$ **11.** $\quad 2\frac{5}{18}$ **12.** $\frac{2}{13} + \frac{3}{13} + \frac{6}{13}$

$\quad +7\frac{4}{9}$ $\quad +4\frac{9}{11}$ $\quad 3\frac{7}{18}$

$\quad\quad\quad\quad\quad\quad\quad\quad\quad\quad +5\frac{11}{18}$

13. $\frac{8}{25} + \frac{9}{25} + \frac{7}{25} + \frac{11}{25}$

5. ADDING FRACTIONS WITH DIFFERENT DENOMINATORS

Fractions having different denominators are called **unlike fractions**. Suppose we wish to add two unlike fractions such as $\frac{1}{2}$ and $\frac{3}{5}$. We must first change these two fractions to equivalent fractions with the same denominator. This **least common denominator** (L.C.D.) is the smallest number that both denominators divide into evenly. In this case, the L.C.D. is 10 (2 × 5), so that

$$\frac{1}{2} = \frac{1 \times 5}{2 \times 5} = \frac{5}{10} \text{ and } \frac{3}{5} = \frac{3 \times 2}{5 \times 2} = \frac{6}{10}$$

Thus, $\frac{1}{2} + \frac{3}{5} = \frac{5}{10} + \frac{6}{10} = \frac{11}{10} = 1\frac{1}{10}$.

RULE: To add unlike fractions, change the fractions to equivalent fractions with the same denominator, known as the *common denominator*.

ILLUSTRATIVE PROBLEM 1: Add $\frac{1}{2}$ and $\frac{2}{3}$.

Solution: The L.C.D. is 6.

$$\frac{1}{2} = \frac{1 \times 3}{2 \times 3} = \frac{3}{6}$$

$$\frac{2}{3} = \frac{2 \times 2}{3 \times 2} = \frac{4}{6}$$

$$\frac{3}{6} + \frac{4}{6} = \frac{7}{6} = 1\frac{1}{6} \quad (answer)$$

Note that, if we had used 12 for the common denominator, then:

$$\frac{1}{2} = \frac{1 \times 6}{2 \times 6} = \frac{6}{12} \text{ and } \frac{2}{3} = \frac{2 \times 4}{3 \times 4} = \frac{8}{12}$$

$$\frac{6}{12} + \frac{8}{12} = \frac{14}{12} = \frac{14 \div 2}{12 \div 2} = \frac{7}{6} = 1\frac{1}{6}$$

The result is the same, but it is simpler to use the *least* common denominator (L.C.D.).

When there are two or more fractions, it is sometimes difficult to find the L.C.D. One procedure is to take the largest denominator and consider its multiples. That is, multiply the denominator by 2, 3, 4, 5, 6, etc., until a product is found that is divisible by all denominators.

ILLUSTRATIVE PROBLEM 2: $\frac{1}{8} + \frac{1}{4} + \frac{2}{5} = ?$

Solution: Consider the multiples of 8:

$$8 \times 2 = 16 \quad 8 \times 3 = 24 \quad 8 \times 4 = 32 \quad 8 \times 5 = 40$$

The first product we reach that is also divisible by 4 and 5 is 40. This is the L.C.D. Thus,

$$\frac{1}{8} = \frac{1 \times 5}{8 \times 5} = \frac{5}{40}$$

$$\frac{1}{4} = \frac{1 \times 10}{4 \times 10} = \frac{10}{40}$$

$$\frac{2}{5} = \frac{2 \times 8}{5 \times 8} = \frac{16}{40}$$

$$\frac{31}{40} \quad (answer)$$

To add mixed numbers with unlike fractions, we may arrange the work as shown in Illustrative Problem 3.

ILLUSTRATIVE PROBLEM 3: $3\frac{3}{4} + 5\frac{2}{3} = ?$

Solution:

$$3 \,\bigg|\, \frac{3}{4} = \frac{9}{12}$$

$$5 \,\bigg|\, \frac{2}{3} = \frac{8}{12}$$

$$8 \,\bigg|\, \quad \frac{17}{12} = 1\frac{5}{12}$$

$$+1\frac{5}{12} \longleftarrow$$

$$9\frac{5}{12} \quad (answer)$$

Here, we are adding the fractional parts of the mixed numbers and then adding this sum to the sum of the whole numbers.

Exercises

In 1 to 15, add the fractions and reduce each answer to *simplest form* (lowest terms for proper fractions; change improper fractions to whole numbers or mixed numbers).

1. $\frac{1}{3} + \frac{1}{6}$

2. $\frac{1}{2} + \frac{1}{3} + \frac{1}{4}$

3. $\frac{5}{12} + \frac{2}{3}$

4. $\frac{1}{6} + \frac{3}{10} + \frac{2}{5}$

5. $\frac{5}{9} + \frac{2}{3} + \frac{1}{6}$

6. $\frac{3}{4} + \frac{7}{16}$

7. $4\frac{5}{8}$
   $3\frac{1}{2}$
   $5\frac{1}{4}$

8. $2\frac{1}{7}$
   $1\frac{3}{4}$

9. $3\frac{5}{8}$
   $7\frac{3}{20}$

10. $4\frac{3}{4}$
    $6\frac{5}{12}$

11. $\frac{1}{2}$
    $7\frac{11}{12}$

12. $12\frac{4}{5}$
    $11\frac{3}{4}$

13. $18\frac{1}{6}$
    $9\frac{2}{3}$
    $5\frac{1}{2}$

14. $5\frac{3}{4}$
    $7\frac{3}{8}$
    $2\frac{5}{16}$

15. $9\frac{7}{8}$
    $4\frac{5}{12}$
    $8\frac{1}{6}$

16. A girl needs $5\frac{1}{3}$ yards of pink ribbon, $4\frac{3}{4}$ yards of blue ribbon, and $3\frac{1}{2}$ yards of yellow ribbon. How many yards of ribbon does she need in all?

17. A carpenter adds an extension of $22\frac{7}{8}$ inches to a shelf that is $35\frac{3}{4}$ inches long. How long is the new shelf?

18. What is the total weight of a can of soup if the soup weighs $12\frac{2}{3}$ ounces and the can weighs $4\frac{4}{5}$ ounces?

6. SUBTRACTING FRACTIONS

The subtraction of fractions is similar to the addition of fractions. Thus, $\frac{5}{7} - \frac{2}{7} = \frac{3}{7}$.

RULE: To subtract like fractions, subtract the numerators and write the difference over the common denominator.

To subtract unlike fractions, we change each fraction to an equivalent fraction with the same common denominator.

ILLUSTRATIVE PROBLEM 1: $\frac{7}{8} - \frac{2}{5} = ?$

Solution: The L.C.D. is 40. Thus,

$$\frac{7}{8} = \frac{7 \times 5}{8 \times 5} = \frac{35}{40}$$

$$\frac{2}{5} = \frac{2 \times 8}{5 \times 8} = \frac{16}{40}$$

$$\frac{7}{8} - \frac{2}{5} = \frac{35}{40} - \frac{16}{40} = \frac{35 - 16}{40} = \frac{19}{40} \quad (answer)$$

When we subtract mixed numbers, we proceed, as in the addition of mixed numbers, to change the fractional parts to equivalent fractions with the same L.C.D.

ILLUSTRATIVE PROBLEM 2: $6\frac{3}{4} - 1\frac{2}{3} = ?$

Solution: The following form is suggested:

$$
\begin{array}{r|ll}
6 & \dfrac{3}{4} & = \dfrac{9}{12} \\[2ex]
-1 & \dfrac{2}{3} & = \dfrac{8}{12} \\[2ex]
\hline
5 & & \dfrac{1}{12}
\end{array}
$$

Answer: The difference is $5\frac{1}{12}$.

RULE: To subtract two mixed numbers, change the fractional parts to equivalent fractions with the same L.C.D., subtract the like fractions, and then subtract the whole numbers.

An added problem arises when the upper fraction is less than the lower fraction. In this case, we overcome the difficulty by borrowing 1 from the whole number in the upper mixed number and adding it to the upper fraction to make it improper.

ILLUSTRATIVE PROBLEM 3: $5\frac{2}{3} - 2\frac{11}{12} = ?$

Solution:

$$
\begin{array}{r|ll}
5 & \dfrac{2}{3} & = \dfrac{8}{12} \\[2ex]
-2 & \dfrac{11}{12} & = \dfrac{11}{12} \\[2ex]
\hline
& &
\end{array}
$$

Now, since we cannot subtract $\frac{11}{12}$ from $\frac{8}{12}$, we borrow 1 from the 5, change the 1 to $\frac{12}{12}$, and add $\frac{12}{12}$ to $\frac{8}{12}$. Thus,

$$
\begin{array}{r|l}
\overset{4}{\cancel{5}} & \dfrac{8}{12} + \dfrac{12}{12} = \dfrac{20}{12} \\[3ex]
-2 & \dfrac{11}{12} \qquad\; = \dfrac{11}{12} \\[3ex]
\hline
2 & \qquad\qquad\quad \dfrac{9}{12}
\end{array}
$$

Answer: The difference is $2\frac{9}{12} = 2\frac{3}{4}$.

ILLUSTRATIVE PROBLEM 4: $\frac{3}{4} + 1\frac{1}{2} - \frac{5}{8} = ?$

Solution:
$$\frac{3}{4} + 1\frac{1}{2} - \frac{5}{8} = \frac{3}{4} + \frac{3}{2} - \frac{5}{8}$$

The L.C.D. is 8. Therefore,

$$\frac{3}{4} = \frac{3 \times 2}{4 \times 2} = \frac{6}{8}$$

$$\frac{3}{2} = \frac{3 \times 4}{2 \times 4} = \frac{12}{8}$$

Thus, $\dfrac{3}{4} + \dfrac{3}{2} - \dfrac{5}{8} = \dfrac{6}{8} + \dfrac{12}{8} - \dfrac{5}{8}$

$$= \frac{18}{8} - \frac{5}{8}$$

$$= \frac{13}{8} \text{ or } 1\frac{5}{8} \quad (answer)$$

Exercises _____

In 1 to 15, subtract. Reduce the answer to lowest terms.

1. $\frac{6}{7} - \frac{2}{7}$ **2.** $\frac{7}{10} - \frac{2}{10}$ **3.** $\frac{7}{8} - \frac{1}{8}$

4. $\quad 6\frac{2}{3}$ **5.** $\quad 6\frac{3}{7}$ **6.** $\quad 9\frac{5}{8}$
$\quad\; -2\frac{1}{3}$ $\quad\; -4\frac{5}{7}$ $\quad\; -4\frac{7}{16}$

7. $\frac{5}{8} - \frac{1}{3}$ **8.** $\frac{7}{8} - \frac{2}{5}$ **9.** $\frac{7}{10} - \frac{1}{3}$ **10.** $\frac{8}{9} - \frac{1}{6}$ **11.** $5 - 2\frac{2}{3}$

12. $4\frac{3}{8} - \frac{5}{8}$ **13.** $\;5\frac{1}{2}$ **14.** $\quad 7\frac{5}{6}$ **15.** $\quad 17$
$\qquad\qquad\quad\; -\frac{3}{4}$ $\quad\;\; -2\frac{4}{9}$ $\quad\; -5\frac{3}{7}$

In 16 to 19, do the indicated adding and subtracting. Reduce the answers to lowest terms.

16. $6 + 3\frac{5}{8} - 2\frac{1}{4}$ **17.** $5\frac{1}{3} + 2\frac{5}{9} - 4\frac{5}{6}$

18. $7\frac{1}{4} + 3\frac{1}{8} - 4\frac{2}{3}$ **19.** $\frac{7}{8} - \frac{3}{16} + \frac{1}{4}$

20. Mrs. Aquino borrowed $\frac{3}{4}$ of a cup of flour and returned $\frac{1}{2}$ of a cup. How much does she still owe?

21. A boy had a board $6\frac{5}{8}$ feet long. He cut off a piece $3\frac{2}{3}$ feet long. How many feet were left?

22. A roast beef weighed $8\frac{2}{5}$ pounds. In the roasting, it lost $2\frac{3}{4}$ pounds. How many pounds remained?

23. Bill lives $4\frac{5}{8}$ miles from school. Tom lives $2\frac{1}{3}$ miles from school along the same straight road as Bill. How much nearer to school does Tom live?

24. A girl has $8\frac{1}{2}$ yards of ribbon and cuts off $3\frac{4}{5}$ yards. How many yards are left?

25. The Dugans bought $3\frac{1}{2}$ tons of coal in the fall and used $1\frac{5}{16}$ tons over the winter. How many tons were left?

---

7. MULTIPLYING FRACTIONS

Just as the product 4×3 means 3 added to itself 4 times, the product $4 \times \frac{2}{3}$ means the fraction $\frac{2}{3}$ added to itself 4 times. Thus,

$$4 \times \frac{2}{3} = \frac{2}{3} + \frac{2}{3} + \frac{2}{3} + \frac{2}{3} = \frac{8}{3} = 2\frac{2}{3}.$$

The same result can be obtained by multiplying the whole number by the numerator of the fraction:

$$4 \times \frac{2}{3} = \frac{4}{1} \times \frac{2}{3} = \frac{4 \times 2}{1 \times 3} = \frac{8}{3} = 2\frac{2}{3}$$

When we multiply whole numbers, the order of multiplication does not matter. Thus, $5 \times 7 = 7 \times 5$. This law of order is known as the **commutative law** or **principle,** and we apply it to fractions as well. Thus,

$$4 \times \frac{2}{3} = \frac{2}{3} \times 4 = \frac{2 \times 4}{3 \times 1} = \frac{8}{3} = 2\frac{2}{3}$$

Note that, in the preceding example, we were able to obtain the product of the two fractions by multiplying the two numerators and multiplying the two denominators to give us the numerator and denominator of the product. It can be shown that this gives us a method for multiplying any two or more fractions.

ILLUSTRATIVE PROBLEM 1: $\frac{3}{4} \times \frac{5}{7} = ?$

Solution:
$$\frac{3}{4} \times \frac{5}{7} = \frac{3 \times 5}{4 \times 7} = \frac{15}{28} \quad (answer)$$

ILLUSTRATIVE PROBLEM 2: $2\frac{1}{2} \times 3\frac{2}{3} = ?$

Solution: Change the mixed numbers to fractions.
$$\frac{5}{2} \times \frac{11}{3} = \frac{55}{6} = 9\frac{1}{6} \quad (answer)$$

As usual, we reduce fractional results to lowest terms. To avoid working with larger numbers, however, it is possible to start this reduction *before* multiplying by dividing out ("cancelling") common factors in numerators and denominators. This process is illustrated in the following problem.

ILLUSTRATIVE PROBLEM 3: $\frac{2}{3}$ of $\frac{3}{7} = ?$

Solution:
$$\frac{2}{3} \times \frac{3}{7} = \frac{6}{21} = \frac{6 \div 3}{21 \div 3} = \frac{2}{7} \quad (answer)$$

Note that we divide the numerator and denominator of $\frac{6}{21}$ by 3, so as to reduce the fraction. We could have done this at the very beginning, like this:

$$\frac{2}{\overset{}{\underset{1}{\cancel{3}}}} \times \frac{\overset{1}{\cancel{3}}}{7} = \frac{2 \times 1}{1 \times 7} = \frac{2}{7}$$

Whenever any number divides evenly into both a numerator and denominator, you may simplify by dividing before multiplying the numerators and denominators. This method is further illustrated in the following problem.

ILLUSTRATIVE PROBLEM 4: $\frac{3}{4} \times \frac{8}{9} = ?$

Solution:
$$\frac{\overset{1}{\cancel{3}}}{\underset{1}{\cancel{4}}} \times \frac{\overset{2}{\cancel{8}}}{\underset{3}{\cancel{9}}} = \frac{2}{3} \quad (answer)$$

Here, we have divided the denominator of the first fraction and the numerator of the second fraction by 4. We have also divided the numerator of the first fraction and the denominator of the second by 3. Thus, the product is already in lowest terms. The same procedure can be applied to three or more fractions.

ILLUSTRATIVE PROBLEM 5: $\frac{1}{3} \times \frac{2}{5} \times \frac{3}{2} = ?$

Solution: $\quad\quad \dfrac{1}{\underset{1}{\cancel{3}}} \times \dfrac{\overset{1}{\cancel{2}}}{5} \times \dfrac{\overset{1}{\cancel{3}}}{\underset{1}{\cancel{2}}} = \dfrac{1}{5}$ *(answer)*

ILLUSTRATIVE PROBLEM 6: $\frac{3}{4} \times \frac{8}{5} \times \frac{15}{16} = ?$

Solution: $\quad\quad \dfrac{3}{4} \times \dfrac{\overset{1}{\cancel{8}}}{\underset{1}{\cancel{5}}} \times \dfrac{\overset{3}{\cancel{15}}}{\underset{2}{\cancel{16}}} = \dfrac{9}{8} = 1\frac{1}{8}$ *(answer)*

Exercises

In 1 to 16, find the products. Reduce to lowest terms by cancellation wherever possible.

1. $12 \times \frac{5}{8}$ **2.** $\frac{4}{9} \times 5$ **3.** $\frac{1}{5} \times \frac{2}{3}$ **4.** $\frac{1}{2} \times \frac{2}{5} \times \frac{1}{3}$

5. $6 \times \frac{3}{4}$ **6.** $\frac{1}{6} \times \frac{4}{3}$ **7.** $\frac{2}{3} \times 6$ **8.** $3\frac{1}{3} \times 10\frac{1}{2}$

9. $12 \times 1\frac{5}{6}$ **10.** $\frac{2}{5} \times 8\frac{3}{4}$ **11.** $\frac{3}{5} \times \frac{7}{12}$ **12.** $4\frac{1}{2} \times \frac{5}{8}$

13. $3\frac{1}{3} \times 2\frac{1}{4}$ **14.** $\frac{1}{3}$ of $\frac{6}{7}$ **15.** $\frac{2}{5}$ of $\frac{1}{2}$ **16.** $24 \times 6\frac{2}{3}$

Note: An alternate solution to exercise 16 is to multiply 24 by 6, multiply 24 by $\frac{2}{3}$, and then add the two partial products, as shown at the right.

$$
\begin{array}{r}
24 \\
\times 6\frac{2}{3} \\
\hline
144 \longleftarrow (24 \times 6 = 144) \\
+16 \longleftarrow \left(\dfrac{\overset{8}{\cancel{24}}}{1} \times \dfrac{2}{\underset{1}{\cancel{3}}} = 16 \right) \\
\hline
160 \quad (answer)
\end{array}
$$

This method is convenient when the whole number is a large number. Use this method to find the products in 17 to 19.

17. $48 \times 3\frac{5}{8}$ **18.** $45 \times 4\frac{2}{5}$ **19.** $54 \times 12\frac{5}{9}$

20. How many feet of lumber are needed to make 15 shelves each $5\frac{2}{3}$ feet long?

21. Tom rides his bike at the rate of $6\frac{1}{4}$ miles per hour. How many miles does he travel in $2\frac{1}{5}$ hours?

22. An inch is $\frac{1}{12}$ of a foot. What fractional part of a foot is $\frac{3}{5}$ of an inch?

23. A *furlong* is a distance that is about $\frac{1}{8}$ of a mile. What fractional part of a mile is:

    a. 2 furlongs?   b. $\frac{2}{3}$ of a furlong?

24. What is the total weight of a dozen chickens, each weighing $4\frac{3}{4}$ pounds?

25. Find the product: $\frac{2}{3} \times \frac{20}{8} \times \frac{9}{5}$

---

8. DIVIDING FRACTIONS

How many $\frac{1}{4}$-inch line segments are there in $2\frac{1}{2}$ inches? This problem is equivalent to dividing $2\frac{1}{2}$ by $\frac{1}{4}$.

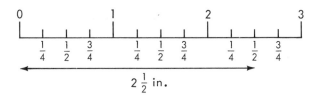

If we look at the $2\frac{1}{2}$-inch line segment in the figure above and count the quarter-inch segments that make it up, we see that there are 10 such smaller segments. Thus, $2\frac{1}{2} \div \frac{1}{4} = 10$ or $\frac{5}{2} \div \frac{1}{4} = 10$. Note that this result can also be obtained by **inverting** the divisor (turning the second fraction upside down) and multiplying the result by the dividend (the first fraction). Thus,

$$\frac{5}{2} \div \frac{1}{4} = \frac{5}{2} \times \frac{4}{1} = \frac{5}{\underset{1}{\cancel{2}}} \times \frac{\overset{2}{\cancel{4}}}{1} = \frac{10}{1} = 10$$

When we invert a fraction, we call the resulting fraction the **reciprocal** of the original fraction. The reciprocal of $\frac{2}{3}$ is $\frac{3}{2}$; the reciprocal of $\frac{5}{1}$ is $\frac{1}{5}$. Note that the product of any fraction and its reciprocal is always 1. For example, $\frac{2}{3} \times \frac{3}{2} = 1$ and $\frac{5}{1} \times \frac{1}{5} = 1$.

We may now state the following rule:

RULE: To divide one fraction by another, *invert* the divisor (the fraction after the division sign) and *multiply*.

Thus, division by a fraction is equivalent to multiplication by the reciprocal of the fraction.

ILLUSTRATIVE PROBLEM 1: Divide 18 by $\frac{3}{5}$.

Solution: $\dfrac{18}{1} \div \dfrac{3}{5} = \dfrac{\overset{6}{\cancel{18}}}{1} \times \dfrac{5}{\underset{1}{\cancel{3}}} = \dfrac{30}{1} = 30$ *(answer)*

ILLUSTRATIVE PROBLEM 2: $\frac{9}{16} \div \frac{3}{10} = ?$

Solution: $\dfrac{9}{16} \div \dfrac{3}{10} = \dfrac{\overset{3}{\cancel{9}}}{\underset{8}{\cancel{16}}} \times \dfrac{\overset{5}{\cancel{10}}}{\underset{1}{\cancel{3}}} = \dfrac{15}{8} = 1\dfrac{7}{8}$ *(answer)*

ILLUSTRATIVE PROBLEM 3: $10 \div 5\frac{3}{4} = ?$

Solution: $\dfrac{10}{1} \div \dfrac{23}{4} = \dfrac{10}{1} \times \dfrac{4}{23} = \dfrac{40}{23} = 1\dfrac{17}{23}$ *(answer)*

ILLUSTRATIVE PROBLEM 4: $2\frac{2}{5} \div 8 = ?$

Solution: $\dfrac{12}{5} \div \dfrac{8}{1} = \dfrac{\overset{3}{\cancel{12}}}{5} \times \dfrac{1}{\underset{2}{\cancel{8}}} = \dfrac{3}{10}$ *(answer)*

Exercises

In 1 to 9, divide.

1. $\frac{3}{8} \div \frac{3}{2}$ 2. $\frac{1}{3} \div \frac{4}{6}$ 3. $14 \div \frac{7}{10}$

4. $3\frac{3}{4} \div \frac{5}{8}$ 5. $\frac{3}{20} \div 6$ 6. $4\frac{2}{3} \div 2\frac{4}{5}$

7. $2\frac{1}{6} \div 2\frac{2}{3}$ 8. $8 \div \frac{4}{9}$ 9. $\frac{4}{9} \div \frac{1}{6}$

10. A $3\frac{3}{4}$-foot length of molding is cut into $\frac{3}{4}$-foot strips. How many strips are there?

11. A woman drove 90 miles in $2\frac{1}{4}$ hours. Assuming a constant rate, how many miles did she drive in 1 hour?

12. A board $9\frac{3}{4}$ feet long is cut into 13 equal pieces. How many feet long is each such piece?

13. If a suit requires $4\frac{1}{5}$ yards of fabric, how many suits can be made with 42 yards of fabric?

14. A man pours $4\frac{1}{2}$ quarts of milk into containers holding $1\frac{1}{2}$ quarts each. How many containers does he fill?

---

9. FINDING FRACTIONAL PARTS

In a class of 30 students, 25 are present. What fractional part of the class is present? Here, we are asking the question "25 is what fractional part of 30?" We can write a fraction with the *total amount* as the *denominator* and the *part* as the *numerator*. Thus,

$$\frac{25}{30} = \frac{25 \div 5}{30 \div 5} = \frac{5}{6}$$

$\frac{5}{6}$ of the class is present.

Check: Does $\frac{5}{6}$ of 30 equal 25?

$$\frac{5}{\overset{}{\underset{1}{\cancel{6}}}} \times \frac{\overset{5}{\cancel{30}}}{1} = \frac{25}{1} = 25 \ \checkmark$$

A useful memory device here is to remember that the number going with the word "is" becomes the numerator and the number going with the word "of" becomes the denominator.

ILLUSTRATIVE PROBLEM 1: 18 is what part of 24?

Solution: $\dfrac{\text{is}}{\text{of}} = \dfrac{18}{24} = \dfrac{18 \div 6}{24 \div 6} = \dfrac{3}{4}$ *(answer)*

ILLUSTRATIVE PROBLEM 2: What part of 80 is 50?

Solution: $\dfrac{\text{is}}{\text{of}} = \dfrac{50}{80} = \dfrac{50 \div 10}{80 \div 10} = \dfrac{5}{8}$ *(answer)*

Exercises

1. 7 is what part of 28?

2. 18 is what part of 20?

3. 16 is what part of 20?

4. What part of 90 is 60?

5. What part of 100 is 73? **6.** What part of 72 is 30?

7. What part of 23 is 18? **8.** What part of 150 is 100?

9. In a class of 28 students, 8 receive A's. What part of the class receives A?

10. In a spelling test of 25 words, Nancy spells 20 correctly. What part of all the words does Nancy spell correctly?

11. In traveling 800 miles, a man does 600 miles by train and 200 miles by car.
  a. What part of the trip does he do by train?
  b. What part of the trip does he drive?

12. In a class of 32 students, 6 are absent.
  a. What part of the class is absent?
  b. What part of the class is present?

---

10. FINDING A NUMBER WHEN GIVEN A FRACTIONAL PART OF IT

ILLUSTRATIVE PROBLEM: $\frac{5}{8}$ of what number is 60?

Solution: In problems of this type, it is convenient to use the equation method we learned in Chapter 3.

Let n = the number.

Then $\frac{5}{8}$ of n = 60, or $\frac{5}{8}n = 60$.

Now, divide both sides of the equation by $\frac{5}{8}$.

$$\frac{\frac{5}{8}n}{\frac{5}{8}} = \frac{60}{\frac{5}{8}}$$

$$n = 60 \div \frac{5}{8} \qquad \textit{Check: Does } \tfrac{5}{8} \text{ of 96 equal 60?}$$

$$n = \frac{\overset{12}{\cancel{60}}}{1} \times \frac{8}{\cancel{5}}_{1} \qquad \frac{5}{\cancel{8}_1} \times \frac{\overset{12}{\cancel{96}}}{1} = \frac{60}{1} = 60 \ \checkmark$$

$$n = 96 \quad (answer)$$

Alternate Solution: Another way of solving this fractional equation is to multiply both sides by the reciprocal of $\frac{5}{8}$, which is $\frac{8}{5}$. Remember that the product of a number and its reciprocal is 1. Thus,

$$\frac{5}{8}n = 60$$

$$\overset{1}{\underset{1}{\cancel{\frac{8}{5}}}} \times \overset{1}{\underset{1}{\cancel{\frac{5}{8}}}}n = \frac{\overset{12}{\cancel{60}}}{1} \times \underset{1}{\cancel{\frac{8}{5}}}$$

$$n = 12 \times 8 = 96 \quad (answer)$$

Exercises

In 1 to 6, solve each equation.

1. $\frac{3}{4}x = 30$ **2.** $\frac{1}{8}y = 14$ **3.** $\frac{7}{5}n = 140$

4. $180 = \frac{6}{7}p$ **5.** $35 = \frac{5}{8}x$ **6.** $80 = \frac{2}{3}s$

7. If $\frac{3}{16}$ of a number is 24, find the number.

8. 31 is $\frac{1}{5}$ of what number?

9. Jane gets $\frac{4}{5}$ of the problems right on a test. If she gets 24 right, how many problems were there on the test?

10. Tom scored 27 points in a basketball game. This was $\frac{3}{8}$ of the total scored by the team. How many points did the team make?

11. Jim delivered $\frac{5}{6}$ of the newspapers he received. If he delivered 40 newspapers, how many did he receive?

12. Solve the equation: $\frac{3}{4}x = \frac{3}{2}$

---

11. SIMPLIFYING COMPLEX FRACTIONS

A *complex fraction* is one that has a fraction in one or both of its terms. Thus,

$$\frac{\frac{3}{5}}{2}, \quad \frac{4}{\frac{5}{8}}, \quad \text{and} \quad \frac{3\frac{1}{3}}{2\frac{1}{2}}$$

are all complex fractions. These are easily simplified if we remember that the fraction line is equivalent to a division sign. Thus,

$$\frac{\dfrac{3}{5}}{\dfrac{3}{4}} \text{ may be written } \frac{3}{5} \div \frac{3}{4} = \frac{\overset{1}{\cancel{3}}}{5} \times \frac{4}{\underset{1}{\cancel{3}}} = \frac{4}{5}$$

RULE: To simplify a complex fraction, rewrite it as the quotient of two fractions, using a division sign in place of the original fraction sign. Then proceed as in the division of fractions.

ILLUSTRATIVE PROBLEM: Simplify the following complex fraction: $\dfrac{3\frac{3}{4}}{7\frac{1}{2}}$

Solution:

$3\dfrac{3}{4} \div 7\dfrac{1}{2}$ (rewrite problem using division sign)

$= \dfrac{15}{4} \div \dfrac{15}{2}$ (change mixed numbers to improper fractions)

$= \dfrac{15}{4} \times \dfrac{2}{15}$ (change the division sign to a multiplication sign and invert the divisor)

$= \dfrac{\overset{1}{\cancel{15}}}{\underset{2}{\cancel{4}}} \times \dfrac{\overset{1}{\cancel{2}}}{\underset{1}{\cancel{15}}}$ (cancel common factors and multiply remaining numerators and remaining denominators)

$= \dfrac{1}{2}$ (*answer*)

Exercises

In 1 to 12, simplify each complex fraction.

1. $\dfrac{\frac{1}{7}}{\frac{7}{8}}$

2. $\dfrac{\frac{1}{3}}{8}$

3. $\dfrac{\frac{4}{5}}{\frac{5}{8}}$

4. $\dfrac{3\frac{1}{2}}{7}$

5. $\dfrac{3\frac{1}{3}}{2\frac{1}{2}}$

6. $\dfrac{4\frac{2}{5}}{7\frac{1}{3}}$

7. $\dfrac{3\frac{2}{3}}{2\frac{2}{5}}$

8. $\dfrac{16\frac{2}{3}}{100}$

9. $\dfrac{\dfrac{7}{8}}{\dfrac{5}{16}}$ 10. $\dfrac{62\dfrac{1}{2}}{100}$ 11. $\dfrac{1\dfrac{2}{3}}{4\dfrac{1}{3}}$ 12. $\dfrac{83\dfrac{1}{3}}{50}$

12 . COMPARING FRACTIONS

Ben lives $\frac{3}{4}$ mile from school and Eli lives $\frac{7}{8}$ mile from the same school. Which one lives farther from the school?

It is difficult to compare the values of the two fractions as they stand. If we change the two fractions to *equivalent* fractions that have the *same denominator*, however, the comparison becomes more apparent.

Thus, $\frac{3}{4} = \frac{3 \times 2}{4 \times 2} = \frac{6}{8}$ and it is obvious that $\frac{7}{8}$ is larger than $\frac{6}{8}$. Therefore, Eli lives farther from the school than Ben.

The basic principle involved here is that, if fractions have the same denominator, the fraction with the largest numerator has the largest value.

PRINCIPLE: To compare the values of two or more fractions:
1. Change the fractions to equivalent fractions with the same denominator.
2. Compare the numerators. The fraction with the largest numerator has the largest value.

ILLUSTRATIVE PROBLEM 1: Which fraction is larger, $\frac{3}{5}$ or $\frac{7}{10}$?

Solution:
$$\frac{3}{5} = \frac{3 \times 2}{5 \times 2} = \frac{6}{10}$$

Answer: Since 7 is larger than 6, it follows that $\frac{7}{10}$ is larger than $\frac{6}{10}$ and, therefore, $\frac{7}{10}$ is larger than $\frac{3}{5}$.

ILLUSTRATIVE PROBLEM 2: Which fraction is smaller, $\frac{5}{8}$ or $\frac{2}{3}$?

Solution: Change both fractions to equivalent fractions with 24 as the common denominator.
$$\frac{5}{8} = \frac{5 \times 3}{8 \times 3} = \frac{15}{24} \qquad \frac{2}{3} = \frac{2 \times 8}{3 \times 8} = \frac{16}{24}$$

Answer: Since $\frac{15}{24}$ is smaller than $\frac{16}{24}$, it follows that $\frac{5}{8}$ is smaller than $\frac{2}{3}$.

ILLUSTRATIVE PROBLEM 3: Arrange the following fractions in order of increasing size: $\frac{1}{3}, \frac{2}{5}, \frac{3}{10}$

Solution: Change each fraction to an equivalent fraction whose denominator is 30.

$$\frac{1}{3} = \frac{1 \times 10}{3 \times 10} = \frac{10}{30} \qquad \frac{2}{5} = \frac{2 \times 6}{5 \times 6} = \frac{12}{30} \qquad \frac{3}{10} = \frac{3 \times 3}{10 \times 3} = \frac{9}{30}$$

Order of increasing size is $\frac{9}{30}, \frac{10}{30}, \frac{12}{30}$.

Answer: $\frac{3}{10}, \frac{1}{3}, \frac{2}{5}$ is the desired order.

Exercises

In 1 to 8, which is the larger fraction?

1. $\frac{5}{9}$ or $\frac{7}{9}$ **2.** $\frac{3}{7}$ or $\frac{4}{7}$ **3.** $\frac{11}{8}$ or $\frac{9}{8}$ **4.** $\frac{17}{21}$ or $\frac{19}{21}$

5. $\frac{7}{8}$ or $\frac{4}{5}$ **6.** $\frac{2}{5}$ or $\frac{3}{10}$ **7.** $\frac{3}{7}$ or $\frac{7}{3}$ **8.** $\frac{5}{12}$ or $\frac{6}{11}$

In 9 to 16, which is the smaller fraction?

9. $\frac{5}{8}$ or $\frac{7}{8}$ **10.** $\frac{8}{9}$ or $\frac{7}{9}$ **11.** $\frac{2}{3}$ or $\frac{3}{4}$ **12.** $\frac{7}{10}$ or $\frac{5}{9}$

13. $\frac{9}{5}$ or $\frac{7}{6}$ **14.** $\frac{5}{6}$ or $\frac{5}{8}$ **15.** $\frac{7}{12}$ or $\frac{8}{11}$ **16.** $\frac{3}{5}$ or $\frac{2}{3}$

In 17 to 19, arrange each set of fractions in order of *increasing* size.

17. $\frac{1}{2}, \frac{1}{3},$ and $\frac{1}{4}$ **18.** $\frac{3}{5}, \frac{7}{10},$ and $\frac{3}{4}$ **19.** $\frac{5}{9}, \frac{5}{6},$ and $\frac{2}{3}$

20. Mr. Albert owns $\frac{3}{4}$ of an acre of land. Mrs. Stone owns $\frac{7}{10}$ of an acre of land. Who owns more land?

21. Walter runs the 100-yard dash in $9\frac{3}{5}$ seconds. Kerry runs the same distance in $9\frac{5}{8}$ seconds. Which one runs the dash in less time?

22. Marge has $\frac{7}{8}$ of a yard of ribbon. Eloise has $\frac{11}{12}$ of a yard of ribbon. Who has more ribbon?

23. One bank pays $6\frac{5}{6}\%$ interest and another pays $6\frac{7}{8}\%$ interest. Which rate of interest is higher?

24. Doris mows $\frac{5}{12}$ of a lawn and Randy mows $\frac{3}{7}$ of the same lawn. Who has mowed more of the lawn?

25. Arrange in order of *decreasing* size: $\frac{5}{8}, \frac{3}{4},$ and $\frac{7}{12}$.

Chapter Review Exercises

In 1 to 12, perform the operation indicated.

1. $\frac{2}{3} \times \frac{1}{2}$ **2.** $\frac{3}{4} - \frac{1}{3}$ **3.** $1\frac{1}{4} + 3\frac{2}{3}$ **4.** $6\frac{1}{5}$
$-3\frac{5}{6}$

5. Add: $6\frac{1}{2}$ **6.** $\frac{3}{8} \div \frac{2}{3}$ **7.** $\frac{5}{6} \times 1\frac{2}{3}$ **8.** $3\frac{1}{8} \div 2\frac{1}{2}$
$5\frac{1}{3}$
$7\frac{3}{8}$

9. $\frac{3}{4} \times 1\frac{1}{8}$ **10.** $2\frac{3}{4} \times 3\frac{1}{7}$ **11.** $5 \div \frac{5}{3}$ **12.** $8 - 1\frac{5}{6}$

13. 25 is what part of 35?   **14.** $4\frac{1}{2} + 1\frac{2}{3} - 3\frac{1}{4} = ?$

15. 65 is $\frac{5}{8}$ of what number?   **16.** 21 is $\frac{3}{4}$ of what number?

17. How many $\frac{3}{4}$-pound bags can be filled from 15 pounds of sugar?

18. A carpenter planes down the thickness of a board from $\frac{7}{8}$ of an inch to $\frac{11}{16}$ of an inch. What was the thickness of the wood removed?

19. Find the sum of $3\frac{3}{32}$, $8\frac{1}{16}$, and $7\frac{5}{64}$.

20. A drainpipe can empty $\frac{1}{3}$ of an oil tank every hour. Another pipe can empty $\frac{1}{2}$ of the tank in the same time.

a. What part of the tank will both pipes empty in an hour?

b. What part will remain in the tank?

21. How much greater is a wire $\frac{3}{8}$ of an inch in diameter than one $\frac{5}{16}$ of an inch in diameter?

22. How many sheets of tin, each $\frac{1}{32}$ of an inch thick, are there in a pile $12\frac{1}{4}$ inches high?

23. Simplify: $\dfrac{5\frac{5}{8}}{10}$

24. Find the average of $8\frac{3}{4}$, $7\frac{1}{2}$, and $4\frac{5}{8}$.

25. Solve the equation: $\frac{3}{4}t = \frac{7}{8}$

CHECK YOUR SKILLS

1. Divide $4\frac{1}{2}$ by $\frac{3}{8}$.   2. Subtract $1\frac{2}{3}$ from $6\frac{3}{4}$.

3. Find the cost of 12 pounds of beef at \$3.15 per pound.

4. What is the Fahrenheit temperature corresponding to 20° Celsius? (Use the formula $F = \frac{9}{5}C + 32$.)

5. What is the cost of 3500 pounds of coal at \$110 per ton? (1 ton = 2000 pounds.)

6. Solve for y: $4y - 9 = 3$.

7. Multiply $4\frac{1}{2}$ by $2\frac{2}{3}$.

8. The sum of three consecutive integers is 90. Find the smallest integer.

9. Write an algebraic expression to represent the total number of cents in q quarters and d dimes.

10. If 3 times a number is subtracted from 7 times the same number, the result is 12. What is the number?

5

Decimal Fractions and Simple Decimal Equations

1. MEANING OF DECIMAL FRACTIONS

A **decimal fraction** is simply a fraction whose denominator is 10, 100, 1000, 10,000, etc. Thus, $\frac{9}{10}$, $\frac{17}{100}$, and $\frac{375}{1000}$ are decimal fractions. Since we use a number system in base 10, we find that decimal fractions often make computation simpler than other fractions.

Instead of writing out the denominators in decimal fractions, we can write the numbers with a **decimal point**. Thus,

$$\frac{9}{10} \text{ becomes } .9 \qquad \frac{17}{100} \text{ becomes } .17 \qquad \frac{375}{1000} \text{ becomes } .375$$

When decimal fractions are written in the *shortened* form with the decimal point, they are usually called just **decimals**. A simple rule for writing a decimal fraction in the shortened form is the following:

RULE: Begin at the right-hand digit of the numerator, count off as many places to the left as there are zeros in the denominator, and place the decimal point to the left of the last digit counted.

Thus, applying the rule to the decimal fraction $\frac{16}{100}$, we start at the 6, count off two places to the left, and get .16. Similarly, $\frac{193}{1000}$ would become .193.

Note that, in some cases, we must write placeholding zeros to the left of the left-hand digit in the numerator.

$$\frac{78}{10,000} = .0078$$

Beginning with the digit 8, we count off four places to the left, adding two zeros as we count, and place the decimal point to the extreme left. Either form is read "seventy-eight ten-thousandths."

Each decimal place in a decimal is given a name. Thus, for the number .37589, we indicate the value of each digit as follows:

	TENTHS	HUNDREDTHS	THOUSANDTHS	TEN-THOUSANDTHS	HUNDRED-THOUSANDTHS
.	3	7	5	8	9

In the number .777, each digit is multiplied by $\frac{1}{10}$ as we move to the right. (The "*place value*" of each 7 is divided by 10.)

$$\text{The first 7 means 7 tenths} = \frac{7}{10} = .7$$

$$\text{The second 7 means 7 hundredths} = \frac{7}{100} = .07$$

$$\text{The third 7 means 7 thousandths} = \frac{7}{1000} = .007$$

If we add these fractions by changing to a common denominator, we obtain:

$$\frac{7}{10} = \frac{700}{1000} = .700$$

$$\frac{7}{100} = \frac{70}{1000} = .070$$

$$\frac{7}{1000} = \frac{7}{1000} = .007$$

$$\textit{Sum:} \quad \frac{777}{1000} = .777$$

Note that we obtain the sum in the shortened decimal form, .777, by simply adding the digit columns of the addends above it. This is one of the advantages of using the shortened decimal form.

From the preceding example, also note that the value of a decimal is not changed by placing zeros after the extreme right-hand digit. Thus, .7, .70, and .700 all have the same value. However, this is not true when zeros are placed between the decimal point and the extreme left-hand digit. Thus, .7, .07, and .007 have different values.

To read a decimal fraction in full, we read both its numerator and denominator. To read .513, we read "five hundred thirteen" (numerator) "thousandths" (denominator). The denominator is always 1 followed by as many zeros as there are decimal places.

A number such as 23.754, which is made up of a whole number and a decimal, is called a **mixed decimal**. Mixed decimals are read the same as mixed numbers. Thus, 112.73 is read "one hundred twelve *and* seventy-three hundredths."

In a number, when a decimal point is not shown, it is always considered to be to the right of the extreme right-hand digit. In the number 237, the decimal point is considered to be to the right of the 7.

Exercises

1. Write the following fractions as decimals:

   a. $\frac{27}{100}$ **b.** $\frac{7}{10}$ **c.** $\frac{64}{100}$ **d.** $\frac{6}{100}$ **e.** $\frac{76}{1000}$

   f. $\frac{9}{1000}$ **g.** $\frac{23}{1000}$ **h.** $\frac{17}{100}$ **i.** $\frac{47}{10,000}$

2. Write the following numbers in decimal form:
   a. seven hundredths
   b. six thousandths
   c. five hundred eighteen thousandths
   d. four hundred eleven ten-thousandths
   e. fifty-four ten-thousandths

3. Write each decimal in exercise **2** as a common fraction.

4. The circle shown has been divided into 10 equal parts. Write as a decimal the part of the circle that is shaded.

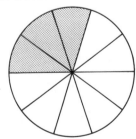

5. A centimeter ruler is shown. Each centimeter is divided into 10 equal parts. Write as a decimal the number of centimeters at *a, b, c,* and *d.*

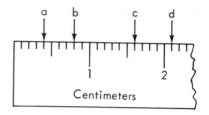

6. Read each of the following numbers:

  a. 30.2   **b.** 23.0   **c.** 27.18   **d.** 6.66

  e. 130.03   **f.** .0804   **g.** 3.007   **h.** .00008

7. In the number 92.09, how many times greater than the place value of the second 9 is the place value of the first 9?

8. Joseph is running a 1000-meter race. Express as a decimal the part of the distance run at the time Joseph has completed 675 meters.

9. A girl buys 100 envelopes in a box. She uses 23 of them. Write as a decimal the part of the box of envelopes she has left.

10. A typist receives a 1000-sheet package of typing paper. He uses 372 of the sheets. Write as a decimal the part of the package that remains.

2.   COMPARING DECIMALS

Which is larger: .79 or .8?

To compare two such decimals, we treat them as we do fractions; that is, we change them to equivalent fractions with the same denominators. Thus,

$$.79 = \frac{79}{100}$$

$$.8 = .80 = \frac{80}{100}$$

Since $\frac{80}{100}$ is greater than $\frac{79}{100}$, we see that .80 or .8 is greater than .79.

Here we can use the fact that writing zeros to the right of the extreme right-hand digit in a decimal does not change its value.

RULE: To compare decimals, change them to equivalent decimals by writing zeros to the right until all of them have the same number of decimal places. Then compare the resulting numbers.

ILLUSTRATIVE PROBLEM: Arrange the following numbers in order of size from smallest to largest:

$$.6 \quad .67 \quad .06 \quad .607$$

Solution: Change all the numbers to three-place decimals:

$$.6 = .600 \quad .67 = .670 \quad .06 = .060 \quad .607 = .607$$

Arrange the three-place decimals in order of size:

$$.060 \quad .600 \quad .607 \quad .670$$

$$\text{or} \quad .06 \quad .6 \quad .607 \quad .67 \quad (answer)$$

Exercises

1. Change each decimal to an equivalent fraction expressed as thousandths.
   a. .3   **b.** .9   **c.** .43   **d.** .05   **e.** .98

2. In each part, choose two decimals having the same value.
   a. .700   .07   .70   .007   **b.** .031   .31   .310   .0031
   c. 4.20   4.02   4.200   4.002

3. In each pair of numbers, choose the larger.
   a. .8 or .82   **b.** .3 or .33   **c.** .1 or .012
   d. .02 or .002   **e.** 2.4 or 2.38   **f.** 5.3 or 5.03
   g. .6 or .66   **h.** .162 or .17   **i.** 3.7 or 3.678

4. Arrange each group of numbers in order of size with the smallest number first.
   a. 3   .3   .03   .003   **b.** .5   .55   .055   .505
   c. .035   .04   .4   .305   **d.** 4.1   .41   .401   4.01
   e. 1.4   1.41   1.414   1.04   **f.** 7.02   7.002   7.19   7.019
   g. 3.2 $\frac{34}{10}$ 3.3 $3\frac{1}{10}$ **h.** 2.5 $\frac{23}{10}$ 2.2 $2\frac{7}{10}$

5. Which of the following numbers are less than 3.25?

$$3.14 \quad 3.3 \quad 3.04 \quad 3.24 \quad 3.254$$

3. ROUNDING OFF DECIMAL FRACTIONS

We may approximate decimals just as we do whole numbers by the process of **rounding off**.

Suppose we want to write the number .257 to the nearest hundredth. Is .257 closer to .25 or to .26? Or, put another way, is .257 closer to .250 or to .260? The number .257 is .007 greater than .250 but .003 less than .260. Since it is closer to .260, we round it off as .26.

Note that, if the number were .253, it would be closer to .25. When rounding off to the nearest hundredth, the deciding digit is the digit in the thousandths place; if it is 5 or more, we increase the hundredths digit by one—if it is less than 5, we drop it and leave the hundredths place unchanged.

RULE: Any decimal may be *rounded off* by first *dropping* all decimal places after the desired place. Then the last figure kept should be *increased by 1* if the first discarded figure is 5 or more. Otherwise, it should be left unchanged.

ILLUSTRATIVE PROBLEM 1: Round off .3243 to the nearest thousandth.

Solution: Keep the first three digits, .324. Since the next digit, 3, is less than 5, drop it. The rounded-off decimal is .324. (*answer*)

ILLUSTRATIVE PROBLEM 2: Round off 3.76 to the nearest tenth.

Solution: Keep the first two digits, 3.7. Because the next digit, 6, is more than 5, increase the digit 7 by 1. The rounded-off decimal is 3.8. (*answer*)

ILLUSTRATIVE PROBLEM 3: Round off 4.1258 to the nearest hundredth.

Solution: Keep 4.12. Since the next digit is 5, increase the digit 2 by 1. The rounded-off decimal is 4.13. (*answer*)

ILLUSTRATIVE PROBLEM 4: Round off 2.98 to the nearest tenth.

Solution: Keep 2.9. Since the next digit is 8, increase the digit 9 by 1. Notice that this results in a change of *two digits*. The rounded-off decimal is 3.0 or 3. (*answer*)

Exercises

1. Round off each number to the nearest *tenth*:
   a. .78
   b. 2.33
   c. 12.95
   d. .549
   e. .083
   f. 2.04

2. Round off each number to the nearest *hundredth*:

   a. .348   **b.** .333   **c.** 2.738

   d. .075   **e.** .3884   **f.** 5.4057

3. Round off each number to the nearest *thousandth*:

   a. .1427   **b.** .5862   **c.** .0456

   d. .00075   **e.** 3.4507   **f.** 8.0095

4. Round off the number 3.1416 to the nearest

   a. thousandth.   **b.** hundredth.

   c. tenth.   **d.** whole number.

5. The length of a rod is 6.745 inches. Round this length off to the nearest *hundredth* of an inch.

6. A high jumper leaped over a bar 5.78 feet above ground. What is this height to the nearest *tenth* of a foot?

7. A boy weighs 63.4 kilograms. Express his weight to the nearest *whole* kilogram.

8. The diameter of the cylinder of an engine measures 5.473 inches. Express this diameter to the nearest *tenth* of an inch.

---

4. CHANGING DECIMALS TO COMMON FRACTIONS

Any decimal may be written as a reduced common fraction. We simply write out the numerator and the denominator (a power of 10) and reduce the fraction to lowest terms, if possible.

EXAMPLES

1. Write .73 as a common fraction in lowest terms.

$$.73 = \frac{73}{100} \quad \text{(This cannot be reduced.)}$$

2. Write .075 as a common fraction in lowest terms.

$$.075 = \frac{75}{1000} = \frac{25 \times 3}{25 \times 40} = \frac{3}{40}$$

3. Write 3.2 as a mixed number in lowest terms.

$$3.2 = 3\frac{2}{10} = 3\frac{1}{5}$$

4. Write $.07\frac{1}{2}$ as a common fraction in lowest terms.

$$.07\frac{1}{2} = \frac{7\frac{1}{2}}{100} = 7\frac{1}{2} \div \frac{100}{1} = \frac{15}{2} \div \frac{100}{1}$$

$$= \frac{\overset{3}{\cancel{15}}}{2} \times \frac{1}{\underset{20}{\cancel{100}}} = \frac{3}{40}$$

Exercises

In 1 to 12, change each decimal to a common fraction in lowest terms.

1. .25   **2.** .5   **3.** .37   **4.** .375

5. .15   **6.** .75   **7.** .80   **8.** .625

9. .05   **10.** .018   **11.** .030   **12.** .0075

In 13 to 20, change each decimal to a fraction in lowest terms.

13. $.12\frac{1}{2}$ **14.** $.33\frac{1}{3}$ **15.** $.16\frac{2}{3}$ **16.** $.87\frac{1}{2}$

17. $.37\frac{1}{2}$ **18.** $.62\frac{1}{2}$ **19.** $.66\frac{2}{3}$ **20.** $.83\frac{1}{3}$

21. Write 3.75 as a mixed number in lowest terms.

22. A sheet of tin is $.06\frac{1}{4}$ of an inch thick. Write this measurement as a common fraction in lowest terms.

23. Tom took .4 of a second to blink his eye. Write this measurement as a common fraction in lowest terms.

5. ADDING AND SUBTRACTING DECIMALS

Since decimals are really fractions, we can add and subtract them in the same way we do fractions. For example,

$$.7 + .2 = \frac{7}{10} + \frac{2}{10} = \frac{9}{10} = .9$$

In the case of decimals, the common denominator for the equivalent fractions is always a power of 10. Adding or subtracting decimals can be accomplished easily by lining up the addends vertically, keeping the decimal points *directly under each other*.

ILLUSTRATIVE PROBLEM 1: $2.5 + 3.57 + 1.02 = ?$

Solution: Line up the decimal points and add:

$$
\begin{array}{r}
2.5 \\
3.57 \\
\underline{1.02} \\
7.09 \quad (answer)
\end{array}
$$

We simply bring down the decimal point and add as we do for whole numbers.

ILLUSTRATIVE PROBLEM 2:   Add \$4.23, \$2.75, and \$6.00.

Solution:

$$
\begin{array}{r}
\$4.23 \\
2.75 \\
\underline{6.00} \\
\$12.98 \quad (answer)
\end{array}
$$

The same principle also works for subtraction of decimals.

ILLUSTRATIVE PROBLEM 3:   Subtract .3 from 4.1.

Solution:

$$
\begin{array}{r}
4.1 \\
\underline{-.3} \\
3.8 \quad (answer)
\end{array}
$$

ILLUSTRATIVE PROBLEM 4:   Subtract .267 from 3.

Solution: Here we need to write three zeros after the decimal point following the number 3. Thus:

$$
\begin{array}{r}
3.000 \\
\underline{-.267} \\
2.733 \quad (answer)
\end{array}
$$

Exercises _____

In 1 to 6, add the decimals.

1. .7	**2.** .39	**3.** 2.73
.5	.04	1.48
4. 2.73	**5.** 3.42	**6.** 2.53
3.5	6.	.97
2.457	4.2	4.25
	8.53	1.404

7. Add: .98 + 4.35 + 35.58

8. Add: 62.2 + 928.6 + 86.4

In 9 to 14, subtract the decimals.

9. .73 −.27	**10.** 2.87 − .08	**11.** 73.0 −14.7
12. 12. − 4.23	**13.** 2.007 −1.4	**14.** \$23.14 − 8.22

15. Subtract: .08 − .032   **16.** From \$12.34 subtract \$5.88.

17. Ann went to the store with a \$5.00 bill. She paid \$2.37 for groceries. What change did she get from the \$5.00 bill?

18. Subtract .14 from 3.2.

19. A man going south drove 347.2 miles one day, 453.7 miles the next day, and 412.6 miles the third day. How many miles did he drive altogether?

20. Peter ran the 100-yard dash in 13.4 seconds, while Bob ran it in 11.9 seconds. How much longer did it take Peter to run the distance?

21. During four weeks in March, Samantha deposited the following amounts in her bank account: \$52.47, \$83.92, \$75.00, and \$121.68. On March 31, she withdrew \$278.47. How much money from her March deposits did she have left in the bank on March 31?

---

6. MULTIPLYING DECIMALS BY WHOLE NUMBERS

If a man earns \$3.75 an hour, how much does he earn in 7 hours? This problem requires multiplying a decimal by a whole number.

$$3.75 \times 7 = 3\frac{75}{100} \times 7 = \frac{375}{100} \times \frac{7}{1} = \frac{2625}{100}$$

$$= 26\frac{25}{100} = \$26.25$$

Note that this result can be obtained by multiplying as we do for whole numbers and then marking off as many decimal places in the product as there are in the original decimal factor.

ILLUSTRATIVE PROBLEM: 3.75 × 7 = ?

Solution: Multiply as for whole numbers.

$$\begin{array}{r} 3.75 \\ \times 7 \\ \hline 2625 \end{array}$$

Mark off 2 decimal places from the right since there are 2 such places in the factor 3.75.

Answer: 3.75 × 7 = 26.25

Exercises _____

In 1 to 8, multiply.

1. .7 ×8	**2.** 2.3 ×6	**3.** 4.7 ×14	**4.** .035 ×8
5. 23.247 ×15	**6.** 18.26 ×23	**7.** $2.85 ×7	**8.** 23 ×.004

9. Find the cost of a dozen hats selling at $6.95 each.

10. A sheet of tin is .18 inch thick. How thick, in inches, is a pile of 40 sheets of tin?

11. A car travels 22.4 miles on a gallon of gas. How many miles does it go on a tankful of 16 gallons?

12. What is the cost of 18 feet of molding at $1.72 per foot?

13. A man bought a television set by paying $50 down and 24 installments of $9.40 each. What was the total cost of the set?

_____

7. MULTIPLYING DECIMALS BY DECIMALS

If we wish to multiply .21 by .7, we may think of these decimals as fractions. Thus,

$$.21 \times .7 = \frac{21}{100} \times \frac{7}{10} = \frac{147}{1000} = .147$$

Since there are two decimal places in the first factor and one in the second factor, there are three decimal places in the product. Thus,

$$\begin{array}{l} .21 \quad \text{(2 decimal places)} \\ \underline{\times .7} \quad \text{(1 decimal place)} \\ .147 \quad \text{(3 decimal places)} \end{array}$$

RULE: To multiply two decimals, multiply as though they were whole numbers. Then count the total number of decimal places in the two factors and mark off that number of places in the product.

Exercises

In 1 to 12, find the products.

1. .4 × .5   **2.** 5.2 × .03   **3.** .57 × .8   **4.** 6.4
                                                                  ×.3

5. 6.85   **6.** .283   **7.**   .08   **8.**   3.8
     ×.23   ×.06   ×1000   ×.0064

9. 5.72   **10.** $3.84   **11.** $8.70   **12.** $212.50
     ×.007   ×1.2   ×.04   ×.07

13. What is the cost of 12.7 gallons of gasoline at $1.20 per gallon?

14. A cubic foot of ice weighs 57.5 pounds. What is the weight in pounds of 10.8 cubic feet of ice?

15. A car travels 20.4 miles on a gallon of gas. How many miles can it travel on 15.3 gallons?

16. A man bought 18.6 yards of ribbon at 24 cents per yard. How much did he pay for the ribbon?

17. A plane flies at an average speed of 387.5 miles per hour. How many miles does it travel in 2.5 hours?

18. A limestone growth in a cave extends downward from the roof at the rate of .008 inch per year. How many inches does it grow in 40.5 years?

19. If a gallon of milk weighs 8.7 pounds, find the weight of 4.25 gallons.

---

8. MULTIPLYING DECIMALS BY 10, 100, 1000, ETC.

Consider the following three multiplication problems:

a.   .375   **b.**   .375   **c.**   .375
       ×10   ×100   ×1000
      3.750   37.500   375.000

Note that in **a**, the effect of the multiplication is to move the decimal point one place to the right; in **b**, two places to the right; and in **c**, three places to the right.

RULE: To multiply a decimal by 10, 100, 1000, etc., move the decimal point to the right as many places as there are zeros in the multiplier.

When the decimal point is at the extreme right of a number, it is usually omitted.

EXAMPLES

1. $12.48 \times 10 = 124.8$ **2.** $.0687 \times 1000 = 68.7$

3. $77 \times 100 = 77. \times 100 = 7700$

Exercises

1. Multiply each number by 10:
   a. .378   **b.** 8.12   **c.** .003   **d.** 83   **e.** $2.18
   f. 2.0083   **g.** .00052   **h.** .07   **i.** 23.04
2. Multiply each of the numbers in exercise **1** by 100.
3. Multiply each of the numbers in exercise **1** by 1000.
4. If a dozen eggs cost $1.08, how much would 10 dozen cost?
5. If gasoline costs $1.094 per gallon, what would be the cost of a 100-gallon drum of gasoline?
6. A merchant buys 1000 doughnuts at $.07 each. How much does the merchant pay for them?

9. DIVIDING DECIMALS BY WHOLE NUMBERS

Consider the examples below.

EXAMPLES

1. $.9 \div 3 = \dfrac{9}{10} \div \dfrac{3}{1} = \dfrac{\overset{3}{\cancel{9}}}{10} \times \dfrac{1}{\underset{1}{\cancel{3}}} = \dfrac{3}{10} = .3 \quad \text{or} \quad 3\overline{).9}^{\,.3}$

2. $2.4 \div 6 = 2\dfrac{4}{10} \div \dfrac{6}{1} = \dfrac{\overset{4}{\cancel{24}}}{10} \times \dfrac{1}{\underset{1}{\cancel{6}}} = \dfrac{4}{10} = .4 \quad \text{or} \quad 6\overline{)2.4}^{\,.4}$

3. $.056 \div 8 = \dfrac{56}{1000} \div \dfrac{8}{1} = \dfrac{\overset{7}{\cancel{56}}}{1000} \times \dfrac{1}{\underset{1}{\cancel{8}}} = \dfrac{7}{1000} = .007$ or $8\overline{\smash{)}.056}^{\,.007}$

Note that, in each case where the division "box" ($\overline{)}$) is used, the decimal point in the quotient is directly above the decimal point in the dividend. It is usually more convenient to use the division box to divide decimals than it is to convert decimals to improper fractions. Here is an example: $45.75 \div 15$

$$
\begin{array}{r}
3.05 \\
15\overline{\smash{)}45.75} \\
45 \\
\hline
75 \\
75 \\
\hline
\end{array}
$$

RULE: To divide a decimal by a whole number, place the decimal point in the quotient directly above the decimal point in the dividend and divide as with whole numbers.

ILLUSTRATIVE PROBLEM 1: Find, correct to the nearest tenth, $2.5 \div 3$.
Solution:

$$
\begin{array}{r}
.83 \\
3\overline{\smash{)}2.50} \\
2\,4 \\
\hline
10 \\
9 \\
\hline
1 \\
\end{array}
$$

In this case, the quotient does not come out evenly. Hence, we add a zero to the dividend to give us a quotient with one more decimal place than is required. Then we round off to the nearest tenth, so that the desired result is .8. (*answer*)

ILLUSTRATIVE PROBLEM 2: Express $3 \div 4$ in decimal form with a remainder of zero.

Solution: Write enough zeros to the right of the decimal point so that the division is exact. Thus,

$$
3 \div 4 = \frac{3}{4} = 4\overline{\smash{)}3.00}
\begin{array}{r}
.75 \ (answer) \\
\end{array}
$$

$$
\begin{array}{r}
2\,8 \\
\hline
20 \\
20 \\
\hline
\end{array}
$$

Exercises _____

In 1 to 9, divide.

1. $3\overline{)6.9}$ **2.** $4\overline{)4.52}$ **3.** $6\overline{).24}$ **4.** $15\overline{)2.25}$ **5.** $35\overline{)3.605}$

6. $4\overline{)\$32.80}$ **7.** $12\overline{).96}$ **8.** $25\overline{)175.50}$ **9.** $463\overline{)41.67}$

In 10 to 15, divide. Write enough zeros after the decimal point in the dividend so that the division is exact.

10. $2\overline{).7}$ **11.** $4\overline{)2.20}$ **12.** $10\overline{)23.7}$

13. $8\overline{).0184}$ **14.** $22\overline{)57.42}$ **15.** $24\overline{)23.4}$

In 16 to 21, find each quotient to the nearest tenth.

16. $3\overline{)7.6}$ **17.** $4\overline{)5.52}$ **18.** $8\overline{)8.7}$

19. $18\overline{)292}$ **20.** $20\overline{)24.45}$ **21.** $16\overline{)4.79}$

22. Write $\frac{3}{5}$ as a decimal by dividing 3 by 5. (Write zeros after the decimal point in the dividend.)

23. a. Write $\frac{3}{8}$ as a decimal, using the method of exercise **22.**
   b. Write $\frac{5}{8}$ as a decimal, using the method of exercise **22.**

24. Mr. Clark used 15 gallons of gasoline to drive 352.5 miles. How many miles per gallon did he average?

25. A sporting goods store sold 9 basketballs for $143.55. What was the cost of each basketball?

26. On a day in January, 11 inches of snow fell in 7 hours. What was the average snowfall per hour to the nearest hundredth of an inch?

_____

10. DIVIDING BY A DECIMAL

Consider $2.44 \div .4$. This problem may be written as the fraction $\frac{2.44}{.4}$. If the divisor (.4) were a whole number, we could divide as in Section 9. We can make .4 a whole number if we remember that we may multiply the numerator and denominator of a fraction by the same number without changing its value.

In this case, we must multiply .4 by 10 to make it a whole number. Thus,

$$\frac{2.44}{.4} = \frac{2.44 \times 10}{.4 \times 10} = \frac{24.4}{4}$$

Now we can divide.

$$\frac{6.1}{4)\overline{24.4}}$$

Hence, 2.44 ÷ .4 = 6.1.

We can check this answer by multiplication.

$$\begin{array}{r} 6.1 \\ \times .4 \\ \hline 2.44 \end{array}$$

If we wish to divide .144 by .12, we must multiply both dividend and divisor by 100 to make the divisor, .12, a whole number. Thus,

$$\frac{.144}{.12} = \frac{.144 \times 100}{.12 \times 100} = \frac{14.4}{12} \qquad \frac{1.2}{12)\overline{14.4}}$$

RULE: To divide by a decimal:
1. Change the divisor to a whole number by multiplying it by a power of 10.
2. Multiply the dividend by the same number.
3. Perform the division with the new dividend and divisor.

ILLUSTRATIVE PROBLEM: 1.645 ÷ .35 = ?

Solution: This time we will accomplish the entire division directly in the division box.

$$.35.)\overline{1.64.5}$$

To change the divisor to a whole number, multiply both the divisor and the dividend by 100. The new place for the decimal point in the dividend is two places to the right of where it had been. We place the decimal point in the quotient directly above the new place in the dividend.

Continue with the division.

$$\begin{array}{r} 4.7 \\ 35)\overline{164.5} \\ \underline{140} \\ 245 \\ \underline{245} \\ 0 \end{array} \text{ (remainder)}$$

Check:
$$\begin{array}{r} 4.7 \text{ (quotient)} \\ \times .35 \text{ (divisor)} \\ \hline 235 \\ \underline{141} \\ 1.645 \text{ (dividend)} \end{array}$$

Answer: 1.645 ÷ .35 = 4.7

Note: To check a division problem, multiply the quotient by the divisor and add the remainder. The result should equal the dividend.

Exercises _____

In 1 to 6, divide.

1. $.7\overline{)\,.056}$ **2.** $.08\overline{)\,\$7.20}$ **3.** $1.6\overline{)\,2.352}$

4. $.012\overline{)\,.156}$ **5.** $1.52\overline{)\,6.536}$ **6.** $2.24\overline{)\,.3584}$

In 7 to 12, find the quotient to the nearest tenth.

7. $.6\overline{)\,.10}$ **8.** $1.2\overline{)\,8.8}$ **9.** $.15\overline{)\,1.28}$

10. $7.7\overline{)\,27.43}$ **11.** $.18\overline{)\,.040}$ **12.** $.25\overline{)\,.163}$

13. If a woman drove 248 miles using 14.5 gallons of gasoline, how many miles per gallon did she average, to the nearest tenth?

14. How many shirts costing $4.75 each can be bought for $95.00?

15. If a student earns $4.25 per hour, how many hours must she work to earn $148.75?

16. An auto travels 243 miles in 4.5 hours. What is its average rate of speed in miles per hour?

17. How many rods .6 foot long can be cut from a wooden rod 9 feet long?

18. A package of paper is 2.8 inches thick. If each sheet is .008 inch thick, how many sheets are there in the package?

19. A boy saves $7.50 per week. In how many weeks will he save $135?

20. If one book costs $4.25, how many such books can be purchased for $119?

---

11. DIVIDING BY 10, 100, 1000, ETC.

In Section 8, we learned that when we *multiply* a number by 10, 100, 1000, etc., we move the decimal point in the number as many places to the *right* as there are zeros in the multiplier. Conversely, when we *divide* a number by 10, 100, 1000, etc., we move the decimal point in the number as many places to the *left* as there are zeros in the divisor.

ILLUSTRATIVE PROBLEM: Divide 256.8 by 100.

Solution: Move the decimal point in 256.8 two places to the left.

$$\frac{256.8}{100} = 2.568 \quad (answer)$$

Exercises

In 1 to 12, divide by 10.

1. 45.12	**2.** 370	**3.** 52	**4.** 7.25
5. .03	**6.** 800	**7.** $8.60	**8.** 945.7
9. .23	**10.** 2.63	**11.** 78,600	**12.** 32.2

13. Divide each of the numbers in exercises **1** to **12** by 100.

14. Divide each of the numbers in exercises **1** to **12** by 1000.

15. A car travels 2237 miles in 10 days. How many miles does it average per day?

16. If 100 shirts cost a retailer $472, how much does each shirt cost him?

17. A roll of 1000 feet of wire costs $380. How much does the wire cost per foot?

18. If a 100-pound bag of potatoes costs $18.00, how much do the potatoes cost per pound?

19. A 100-gallon tank of oil is filled for $121.40. What is the cost per gallon?

20. If 100 flower bulbs cost $14, how much does each bulb cost?

---

12. SOLVING SIMPLE DECIMAL EQUATIONS

If we wish to solve an equation with decimals, such as $.3x = 2.1$, we may proceed by methods used in Chapter 2.

$$\frac{.3x}{.3} = \frac{2.1}{.3}$$

$$x = \frac{21}{3} = 7$$

An alternate method is to clear the equation of decimals by multiplying both sides by 10, 100, 1000, or a greater power of 10. In this case, we multiply both sides by 10.

$$10 \times .3x = 10 \times 2.1$$

$$3x = 21$$

$$\frac{3x}{3} = \frac{21}{3}$$

This method of solving decimal equations by clearing of decimals usually makes the work easier.

$$x = 7$$

ILLUSTRATIVE PROBLEM: Solve: $.05y = .725$

Solution: Multiply both sides by 100.

$$100 \times .05y = 100 \times .725$$

$$5y = 72.5$$

$$y = \frac{72.5}{5} = 14.5 \quad (answer)$$

Generally, we multiply both sides of an equation by that power of 10 that eliminates the decimal from the coefficient of the unknown.

Exercises

In 1 to 10, solve the equations.

1. $.75t = 40$ **2.** $1.50m = 72$ **3.** $.5p = 30$ **4.** $.18k = 2.7$

5. $.06y = 7.2$ **6.** $2.5x = 1.25$ **7.** $3.14d = 9.42$ **8.** $.03z = 48$

9. $1.5r - r = .9$ **10.** $2.3r + .9r = 64$

In 11 to 15, write an equation and then solve it.

11. $.4$ of what number is 76?

12. $.35$ of a certain number is 70. What is the number?

13. Bill bought a football for $12, which was .8 of the regular price. What was the regular price?

14. The selling price of a suit is 1.2 of its cost. If the selling price is $84, what is the cost of the suit?

15. The freshman class of a school is made up of 180 students. If this is .36 of the total number of students in the school, how many students are enrolled in the school?

Chapter Review Exercises

1. Write each of the following fractions as a decimal:
   a. $\frac{3}{10}$ **b.** $\frac{7}{20}$ **c.** $\frac{8}{25}$ **d.** $\frac{17}{50}$

2. Write each of the following numbers as a common fraction or a mixed number in lowest terms.
   a. .65   **b.** 4   **c.** 2.75   **d.** 1.375

3. A cubic foot of water weighs 62.5 pounds. What is the weight, in pounds, of 10 cubic feet of water?

4. John bought $3.87 worth of fruit. What change did he receive from a $5.00 bill?

5. Arrange the following numbers from the smallest to the largest:

   .3 .03 .33 3.3 .003

6. The noon temperatures for five successive days were 63.2°, 65.5°, 61.3°, 59.8°, and 59.7°. What was the average noon temperature for the five days?

7. Divide 23.24 by 2.8 and check.

8. Change $\frac{5}{7}$ to a decimal. Round off to the nearest hundredth.

9. Mr. Jones spends $450 per year for subway fares. What is his average monthly expense for subway fares?

10. How many sheets of tin each .12 inch thick are needed to make a pile 15 inches high?

11. Jane earns $4.35 an hour working after school. How much, to the nearest cent, does she earn in a week if she works $24\frac{1}{2}$ hours?

12. .7 of a certain number is .035. Write an equation for this statement and solve it to find the number.

CUMULATIVE REVIEW

1. Solve for p: $0.2p + 3 = 5$

2. Express in terms of x and y the total value, in dollars, of $3x$ books sold at y dollars each.

3. If $2x + 7$ represents an odd number, represent the next larger odd number.

4. Divide 241.67 by 3.4, and calculate the answer to the nearest tenth.

5. Marian bought a stereo for $302. If she has 12 months to pay, what is the amount, to the nearest cent, of each monthly payment?

6. Solve for x: $7x + 6 = 62$

7. Represent, in terms of m, the number of seconds in m minutes and 8 seconds.

8. Change $86°$ Fahrenheit to its equivalent Celsius temperature. [Use $C = \frac{5}{9}(F - 32)$.]

9. If one liter is equal to 1.06 quarts, how many quarts are there in 500 liters?

10. A man earns a total of $237.50 for a 38-hour week. What is his hourly rate of pay?

Answers to Cumulative Review

If you get an incorrect answer, refer to the chapter and section shown in brackets for review. For example, [5-2] means chapter 5, section 2.

1. 10 [5-12]   2. $3xy$ [3-1]   3. $2x + 9$ [3-3]

4. 71.1 [5-10]   5. $25.17 [5-9]   6. 8 [2-5]

7. $60m + 8$ [3-1]   8. $30°C$ [1-5]   9. 530 [5-6]

10. $6.25 [5-9]

6

Ratio and Proportion

1. MEANING OF RATIO

If Jim is 18 years old and his mother is 54 years old, we may compare their ages by saying that Jim's mother is 36 years older than he is. This is a comparison by subtraction.

Another way of comparing their ages is to say that Jim's mother is 3 times as old as he is. This is a comparison by division: $54 \div 18 = 3$.

When we compare two numbers by division, we say that we are finding their **ratio;** thus, we say that the ratio of the mother's age to Jim's age is 3 to 1. We are considering the fraction $\frac{54}{18}$ and reducing it to $\frac{3}{1}$.

Thus, a ratio is a comparison of two quantities by division. We may write the ratio 3 to 1 as $3 \div 1$ or $\frac{3}{1}$ or $3:1$. What we are, in effect, saying here is that, for every 3 years in the mother's age, there is one year in Jim's age. Note that the order here is important. We may reverse the order by saying that the ratio of Jim's age to his mother's age is $1:3$. This means that Jim's age is $\frac{1}{3}$ of his mother's age. The numbers 1 and 3 are called the **terms** of the ratio.

Since a ratio is really a fraction, we may use the rules for reducing fractions to find equivalent ratios. Thus, in the example above, we can say that

$$\frac{54}{18} = \frac{27}{9} = \frac{3}{1} \quad \text{or} \quad 54:18 = 27:9 \text{ (read ``54 is to 18 as 27 is to 9'')}$$

$$\text{and } 27:9 = 3:1$$

We see that the ratio $54:18$ is equivalent to the ratio $27:9$ or the ratio $3:1$.

Likewise, if we start with a ratio, say $\frac{3}{5}$, we may multiply the terms by any number, for example 4, and get an equivalent ratio, $\frac{12}{20}$.

We may now state the following:

PRINCIPLE: If both terms of a ratio are divided or multiplied by the same number (not zero), an equivalent ratio is obtained.

If we wish to compare 1 yard to 1 foot by means of a ratio, we have to change both lengths to the same units. Thus, since 1 yard = 3 feet, the ratio is $3:1$.

The ratio of 2 pounds to 12 ounces is $\frac{32}{12} = \frac{8}{3}$ or $8:3$.

RULE: To find the ratio of two quantities in different *units* of measure, convert both quantities to the same unit. Then reduce the resulting ratio to lowest terms.

PRINCIPLE: When a ratio compares different *kinds* of measure, it is called a **rate**.

When we say that a car travels 90 miles in 2 hours, we can compare the number of miles with the number of hours by saying that the ratio of these two quantities is $90:2$ or $45:1$. We usually state the latter by saying that the rate of the car is 45 miles per hour.

Thus, a rate usually indicates a ratio where the unit following the word "per" is understood to be 1. If we say that milk is selling for 50¢ per quart, we mean that the ratio of the price to the number of quarts is $50:1$.

If we buy 9 bars of chocolate for 3 children, we say that the ratio is $9:3$ or $3:1$, meaning 3 bars per child.

ILLUSTRATIVE PROBLEM 1: Mary has 3 quarters and Anne has 5 dimes.

a. What is the ratio of Mary's amount to Anne's?

b. What is the ratio of Anne's amount to the total that they both have?

Solution:

a. $\dfrac{3 \text{ quarters}}{5 \text{ dimes}} = \dfrac{3 \times 25}{5 \times 10} = \dfrac{75}{50} = \dfrac{3}{2}$ *(answer)*

b. $\dfrac{50}{75 + 50} = \dfrac{50}{125} = \dfrac{2}{5}$ *(answer)*

ILLUSTRATIVE PROBLEM 2: A rectangular room has dimensions of 9 feet by 12 feet.

a. What is the ratio of its width to its length?

b. What is the ratio of its length to its width?

c. What is the ratio of its length to its perimeter?

Solution:

a. $\dfrac{9}{12} = \dfrac{3}{4}$ *(answer)* **b.** $\dfrac{12}{9} = \dfrac{4}{3}$ *(answer)*

c. perimeter $= 2(9 + 12) = 2 \cdot 21 = 42$

$$\frac{\text{length}}{\text{perimeter}} = \frac{12}{42} = \frac{2}{7} \quad (answer)$$

Exercises

1. Write the following ratios as fractions and reduce to lowest terms:
   a. 5 pounds to 15 pounds **b.** $16 to $12 **c.** 3:9 **d.** 15:21

2. In the ratios in exercise **1,** write the ratio of the second quantity to the first as a fraction and reduce to lowest terms.

3. What is the ratio of the lengths of two lines, one 12 inches long and the other 15 inches long?

4. A baseball team won 10 games and lost 12. What is the ratio of the games won to the games played?

5. In each of the following, give the ratio of the first quantity to the second:
   a. 1 pound to 2 ounces   **b.** 1 dollar to 1 quarter
   c. 1 kilogram to 100 grams   **d.** 1 hour to 20 minutes
      (1 kilogram = 1000 grams)
   e. 1 pint to 1 gallon   **f.** 1 dime to 1 quarter

6. A living room is 15 feet wide and 20 feet long. Give the ratio of the
   a. length to the width.   **b.** width to the length.
   c. length to the perimeter.   **d.** perimeter to the width.

7. A class consists of 20 boys and 12 girls. Give the ratio of
   a. the number of boys to the number of girls.
   b. the number of girls to the number of boys.
   c. the number of boys to the total number of students.
   d. the number of girls to the total number of students.

8. Express the following ratios in reduced form:

    a. $15:18$ **b.** $\dfrac{2}{3}:\dfrac{5}{3}$ **c.** $125:175$ **d.** $.4:1.0$ **e.** $\dfrac{1}{5}:2$

    f. $\dfrac{3}{4}:\dfrac{5}{8}$ **g.** $1\dfrac{1}{2}:2\dfrac{3}{4}$ **h.** $4.0:.75$ **i.** $.23:.023$

9. A student did 16 problems correctly on a 20-problem test.

    a. What is the ratio of the number right to the total number of problems?

    b. What is the ratio of the number wrong to the total number of problems?

    c. What is the ratio of the number right to the number wrong?

10. Express the following ratios in reduced form:

    a. 40 minutes to 2 hours   **b.** 1 meter to 20 centimeters
                                                  (1 meter = 100 centimeters)

    c. 2 miles to 440 yards   **d.** 3 days to 1 week

    e. 3 pounds to 8 ounces   **f.** 1 yard to 8 inches

    g. 2 yards to 15 feet   **h.** 250 grams to 1 kilogram

11. A certain alloy is made by mixing 12 pounds of copper with 20 pounds of nickel.

    a. What is the ratio of the weight of copper to the weight of nickel?

    b. What is the ratio of the weight of copper to the weight of the total alloy?

    c. What is the ratio of the weight of nickel to the weight of the total alloy?

12. A cake recipe calls for $\frac{2}{3}$ cup of milk to $1\frac{1}{2}$ cups of flour. What is the ratio of the number of cups of milk to the number of cups of flour?

13. A 12-ounce bottle of tincture of iodine contains $\frac{2}{3}$ ounce of pure iodine. What is the ratio of the weight of pure iodine to the weight of the total mixture?

14. Express each of the following phrases as a ratio in simplest form:

    a. 200 miles in 4 hours   **b.** 12 cookies for 3 girls

    c. 80¢ for 4 bars of candy   **d.** 1 mile in 4 minutes

    e. 100 yards in 10 seconds   **f.** $1.25 for 5 pounds

15. A particular rectangle has its length and width in the ratio of $3:2$. Which of the following rectangles have the same ratio of length to width? Length : width is

    a. $18:15$ **b.** $20:14$ **c.** $27:18$ **d.** $4\frac{1}{4}:2\frac{1}{2}$ **e.** $3\frac{3}{4}:2\frac{1}{2}$ **f.** $8.1:5.4$

2. SOLVING RATIO PROBLEMS USING EQUATIONS

Consider the following problem:

In a certain class, the ratio of the number of boys to the number of girls is 5:3. If there are 32 students in the class, how many boys and how many girls are there?

Although the problem can be done by an arithmetic approach, the use of equations can give us a very direct way of solving many different types of such problems.

Since the ratio of boys to girls is 5:3, we may represent the number of boys as $5x$ and the number of girls as $3x$, where x, in this case, is any positive integer. Note that our aim is not merely to find x, but to find $5x$ and $3x$. The form of the solution follows:

Solution: Let $5x$ = the number of boys

and $3x$ = the number of girls. Then

$$5x + 3x = 32$$
$$8x = 32$$
$$x = 4, 5x = 20, \text{ and } 3x = 12$$

Check:

$$\text{Is } \frac{20}{12} = \frac{5}{3}? \quad \frac{20}{12} = \frac{5 \cdot 4}{3 \cdot 4} = \frac{5}{3}$$

Is $20 + 12 = 32$? $32 = 32$ ✔

Answer: There are 20 boys and 12 girls in the class.

Remember that finding x is not the final answer. It is merely the common multiplier. Once x is found, we must determine $5x$ and $3x$ as our desired answers.

If Bill's height is 60 inches, Maria's height is 50 inches, and Don's height is 40 inches, then the ratio of Bill's height to Maria's height is 60:50 and the ratio of Maria's height to Don's height is 50:40. These two ratios can be combined and written in the following shortened form:

$$60:50:40 = 6(10):5(10):4(10) = 6:5:4$$

The ratio of the numbers p, q, and r, where none of these numbers is equal to zero, may be written $p:q:r$.

To represent three numbers in the ratio 3:4:5, for example, we may write these three numbers as $3x$, $4x$, and $5x$, where x does not equal zero.

Exercises

1. Find two numbers in the ratio of $3:5$ whose sum is 96.

2. Divide a 35-inch line segment into two segments whose lengths are in the ratio $2:3$. How many inches are there in the length of each segment?

3. Jim and his younger brother divide a profit of \$78 in the ratio of $8:5$. How much does each get?

4. Two numbers are in the ratio of $7:4$ and their difference is 42. Find the numbers.

5. Two numbers are in the ratio of $7:5$ and their difference is 16. Find the numbers.

6. The three angles of a triangle are in the ratio $3:4:5$. The sum of the angles is $180°$. Find the number of degrees in each angle. (*Hint:* Let the angles be $3x$, $4x$, $5x$.)

7. The sides of a triangle are in the ratio $5:6:7$. If the perimeter of the triangle is 126 inches, find the length of each side.

8. In a certain high school, the ratio of freshmen to sophomores is $6:5$. If there are 880 students in both classes combined, how many students are there in each class?

9. Two business partners divide a profit in the ratio of $11:10$. If the profit one year is \$10,500, how much does each partner receive?

10. Two numbers are in the ratio of $7:10$. If 8 is added to their sum, the result is 93. Find the numbers.

11. The ratio of the length to the width of a rectangle is $8:7$. If the perimeter of the rectangle is 480 feet, find the length and width in feet.

12. In a basketball game, Arlene and Tina scored points in the ratio of $9:5$. If Arlene scored 12 more points than Tina, how many points did each make?

3. PROPORTION

The ratio $\frac{4}{6}$ is equal to the ratio $\frac{2}{3}$. We may thus write the equation $\frac{4}{6} = \frac{2}{3}$. Such an equation is called a **proportion**. A proportion is an equation stating that two ratios are equal.

The preceding proportion may also be written as $4:6 = 2:3$. Recall that we read it as "4 is to 6 as 2 is to 3."

In general, we may write any proportion in the form

$$\frac{p}{q} = \frac{r}{s} \text{ or } p:q = r:s \quad \text{where} \quad q \neq 0 \quad \text{and} \quad s \neq 0$$

There are four terms in this proportion, namely, p, q, r, and s. The first and fourth terms are called the **extremes** and the second and third are called the **means:**

Note that, in the proportion $15:20 = 3:4$, the product of the means, 20×3, equals 60 and the product of the extremes, 15×4, also equals 60.

Likewise, in the proportion, $3:8 = 9:24$, the product of the means, 8×9, and the product of the extremes, 3×24, both equal 72.

These examples illustrate a most important principle of proportions, namely:

PRINCIPLE: In a proportion, the product of the means is equal to the product of the extremes.

By treating the general proportion $\frac{p}{q} = \frac{r}{s}$ as a fractional equation, we can show that the principle just stated is true. That is, $ps = qr$.

One way of testing to see if two ratios are equal is to take the products of the means and the extremes (*cross-multiply*). If these products are not equal, the two ratios do not form a proportion.

ILLUSTRATIVE PROBLEM 1: Solve for p: $\dfrac{3}{16} = \dfrac{p}{80}$

Solution: Since, in a proportion, the product of the means equals the product of the extremes, we may cross-multiply and obtain:

$16p = 3 \cdot 80$

$16p = 240$

$p = 15 \quad (answer)$

Check: $\dfrac{3}{16} \overset{?}{=} \dfrac{15}{80}$

Show that the two ratios are equal by reducing fractions.

$$\frac{3}{16} = \frac{3}{16} \checkmark$$

ILLUSTRATIVE PROBLEM 2: Do the ratios $\frac{4}{5}$ and $\frac{8}{9}$ form a proportion?

Solution: product of extremes = 4 · 9 = 36
            product of means = 5 · 8 = 40

Answer: Since these products are not equal, the two fractions do not form a true proportion, and $\frac{4}{5} \neq \frac{8}{9}$.

ILLUSTRATIVE PROBLEM 3: If an automobile runs 66 miles on 3 gallons of gas, how many miles will it run on 7 gallons of gas?

Solution: Let x = the number of miles run on 7 gallons of gas. Then

$$\frac{66}{3} = \frac{x}{7}$$

$$3x = 462 \quad \text{(cross-multiply)}$$

$$x = 154 \text{ miles} \quad (answer)$$

ILLUSTRATIVE PROBLEM 4: Solve for t: $\dfrac{40}{36} = \dfrac{10}{3t}$

Solution:

40 · 3t = 36 · 10 (cross-multiply)

120t = 360

$t = 3$ *(answer)*

Check:

$$\frac{40}{36} \overset{?}{=} \frac{10}{3 \cdot 3}$$

$$\frac{40}{36} \overset{?}{=} \frac{10}{9}$$

Show that the two ratios are equal by reducing fractions.

$$\frac{10}{9} = \frac{10}{9} \quad \checkmark$$

Exercises

In 1 to 9, solve the proportions and check.

1. $\dfrac{6}{3} = \dfrac{12}{x}$

2. $\dfrac{4}{5} = \dfrac{9}{b}$

3. $\dfrac{7}{5} = \dfrac{q}{6}$

4. $\dfrac{p}{4} = \dfrac{15}{12}$

5. $\dfrac{24}{66} = \dfrac{3}{r}$

6. $\dfrac{25}{t} = \dfrac{5}{2}$

7. $6:5 = m:45$

8. $15:2k = 5:12$

9. $16:4y = 3:12$

In 10 to 15, state whether or not the given ratios form a true proportion. (Answer *yes* or *no*.)

10. $\dfrac{2}{7}, \dfrac{4}{49}$

11. $\dfrac{6}{10}, \dfrac{9}{15}$

12. $\dfrac{p}{4p}, \dfrac{2}{8}$

13. $\dfrac{5}{9}, \dfrac{9}{5}$

14. $\dfrac{5}{4}, \dfrac{10}{8}$

15. $\dfrac{21}{27}, \dfrac{28}{36}$

16. The speeds of two cars are in the ratio 3 to 5. If the slower car goes 36 miles per hour (mph), what is the speed of the faster car?

17. If 12 typewriters cost $1020, how much will 9 cost at the same rate?

18. If 1 inch on a map represents 50 miles, how many inches on that map represent 540 miles?

19. If 6 limes cost 80 cents, find the cost of 15 limes at the same rate.

20. If pens cost $1.08 a dozen, how much should 5 such pens cost?

21. The weight of 80 feet of telephone wire is 5 pounds. Find the weight of 360 feet of this wire in pounds.

22. A student received $90 for working 14 hours. At the same rate of pay, how many hours must the student work to earn $150?

23. A boy travels 80 miles on his motorcycle in 3 hours. At the same rate, how many miles will he go in 5 hours?

24. Two numbers are in the ratio 8:5. If the smaller number is 75, what is the larger number?

25. The ratio of the length of a rectangle to its width is 7:4. If the width is 68 inches, how many inches are in the length?

26. Marsha owns a house worth $72,000 and pays $5400 in taxes. At this rate, how much tax should be paid on a house worth $96,000?

27. A car travels 240 miles in $4\frac{3}{4}$ hours. At this rate, how far will it travel in 6 hours?

28. A picture 3 inches wide by 5 inches long is to be enlarged so that its width is $6\frac{1}{2}$ inches. What will be the length in inches of the enlarged picture?

29. A board is cut into two pieces having the ratio 3:5. If the longer piece is 12 feet, how many feet are in the shorter piece?

30. If 3 cans of peas sell for $1.77, how much would 8 such cans cost?

31. It took a typist 32 minutes to type 6 pages of a report. At this rate, how many minutes will it take her to type the remaining 27 pages of the report?

4. SCALE DRAWINGS

In making floor plans for an apartment or a house, it is common practice for architects to draw the rooms to **scale.** For example, in the figure, a $16' \times 20'$ living room is drawn to a scale of $\frac{1}{8}$ inch to 1 foot. To find the size of the drawing, we can first determine the number of inches in the width, x, of the rectangle by solving the proportion:

Living Room
16' × 20'

Scale: $\frac{1}{8}'' = 1'$

$$\frac{1}{8} : x = 1 : 16 \quad \text{or} \quad \frac{\frac{1}{8}}{x} = \frac{1}{16}$$ (Note that $\frac{1}{8}'' = \frac{1}{8}$ inch and $1' = 1$ foot.)

Cross-multiplying gives $x = \frac{1}{8} \cdot 16 = 2''$ (width of rectangle).

To determine the number of inches in the actual length, y, of the drawing, we must solve the proportion:

$$\frac{1}{8} : y = 1 : 20 \quad \text{or} \quad \frac{\frac{1}{8}}{y} = \frac{1}{20}$$

Cross-multiplying gives $y = \frac{1}{8} \cdot 20 = 2\frac{1}{2}''$ (length of rectangle).

Thus, a rectangle $2'' \times 2\frac{1}{2}''$ will represent the living room on the floor plan.

Remember that the first number on the scale indicates the drawing dimension and the second number represents the actual dimension of the figure or object. In the figure just discussed, the drawing was a reduction of the actual living room. In some cases, the drawing may be an enlargement of the actual object or figure. For example, a drawing of an ant in a biology book may require an enlargement of 8 times its actual size for details to be seen. In this case, the scale would read $8:1$ or $1'' = \frac{1}{8}''$. Remember again that the first number is the drawing size and the second number is the actual size. Drawings showing enlargements are very common in making diagrams of small machine tools. When drawings are made to actual size, the scale in this case is simply $1:1$.

In making scale drawings, we are constructing figures that have the same shape as the actual figures but are reduced or enlarged in size in a proportionate manner. Figures reduced or enlarged in this way are said to be **similar** to the actual figures.

A map of a particular region is merely a scale drawing showing a considerable reduction of the region. Usually, in one corner of the map, the scale is shown. In the figure, the scale is shown to be 1 in. = 100 mi. If this is the scale of a New York State map and the straight-line distance between Albany and New York City measures $1\frac{1}{4}$ inches, what is the actual distance in miles?

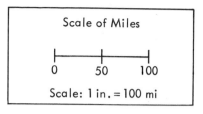

Scale of Miles

0   50   100

Scale: 1 in. = 100 mi

We form the proportion: $\dfrac{1\text{ in.}}{100\text{ mi}} = \dfrac{1\frac{1}{4}\text{ in.}}{x\text{ mi}}$

Cross-multiplying, we obtain

$$1 \cdot x = 100 \cdot 1\frac{1}{4}, \text{ or } x = 125 \text{ miles}$$

ILLUSTRATIVE PROBLEM 1: In a photograph, the height of a boy is $1\frac{1}{2}$ inches. If the boy's height is actually 4 feet 6 inches, find the number of feet represented by 1 inch.

Solution: Using x to represent the unknown, we can write that 1 inch represents a height of x feet. Using the information given, we know that $1\frac{1}{2}$ inches represent $4\frac{1}{2}$ feet.

Form the proportion: $\dfrac{1\frac{1}{2}''}{4\frac{1}{2}'} = \dfrac{1''}{x}$

Cross-multiply: $1\frac{1}{2}x = 4\frac{1}{2}$

$$\frac{3}{2}x = \frac{9}{2}$$

Multiply both sides by 2: $3x = 9$

$$x = 3$$

Answer: 1 inch represents 3 feet.

ILLUSTRATIVE PROBLEM 2: If $\frac{3}{4}$ inch represents 20 yards, how many inches long must a scale length be to represent a football field 100 yards long?

Solution:
$$\frac{\frac{3}{4}}{20} = \frac{x}{100}$$

Cross-multiply:
$$20x = \frac{3}{4} \cdot 100$$

$$20x = 75$$

Divide both sides by 20:
$$x = 3\frac{3}{4} \text{ in.}$$

Answer: $3\frac{3}{4}$ inches represent 100 yards.

ILLUSTRATIVE PROBLEM 3: The measured distance on a map between Pittsburgh and New York City is $3\frac{1}{2}$ inches. If the scale of the map is 1 inch to 70 miles, what is the actual distance in miles between these two cities?

Solution:
$$\frac{1 \text{ in.}}{70 \text{ mi}} = \frac{3\frac{1}{2} \text{ in.}}{x \text{ mi}}$$

Cross-multiply:
$$x = 70 \cdot 3\frac{1}{2}$$

$$x = 245 \text{ mi}$$

Answer: The actual distance between Pittsburgh and New York City is 245 miles.

Exercises

1. Using a scale of $\frac{1}{8}$ in. = 1 ft, represent each of the following measurements in inches.
   a. 40 feet
   b. 30 feet
   c. 26 feet
   d. 120 feet
   e. 33 feet
   f. 58 feet

2. The scale used in making a scale drawing of a house is $\frac{1}{2}$ in. = 1 ft. Give the actual dimensions in feet of the rooms whose dimensions on the drawing are as follows:

a. 4 in. by 6 in.

b. $8\frac{1}{2}$ in. by 12 in.

c. $5\frac{1}{4}$ in. by $7\frac{1}{2}$ in.

d. $6\frac{3}{4}$ in. by 10 in.

3. a. Two towns are $4\frac{3}{4}$ inches apart on a map drawn to a scale of 1 in. = 50 mi. What is the distance in miles between towns?

b. Still considering this map, find how long (in inches) a line must be to represent a distance of 425 miles.

4. In the following, tell whether the scale drawing will be a *reduction*, an *enlargement*, or the *same size* as the object if the scale is:

a. $\frac{1}{2}$ in. = 10 ft

b. $\frac{1}{8}$ in. = 100 mi

c. 1 in. = $\frac{1}{8}$ in.

d. 10:10

e. 8:5

f. 1000:1

5. A U.S. dollar bill is $6\frac{1}{8}$ inches long and $2\frac{5}{8}$ inches wide. If we draw it to a scale of $1'' : 3''$, how many inches are there in the length and width of the drawing?

6. Two towns are $8\frac{3}{4}$ inches apart on a map drawn to a scale of 1 inch = 60 miles. Find the distance in miles between the towns.

7. In the scale at the right, how many miles does 1 inch represent? (*Hint:* Each square is $\frac{1}{4}$ inch on a side.)

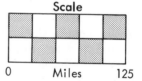

Scale

0   Miles   125

8. A map of Europe is drawn to the scale of 1 inch = 400 miles. Find the distance in miles represented on the map by each of the following lengths:

a. 3 inches

b. $4\frac{1}{2}$ inches

c. $2\frac{3}{4}$ inches

d. $4\frac{7}{8}$ inches

e. $3\frac{5}{16}$ inches

f. $\frac{5}{8}$ inch

9. Considering the scale in exercise **8**, find the number of inches on the map that would represent a distance of

a. 250 miles.

b. 840 miles.

c. 1000 miles.

d. 325 miles.

e. 420 miles.

f. 1400 miles.

10. Find the scale of a drawing if

a. a man 6′ tall is represented by a line 4″ long.

b. a bee $\frac{1}{2}''$ long is represented by a line 2″ long.

c. two towns 150 miles apart are separated on a map by a distance of 6 inches.

d. an auto 10 ft long is represented by a figure $2\frac{1}{2}$ in. long.

11. Using the scale 1 inch = 8 feet, how many inches would you use to represent:

    a. a horse 6 feet high?   **b.** a piano 5 ft 4 in. long?

12. In making a model airplane, a boy used a scale of 1:80. If the length of the plane is actually 100 feet, how long should the model plane be?

13. A pole is 20 ft high. A wire is stretched from the top of the pole to a point on the ground 15 ft from the base of the pole. Using a scale of 1 in. = 5 ft, make a scale drawing of the pole and wire. Measure the number of inches in the line in the drawing that represents the wire. What is the length of the wire in feet?

---

5. SOME OTHER USES OF RATIO AND PROPORTION

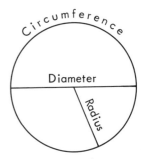

In a circle, it is clear that the diameter is twice the radius, or the ratio of the diameter to the radius is 2:1. The **circumference** of a circle is the distance around the circle. It is very much like the perimeter of a polygon. If we measure the circumference of any circle and its diameter, we find that the ratio of the circumference, C, of the circle to its diameter, d, is always slightly more than 3. More precise measurement indicates that this ratio is about $3\frac{1}{7}$ or 3.14. The ancient Greeks represented this ratio by the Greek letter π, pronounced "pie." Thus, we may write

$$\frac{C}{d} = \pi = \frac{22}{7} = 3\frac{1}{7} = 3.14 \quad (approximately)$$

We may also write $C = \pi d$.

ILLUSTRATIVE PROBLEM 1: Find the length of the circumference of a circle of diameter 14 inches. Use $\pi = \frac{22}{7}$.

Solution: $C = \pi d$

$$C = \frac{22}{7} \cdot 14 = 44 \text{ inches} \quad (answer)$$

The ancient Greeks also decided, after studying many rectangles, that the rectangle that "looks best" is one whose length and width are in the ratio 3:2. Many of the rectangles used in ancient Greek art and architecture maintain this ratio.

ILLUSTRATIVE PROBLEM 2: An enlarged picture is $4\frac{1}{2}$ feet long. How many feet should its width be to make it "look best"?

Solution: Let x = the number of feet in the width. Then

$$\frac{4\frac{1}{2}}{x} = \frac{3}{2}$$

$$3x = 9$$

$$x = 3 \text{ feet} \quad (answer)$$

Ratios are also used to determine the standing of teams in various sports. For example, if a baseball team wins 50 games out of 75 games played, we say its record is $\frac{50}{75} = \frac{2}{3} = .667$. The ratio is converted to its decimal form carried out to the nearest thousandth.

Exercises

1. Find in inches the circumference of a circle whose diameter is 21 inches.

2. The width of a rectangle is 24 inches. What must be its length, in inches, if it is to "look best"?

3. What is the radius, in inches, of a circle whose circumference is 176 inches?

4. A baseball team won 80 games and lost 60 games one season. What was its record?

5. What is the ratio of the circumference of a circle to its radius? Write a formula for the circumference of a circle in terms of its radius.

6. A team plays 120 games one season. How many games must it win to achieve a record of .875?

7. A rectangle has a perimeter of 75 feet. What must be its dimensions, in feet, in order that the ratio of the length to the width is $3:2$?

8. A circle has a diameter of 3 ft 6 in. What is its circumference in feet?

9. After winning 45 games, a team's record is .625. How many games did the team play?

Chapter Review Exercises

1. The length of a rectangle is 14 inches and its width is 8 inches.
   a. What is the ratio of the length to the width?
   b. What is the ratio of the width to the length?
   c. What is the ratio of the length to the perimeter of the rectangle?

2. The distance from New York to Buffalo is 400 miles. How many inches would represent this distance on a map whose scale is 1 inch = 60 miles?

3. Find the value of y in each of the following proportions:
   a. $2:3 = 16:y$
   b. $3:4 = 5:y$
   c. $12:18 = y:\frac{1}{2}$
   d. $y:\frac{2}{3} = 1\frac{1}{4}:2\frac{1}{2}$

4. If a recipe for 3 dozen cookies calls for 2 cups of sugar, how much sugar is needed for 5 dozen cookies?

5. Two circles have diameters of $8''$ and $10''$. What is the ratio of their
   a. diameters?   b. radii?   c. circumferences?

6. Consider the two rectangles shown below:

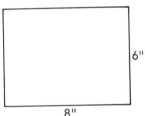

   a. What is the ratio of their lengths?
   b. What is the ratio of their widths?
   c. What is the ratio of their perimeters?

7. If $1\frac{1}{2}$ inches on a map represents 120 miles, what distance, in miles, does $2\frac{7}{8}$ inches represent?

8. Two boys divide a profit of $36 in the ratio of $5:3$. How much does each boy get?

9. A 1-pound object on the earth weighs .16 pound on the moon.
   a. At this rate, what would a 180-pound earthman weigh on the moon?
   b. If a rock on the moon weighs 8 pounds, how much would it weigh on earth?

10. If 5 pounds of potatoes cost 67¢, what will 12 pounds of potatoes cost at the same rate?

11. A woman uses 14 gallons of gasoline for a trip of 340 miles. At this rate, how many gallons will she use on a 510-mile trip?

12. A team wins 80 games and loses 40 games in one season. What is its record?

13. If 45 feet of copper wire weigh 9 pounds, what will 60 feet of the same wire weigh?

14. A typist found that it took him 18 minutes to type 4 pages of a report. At this rate, how long will it take him to type the entire 30-page report?

15. The sides of a triangle are in the ratio 8:9:10. The perimeter of the triangle is 135 inches. Find the length in inches of the shortest side.

CHECK YOUR SKILLS

1. Divide 1366.56 by 15.6 and multiply the quotient by 14.

2. How many minutes are there in $3\frac{1}{4}$ hours?

3. Subtract $9\frac{5}{8}$ from 20.

4. At $.42 a liter, what will 109 liters of milk cost?

5. Solve for y: $3y + 10 = 19$

6. The ratio of two numbers is 1:5 and their sum is 24. What is the smaller number?

7. Using the formula $A = \frac{1}{2}bh$, what is the value of A when $b = 5.6$ and $h = 18.4$?

8. Solve for w: $\dfrac{w}{6} = \dfrac{7}{2}$ 9. Solve for x: $.05x - 4 = 6$

10. Express in terms of x the perimeter of a rectangle whose width is represented by x and whose length is represented by $2x$.

7

Percentage

1. MEANING OF PERCENT

As early as the 15th century, businesspeople made use of certain decimal fractions so much that they gave them the special name **percent.** Percent means "per hundred" or "by the hundred." The symbol for percent is %. Thus, if 20 students are absent in a freshman class of 100, we may say that 20% are absent. Since 80 students would then be present, we may say that 80% of the class is present.

Thus, we say that *percent* means *hundredths.* For example, 73% is $\frac{73}{100}$, or .73. Also, 1% is 1 out of 100, which is $\frac{1}{100}$, or .01. If a student answers 87 questions correctly in a test of 100 questions, we say his score is 87%, which is equal to $\frac{87}{100}$ written as a fraction, or .87 written as a decimal. A score of 100% would mean he answered 100 correctly out of 100 questions.

$$\text{Definition: } n\% = \frac{n}{100}$$

Here are some questions about the meaning of percent. Try to answer them before looking at the answers to the right.

1. Write seven hundredths as a percent.   7%   (*answer*)

2. Write $\frac{19}{100}$ as a percent.   19%   (*answer*)

3. Write 83% as a common fraction. $\dfrac{83}{100}$ (*answer*)

4. Write $6\frac{1}{2}\%$ as a fraction. $\dfrac{6\frac{1}{2}}{100}$ (*answer*)

5. In a class of 25 students, 100% are present. How many students are present?   25   (*answer*)

ILLUSTRATIVE PROBLEM: If 17% of a class is absent, what percent of the class is present?

Solution: Since the entire class is represented by 100%, subtract 17% from 100%. Thus, 100% − 17% = 83% (*answer*).

Exercises

1. Write each percent as a fraction with denominator 100:
   a. 17%   **b.** 8%   **c.** 35%   **d.** 42%
   e. 88%   **f.** 63%   **g.** $12\frac{1}{2}$%   **h.** $9\frac{1}{4}$%

2. Write as a percent:
   a. 9 hundredths   **b.** 27 hundredths
   c. 49 hundredths   **d.** 73 hundredths
   e. $\frac{18}{100}$ **f.** $\frac{37}{100}$ **g.** $\frac{54}{100}$ **h.** $\frac{68}{100}$

3. Anne earned $100 last week. She spent $28 for lunches, $12 for transportation, $33 for clothes, $15 for books, and put the rest in the bank.
   a. What percent of her earnings did she spend for lunches?
   b. What percent did she spend for transportation?
   c. What percent did she spend for clothes?
   d. For books?
   e. What percent did she put in the bank?

4. In a spelling test of 27 words, Jack got 100% right. How many did he get right?

5. In a class of 100 sophomores, 42 were boys. What percent of the class was made up of girls?

6. A tie made of polyester and silk is 60% polyester. What percent is silk?

7. A baseball player one season got hits 32% of the times he was at bat. What percent of the times at bat did he not get hits?

8. In his will, Mr. Scott left 55% of his estate to his wife, 12% to his son, 12% to his daughter, and the rest to charity. What percent went to charity?

9. In a crate of 100 oranges, 8% are spoiled. What percent of the oranges are not spoiled?

10. A woman bought a $100 armchair and had to pay an additional 7% of the cost as sales tax. How many dollars did she pay as sales tax?

2. CHANGING PERCENTS TO DECIMALS AND DECIMALS TO PERCENTS

Using our definition of percent, we can change percents to decimals as shown in the following examples.

EXAMPLES

1. $37\% = \dfrac{37}{100} = .37$ **2.** $8\% = \dfrac{8}{100} = .08$

3. $7.4\% = \dfrac{7.4}{100} = \dfrac{7.4 \times 10}{100 \times 10} = \dfrac{74}{1000} = .074$

4. $.5\% = \dfrac{.5}{100} = \dfrac{.5 \times 10}{100 \times 10} = \dfrac{5}{1000} = .005$

5. $\left. \begin{array}{l} 6\frac{1}{2}\% = \dfrac{6\frac{1}{2}}{100} = .06\frac{1}{2} \\[4mm] 6.5\% = \dfrac{6.5}{100} = .065 \end{array} \right\}$ (same value)

From these examples, we form the following general rule:

RULE: To change a percent to a decimal, drop the percent sign and move the decimal point two places to the *left*.

It then follows that, if we want to write a decimal as a percent, we simply reverse the procedure:

RULE: To change a decimal to a percent, move the decimal point two places to the *right* and write the % sign.

Using this rule, we can change decimals and fractions to percents as shown below.

EXAMPLES

6. $.41 = 41\%$ **7.** $.03 = 3\%$ **8.** $.045 = 4.5\%$

9. $.007 = .7\%$ **10.** $.37\frac{1}{2} = 37\frac{1}{2}\%$ **11.** $1.20 = 120\%$

12. $1\frac{1}{2} = 1.50 = 150\%$ **13.** $2\frac{1}{4} = 2.25 = 225\%$

Note that a mixed number or a mixed decimal, as in examples **11, 12,** and **13,** changes to a percent that is greater than 100%.

Exercises

In 1 to 16, change each percent to a decimal.

1. 32%   **2.** 25%   **3.** 9%   **4.** 1%

5. 73%   **6.** 2.5%   **7.** $9\frac{1}{2}\%$ **8.** 100%

9. $5\frac{3}{4}\%$ **10.** .63%   **11.** 125%   **12.** .2%

13. $\frac{1}{2}\%$ **14.** 7.25%   **15.** $66\frac{2}{3}\%$ **16.** 5.3%

In 17 to 32, change each decimal to a percent.

17. .75   **18.** .47   **19.** .06   **20.** .32

21. .8   **22.** .035   **23.** $.07\frac{1}{2}$ **24.** .009

25. .363   **26.** .897   **27.** 1.00   **28.** 1.25

29. 2.1   **30.** $.63\frac{1}{3}$ **31.** .0065   **32.** .6

33. In each part, choose two numbers that have the same value.
   a. .17, 17, 17%   **b.** 40%, 4%, .04   **c.** 150%, 1.5, 15

34. The "cost of living" in one year increased 11%. At this rate, how many cents more would you pay for an item that cost $1.00 last year? $8.00 last year?

35. A team won .7 of the games it played last year.
   a. What percent of games played did it win?
   b. What percent of games played did it lose?

36. A certain bank pays $6\frac{1}{2}\%$ interest on all money deposited for a year. Write this rate of interest as a decimal fraction.

37. A salesman receives $12\frac{1}{2}¢$ for every dollar of merchandise he sells. Write this rate of commission as a percent.

38. A family spends .275 of its income for rent.
   a. What percent of its income does it spend for rent?
   b. What percent of its income is left?

3. CHANGING PERCENTS TO FRACTIONS AND FRACTIONS TO PERCENTS

ILLUSTRATIVE PROBLEM 1: Phillip received a mark of 80% on a short-answer test. What fractional part of the questions did he answer correctly?

Solution: $$80\% = \frac{80}{100} = \frac{4}{5}$$

Answer: Phillip answered $\frac{4}{5}$ of the questions correctly.

RULE: To change $n\%$ to a fraction, write the fraction $\dfrac{n}{100}$ and reduce it to lowest terms.

EXAMPLES

1. $75\% = \dfrac{75}{100} = \dfrac{3}{4}$

2. $7\frac{1}{2}\% = \dfrac{7\frac{1}{2}}{100} = 7\frac{1}{2} \div \dfrac{100}{1} = \dfrac{\overset{3}{\cancel{15}}}{2} \times \dfrac{1}{\underset{20}{\cancel{100}}} = \dfrac{3}{40}$

ILLUSTRATIVE PROBLEM 2: In a class of 25 students, 20 students are present. What percent of the class is present?

Solution: $$\frac{20}{25} = \frac{20 \times 4}{25 \times 4} = \frac{80}{100} = 80\%$$

Answer: 80 percent of the class is present.

RULE: To change a fraction to a percent, change the fraction to an equivalent fraction with a denominator of 100. This fraction is then easily changed to a percent.

Another way of changing a fraction to a percent is to change the fraction to a two-place decimal by dividing the numerator by the denominator. This hundredth-place decimal can then be written as a percent. For example, let us change $\frac{1}{8}$ to a percent:

$$\frac{1}{8} = 1 \div 8 = 8)\overline{1.00}^{\quad .12\frac{1}{2}} = 12\frac{1}{2}\%$$

$$\frac{8}{}$$
$$20$$
$$\underline{16}$$
$$\frac{4}{8} = \frac{1}{2}$$

This latter method is particularly useful when the denominator of the given fraction is not a divisor of 100.

ILLUSTRATIVE PROBLEM 3: Change $\frac{3}{4}$ to a percent.

Solution: $\dfrac{3}{4} = \dfrac{3 \times 25}{4 \times 25} = \dfrac{75}{100} = 75\%$ *(answer)*

ILLUSTRATIVE PROBLEM 4: Change $\frac{5}{8}$ to a percent.

Solution: $8)\overline{5.00}^{\quad .62\frac{1}{2}} = 62\frac{1}{2}\%$ *(answer)*

Exercises

In 1 to 16, change each percent to a fraction in *lowest terms* or to a mixed number.

1. 15%	**2.** 65%	**3.** 70%	**4.** 85%
5. 18%	**6.** 6%	**7.** 132%	**8.** 125%
9. $66\frac{2}{3}\%$	**10.** $37\frac{1}{2}\%$	**11.** 12.5%	**12.** $83\frac{1}{3}\%$
13. $6\frac{1}{4}\%$	**14.** $87\frac{1}{2}\%$	**15.** $33\frac{1}{3}\%$	**16.** $8\frac{1}{3}\%$

In 17 to 28, change each fraction to a percent.

17. $\frac{1}{3}$	**18.** $\frac{3}{4}$	**19.** $\frac{1}{2}$	**20.** $\frac{3}{8}$
21. $\frac{1}{6}$	**22.** $\frac{5}{6}$	**23.** $\frac{1}{4}$	**24.** $\frac{7}{8}$
25. $\frac{21}{25}$	**26.** $\frac{12}{18}$	**27.** $\frac{15}{24}$	**28.** $\frac{5}{12}$

29. In each part, choose two numbers that have the same value.

    a. $\frac{1}{4}$, 4%, 25%   **b.** $\frac{3}{5}$, 60, 60%   **c.** 150, $1\frac{1}{2}$, 150%

30. A radio is being sold for "20% off" the marked price.

    a. By what fractional part of the marked price is the price of the radio being reduced?

    b. What fractional part of the marked price must you pay for the radio?

31. In running for mayor of a town, a woman receives $\frac{3}{5}$ of the votes cast. She has one opponent.

    a. What percent of the total vote does she receive?

    b. What percent does her opponent receive?

32. Joan saved $33\frac{1}{3}\%$ of her earnings. What fractional part of her earnings did she spend?

33. Pick the largest number: $\frac{3}{5}$, 58%, .61

34. A team has won $\frac{5}{8}$ of the games played.

    a. What percent of its games has it won?

    b. What percent of its games has it lost?

35. Which of the following numbers are equal to the fraction $\frac{3}{8}$?

$$.37\tfrac{1}{2} \qquad 37\tfrac{1}{2} \qquad 37\tfrac{1}{2}\% \qquad .37\tfrac{1}{2}\% \qquad \tfrac{3}{8}\% \qquad 37.5$$

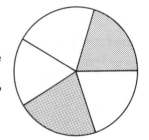

36. a. What percent of the circle shown at the right is shaded?

    b. What percent of the circle is unshaded?

37. Mr. Jones used $\frac{7}{8}$ of the oil in his tank.

    a. What percent of the oil did he use?

    b. What percent of the oil remains in the tank?

Following are some commonly used percents and their fractional equivalents. Learn them.

Percent	Fractional equivalent	Percent	Fractional equivalent
$10\% =$	$\frac{1}{10}$	$12\frac{1}{2}\% =$	$\frac{1}{8}$
$20\% =$	$\frac{1}{5}$	$25\% =$	$\frac{1}{4}$
$30\% =$	$\frac{3}{10}$	$37\frac{1}{2}\% =$	$\frac{3}{8}$
$40\% =$	$\frac{2}{5}$	$62\frac{1}{2}\% =$	$\frac{5}{8}$
$50\% =$	$\frac{1}{2}$	$75\% =$	$\frac{3}{4}$
$60\% =$	$\frac{3}{5}$	$87\frac{1}{2}\% =$	$\frac{7}{8}$
$70\% =$	$\frac{7}{10}$	$16\frac{2}{3}\% =$	$\frac{1}{6}$
$80\% =$	$\frac{4}{5}$	$33\frac{1}{3}\% =$	$\frac{1}{3}$
$90\% =$	$\frac{9}{10}$	$66\frac{2}{3}\% =$	$\frac{2}{3}$
$100\% =$	1	$83\frac{1}{3}\% =$	$\frac{5}{6}$

4. FINDING A PERCENT OF A NUMBER

Angelo buys a radio marked $48 and gets 25% off the marked price. In order to find out how much he saves, we can change 25% to .25 and multiply this decimal by $48. We see that he gets $12 off the marked price:

$$\begin{array}{r} \$48 \\ \times .25 \\ \hline 2\ 40 \\ 9\ 6 \\ \hline \$12.00 \end{array}$$

It is also possible to do this problem by changing 25% to the fraction $\frac{1}{4}$ and then multiplying $\frac{1}{4}$ by $48. Thus, $\dfrac{1}{\overset{}{\underset{1}{4}}} \times \overset{12}{\cancel{48}} = \$12$.

EXAMPLES

1. Find 28% of 45.

$28\% = .28$

$$\begin{array}{r} 45 \\ \times .28 \\ \hline 3\ 60 \\ 9\ 0 \\ \hline 12.60 = 12.6 \end{array}$$

2. Find $37\frac{1}{2}\%$ of $32.

$$37\frac{1}{2}\% = \frac{3}{8}$$

$$\dfrac{3}{\overset{}{\underset{1}{8}}} \times \overset{4}{\cancel{32}} = \$12$$

3. Find $\frac{1}{4}$% of $1200.

$$\frac{1}{4}\% = .00\frac{1}{4}$$

$$\begin{array}{r} 1200 \\ \times .00\frac{1}{4} \\ \hline \$3.00 \end{array}$$

4. Find 7% of $12.40 to the nearest cent.

$$7\% = .07$$

$$\begin{array}{r} 12.40 \\ \times .07 \\ \hline .8680 = \$.87 \end{array}$$

RULE: To find a percent of a number, change the percent to an equivalent decimal or fraction and multiply this decimal or fraction by the number.

Note that, in example **2,** it is simpler to multiply 32 by $\frac{3}{8}$ than to multiply 32 by .375. Use this method when you can find a common fraction equivalent to the given percent.

Exercises

In 1 to 30, find the indicated percent.

1. 25% of 740   **2.** 6% of 50   **3.** $12\frac{1}{2}$% of 88

4. 16% of 85   **5.** 2.7% of 30   **6.** 82% of 94

7. $5\frac{1}{2}$% of $780   **8.** $33\frac{1}{3}$% of 63   **9.** $8\frac{1}{4}$% of $2400

10. 30% of 110   **11.** $62\frac{1}{2}$% of 96   **12.** 60% of 75

13. $66\frac{2}{3}$% of $360   **14.** 75% of 84   **15.** 80% of $65

16. $87\frac{1}{2}$% of 400   **17.** 12% of 320   **18.** 9% of 1400

19. $\frac{1}{2}$% of 1600   **20.** $3\frac{1}{2}$% of 600   **21.** 120% of 85

22. 1% of $820   **23.** $83\frac{1}{3}$% of 720   **24.** 90% of 380

25. $37\frac{1}{2}$% of 640   **26.** 8.4% of $300   **27.** 4.5% of 40

28. $\frac{1}{3}$% of 630   **29.** 150% of 78   **30.** $112\frac{1}{2}$% of 16

31. A man earning $325 per week saves 20% of his salary. How much does he save each week?

32. There are 2400 students in a school. On a certain day, 94% are present.
   a. How many students are present that day?
   b. How many students are absent that day?

33. 70% of the weight of the human body is water. If Anne weighs 110 pounds, how much of her weight (in pounds) is water?

34. A boy pays an 8% sales tax on a $60 radio. How much tax does he pay?

35. A roast beef loses 25% of its weight in cooking. How much will a 6-pound roast weigh after it is cooked?

36. In a shipment of 1100 oranges, about 4% are spoiled.
   a. How many oranges are unspoiled?
   b. How many oranges are spoiled?

37. A team lost 18% of the 50 games it played one season.
   a. How many games did it win?
   b. How many games did it lose?

38. Paul receives a $12\frac{1}{2}$% commission on whatever he sells. If Paul's sales are $1680 one week, how much is his commission?

---

5. FINDING WHAT PERCENT ONE NUMBER IS OF ANOTHER

In a class of 30 students, 6 are absent. What percent of the class is absent?

We already know how to find what fractional part 6 is of 30: it is $\frac{6}{30} = \frac{1}{5}$. We then know from our table that $\frac{1}{5} = 20\%$. Or, we can convert $\frac{1}{5}$ to a percent by division.

To avoid difficulty, remember that the number with the word "of" is in the denominator of the fraction formed and the number with the word "is" is in the numerator.

ILLUSTRATIVE PROBLEM: Find, to the nearest whole percent, what percent of 7 is 3.

Solution: $\dfrac{\text{is}}{\text{of}} = \dfrac{3}{7}$

$7 \overline{)3.000} .428 = .43 = 43\%$ (to nearest whole percent)

Answer: To the nearest whole percent, 3 is 43% of 7.

Exercises _____

1. 12 is what percent of 60?   2. 32 is what percent of 40?
3. 50 is what percent of 150?   4. What percent of 20 is 18?
5. What percent of 40 is 28?   6. What percent of 40 is 15?
7. 18 is what percent of 18?   8. What percent of $30 is $1.50?
9. What percent of $17 is $.34?   10. 1.5 is what percent of 9?
11. What percent of 12 is 15?   12. 100 is what percent of 250?

13. Answer, correct to the *nearest percent:*
    a. What percent of 28 is 4?   **b.** 35 is what percent of 120?
    c. 6 is what percent of 90?   **d.** $14 is what percent of $78?
    e. What percent of 9 is 4?   **f.** 15 is what percent of 48?

14. Nancy got 4 questions wrong on a test of 25 questions.
    a. What percent of the questions did she get wrong?
    b. What percent did she get right?

15. A class is made up of 12 girls and 18 boys.
    a. What percent of the class consists of girls?
    b. What percent consists of boys?

16. A $35 jacket is reduced to $28. What percent of the original cost is the amount of the reduction?

17. 9 inches is what percent of a foot?

18. A pint is what percent of a gallon?

19. A 150-pound man on earth weighs 24 pounds on the moon. What percent of his weight on earth is his weight on the moon?

20. A man earning $300 per week in salary is given an increase to $360 per week. What percent of his original salary is the increase?

21. In a class of 32 students, 4 are absent.
    a. What percent of the class is absent?
    b. What percent is present?

22. A team won 24 games out of a total of 33 played in one season.
    a. What percent of its games played did it win (correct to the nearest percent)?
    b. What percent did it lose?

23. In 15 quarts of milk, there is 1 quart of butterfat. What is the percent of butterfat?

24. A woman bought a $48.00 tire and paid a total of $51.84, including sales tax.
   a. How much is the sales tax?
   b. What percent of the price before the tax is the sales tax?

25. The population of a town was 2000 and grew to 5000 ten years later.
   a. How much is the increase in population?
   b. By what percent has the original population increased?

26. A clothing dealer buys a jacket for $40 and sells it for $90.
   a. How much is his profit on the sale?
   b. What percent of his cost is his profit?

---

6. FINDING A NUMBER WHEN A PERCENT OF IT IS KNOWN

Twenty-five percent of a number is 21. Find the number.

For this type of problem, it is frequently convenient to use equations. Let x represent the number. Then:

Method 1

$$25\%x = 21$$
$$.25x = 21 \quad (25\% = .25)$$
$$25x = 2100 \quad \text{(multiply both sides by 100)}$$
$$\frac{25x}{25} = \frac{2100}{25} \quad \text{(divide both sides by 25)}$$
$$x = 84$$

Method 2

$$25\%x = 21$$
$$\frac{1}{4}x = 21 \quad \left(25\% = \frac{1}{4}\right)$$
$$\frac{1}{4}x \cdot 4 = 21 \cdot 4 \quad \text{(multiply both sides by 4)}$$
$$x = 84$$

Check: Is 25% of 84 equal to 21?
$$\tfrac{1}{4} \times 84 = 21 \;\checkmark$$

Answer: The number is 84.

ILLUSTRATIVE PROBLEM: Carl spent $153.00 for a suit, which was 85% of the original price. What was the original price?

Solution: Let x = the original price. Then

$$.85x = \$153$$

$$85x = 15300 \quad \text{(multiply both sides by 100)}$$

$$\frac{85x}{85} = \frac{15300}{85}$$

$$x = 180$$

Check: Does 85% of $180 yield $153.00?

$$
\begin{array}{r}
180 \\
\times\ .85 \\
\hline
900 \\
14400 \\
\hline
\$153.00 \quad \checkmark
\end{array}
$$

Answer: $180.00

Exercises

1. 40% of what number is 32?   **2.** 30 is 20% of what number?

3. 50% of a certain number is 13. What is the number?

4. If $32 is 15% of a certain amount, what is the amount? (Answer to the nearest cent.)

5. If 18 is $33\frac{1}{3}$% of a number, find the number.

6. 7% of what number is 28?   **7.** 24 is $37\frac{1}{2}$% of what number?

8. 35 is $62\frac{1}{2}$% of what number?   **9.** 75% of what number is 120?

10. $83\frac{1}{3}$% of what number is 65?

11. 120% of some number is 35. Find the number.

12. David bought a camera for $187.50, which was 75% of the original price. What was the original price?

13. How much must a salesman sell per week to earn $160, if he is paid 8% commission on sales?

14. The selling price of a radio is 130% of its cost. If the selling price is $65, what is the cost of the radio?

15. The graduating class of a school consists of 180 students. If this is 16% of the school's enrollment, what is the total enrollment of the school?

16. A family spends $450 per month for rent, which is 24% of their monthly income. What is their monthly income?

17. A team won 48 games, which was $66\frac{2}{3}\%$ of the games it played that season.

    a. How many games did it play?

    b. How many games did it lose?

18. Nancy saved $3.25, which was $62\frac{1}{2}\%$ of her allowance for the week. What was her weekly allowance?

---

7. DISCOUNT PROBLEMS

A suit marked $160 is sold at a 25% discount. How much does the buyer pay for the suit?

The buyer figures that the amount of the discount (or reduction) is 25% of $160, which is .25 × 160 = 40. He then subtracts $40 from $160 and arrives at $120, which he pays for the suit.

The original price, $160, is called the *list price* or *marked price*. The percent of reduction, 25%, is called the *rate of discount*. The amount of reduction, $40, is called the *discount*. The reduced price paid by the buyer, $120, is called the *net price* or *sale price*. To summarize:

Discount = List price × Rate of discount

Net price = List price − Discount
(Sale price)   (Marked   (Amount of
             price)   reduction)

Another way of solving this problem is to think of the original price as 100% and the net price as 100% − 25% = 75% of the original price. Thus, the net price is 75% of $160 = .75 × $160 = $120.

ILLUSTRATIVE PROBLEM 1: A $360 TV set is offered at a discount of 40%. What is the sale price of the set?

Solution:

Method 1	Method 2
$360 × .40 = $144 (discount)	100% − 40% = 60%
$360 − $144 = $216 (*answer*)	$360 × .60 = $216 (*answer*)

ILLUSTRATIVE PROBLEM 2:   Tom paid $45 for a jacket marked at $60. What was the rate of discount?

Solution:

$60   (marked price)
−$45   (sale price)
$15   (discount)

   We must now find what percent the discount, $15, is of the marked price, $60.

Rule: Rate of discount $= \dfrac{\text{Amount of discount}}{\text{Marked price}}$.

Express the fraction as a percent.

$$\frac{\text{Amount of discount}}{\text{Marked price}} = \frac{\$15}{\$60} = \frac{1}{4} = 25\% \quad (answer)$$

ILLUSTRATIVE PROBLEM 3:   A dealer buys a sofa, listed at $350, at a discount of 20%. He decides to pay cash for it and gets a further discount of 2%. What was the net price?

Solution: Such discounts are called **successive discounts** and are common in the business world. We figure successive discounts step by step:

Step 1. $350   (list price)
             ×.20   (first rate of discount)
           $70.00   (first discount)

Step 2. $350   (list price)
             −70   (first discount)
            $280   (price after first discount)

Step 3. $280   (price after first discount)
             ×.02   (rate of second discount)
            $5.60   (second discount)

Step 4. $280.00   (price after first discount)
           −5.60   (second discount)
         $274.40   (*answer*)

   Note that the second discount is applied to the price *after* the first discount has been considered. Two successive discounts of 20% and 2% are not equivalent to a single discount of 22%. For a single discount of 22%, the net price is $273, as shown.

$350 $350
×.22   −77
 700     $273
 700
$77.00

ILLUSTRATIVE PROBLEM 4:   Tom buys a bicycle for $105 after receiving a 30% discount. What was the original (marked) price of the bicycle?

Solution: We may use equations to solve such problems. If the rate of discount was 30%, then Tom paid 100% − 30% = 70% of the original price.

$$\text{Let } x = \text{the original price in dollars.}$$

$$\text{Then } .70x = \text{the amount Tom paid.}$$

$$\text{Thus: } .70x = \$105$$

$$70x = \$10,500 \text{ (multiply by 100)}$$

$$\frac{70x}{70} = \frac{\$10,500}{70}$$

$$x = \$150 \quad (answer)$$

Exercises

1. Mrs. Gomez bought a $40 dress at a discount of 15%. How much did she pay for the dress?

2. A shirt listed at $12 is sold at a 20% discount.
   a. How much is the discount?   **b.** How much is the sale price?

3. Mr. Gold buys a $5600 used car at a discount of 12%.
   a. How much is the discount?   **b.** How much does he pay?

4. A man buys a $40 jacket for $30.
   a. How much was the discount?
   b. What was the rate of discount?

5. A $30 toaster is sold for $25.
   a. How much was the discount?
   b. What was the rate of discount?

6. A steam iron marked $40 is sold at a discount of 18%.
   a. How much is the discount?   **b.** How much is the sale price?

7. Two stores were selling the same model camera. In shop A, the sign said "$80 less 20%." In shop B, the sign said "$88 less 25%."
   a. Which store made the better offer?
   b. How much better is it?

8. A $40 tennis racket is sold for $36. What is the rate of discount?

9. A teacher buys textbooks listed at $5.40 each at a 10% discount. How much does she pay for each book?

10. A dealer receives a bill for $480. He pays within 10 days and receives a 3% discount. How much does he pay?

11. An auto dealer offers a discount of 15% on a $3000 used car. He offers a further 3% discount for cash payment. What is the least amount for which the car can be purchased?

12. If you received a $3 discount on a bill for $25, what rate of discount did you receive?

In 13 to 18, find the *rate of discount.*

	List price	Net price
13.	$80	$50
14.	$125	$110
15.	$42.50	$30.00
16.	$120	$110
17.	$245	$195
18.	$58.40	$52.56

19. A chair listed at $240 is sold with successive discounts of 30% and 5%. What was the sale price?

In 20 to 22, find the *net price.*

	List price	Rates of successive discount
20.	$400	20% and 10%
21.	$80	10% and 4%
22.	$480	$12\frac{1}{2}$% and 5%

23. A man can buy a $200 motor from dealer A at a 15% discount. He can get the same $200 motor from dealer B with successive discounts of 10% and 5%.

    a. Which is cheaper?   **b.** How much cheaper is it?

24. Mrs. White buys a chair for $54 after a 10% discount. What was the list price of the chair?

25. Mr. Roberts pays $68 for a lawn mower after a 15% discount. What was the list price of the mower?

In 26 to 28, find the *list price*.

	Net price	Rate of discount
26.	$51.80	30%
27.	$11.50	8%
28.	$150.00	$37\frac{1}{2}\%$

29. Mr. Finney bought a $160 power saw at a sale and got a 15% discount.

a. What was the net price of the saw?

b. If he paid a 7% sales tax, what was the total amount of the bill?

30. Maria paid $36.00 for a baseball uniform after receiving a 20% discount. What was the marked price of the uniform?

---

8. INTEREST FORMULA

When people in business borrow money for a period of time, they must pay for the use of this money. If Mr. A borrows $1000 from Mr. B for a period of a year, and then pays him back $1060, he pays $60 for the use of the $1000 for a year. This $60 is called the **interest** on the loan.

Likewise, when you put money in the bank, you receive interest for "lending" it to the bank. (The bank, in turn, lends it out to others.) If you put $100 in the bank at an annual rate of interest of 6%, you get back $106 at the end of the year. This is the original $100 plus 6% of $100, which is $6. The $100 is called the **principal;** the 6% is the **rate of interest;** the $6 is the **interest;** the $106 is the **amount** (meaning the amount returned).

The rate of interest is usually stated **annually,** that is, for a one-year period. If the loan just described were for $\frac{1}{2}$ year, the interest would be $\frac{1}{2}$ of $6, or $3. If the loan were for 2 years, then the interest would be $2 \times \$6$, or $12. Thus, we see that we must multiply the annual interest by the time in years to obtain the interest for any loan.

We may summarize the relationship as follows:

Interest = Principal × Rate of interest × Time in years

As a formula, we may write

$$I = PRT$$

where I is the interest, P is the principal, R is the rate of interest, and T is the time in years.

ILLUSTRATIVE PROBLEM 1:

a. Find the interest on $650 at 8% for 9 months.

b. Find the amount.

Solution:

a. $I = PRT$

Since 9 months = $\frac{9}{12} = \frac{3}{4}$ year, we substitute $P = 650$, $R = \frac{8}{100}$, and $T = \frac{3}{4}$:

$$I = 650 \times \frac{\overset{2}{\cancel{8}}}{100} \times \frac{3}{\underset{1}{\cancel{4}}} = \frac{3900}{100} = \$39 \quad (answer)$$

b. Note that to find the *amount* we must add the principal to the interest. As a formula, $A = P + I$, where A is the amount.

$$A = 650 + 39 = \$689 \ (answer)$$

ILLUSTRATIVE PROBLEM 2: What principal will yield an interest of $22.50 if invested at 5% annually for 6 months?

Solution: $I = PRT$

Since 6 months = $\frac{1}{2}$ year, we substitute $I = 22.50$, $R = \frac{5}{100}$, $T = \frac{1}{2}$:

$$22.50 = P \times \frac{5}{100} \times \frac{1}{2}$$

$$22.50 = \frac{5}{200}P$$

$$22.50 = \frac{1}{40}P \ \text{ or } \ \frac{1}{40}P = 22.50$$

Multiplying both sides by 40, we have

$$P = 40 \times 22.50 = \$900 \quad (answer)$$

ILLUSTRATIVE PROBLEM 3: At what annual rate of interest will a bond of $450 yield an interest payment of $67.50 in $2\frac{1}{2}$ years?

Solution:

$$I = PRT$$

Substitute: $67.50 = \overset{225}{\cancel{450}} \times R \times \dfrac{5}{\underset{1}{\cancel{2}}}$

$$\frac{67.50}{1125} = \frac{1125R}{1125}$$

$$R = \frac{67.50}{1125} = .06$$

$$R = 6\% \quad (answer)$$

Exercises

In 1 to 5, find the interest on the loans.

1. $600 at 5% for $2\frac{1}{2}$ years   **2.** $350 at 4% for 2 years

3. $1200 at $4\frac{1}{2}$% for 9 months   **4.** $1500 at 6% for 6 months

5. $2400 at 5% for 1 year 6 months

6. Mr. McCrea borrowed $1200 at 11% interest.
  a. How much interest did he pay at the end of 2 years?
  b. What is the total *amount* he returned?

7. What principal yields an interest of $80 after 2 years at an annual rate of interest of 6%?

8. Sue's father borrowed some money at 11% annual interest for 9 months. When he repaid the principal, he paid $55.00 in interest.
  a. How much did he borrow?
  b. What is the total *amount* he returned?

In 9 to 15, find the missing item.

	Principal	Rate	Time	Interest
9.	?	7%	2 yr.	$14
10.	$325	?	1 yr. 6 mo.	$39
11.	$375	6%	?	$56.25
12.	$1200	?	$2\frac{1}{2}$ yr.	$165
13.	$1560	$5\frac{1}{2}$%	2 yr.	?
14.	?	$7\frac{1}{2}$%	$1\frac{1}{2}$ yr.	$87.75
15.	$4000	?	2 yr. 6 mo.	$750

16. Mrs. Perkus paid $85.50 interest for a 2-year loan at 9% annual interest. How much did she borrow?

17. If you paid $42 interest for a $175 loan for 1 yr. 6 mo., what was the rate of interest?

18. If you borrow $450 at 7% interest, for what period of time would you have the money if the interest is $63?

19. How much must you deposit in a bank at $5\frac{1}{2}$% annual interest in order to earn an interest payment of $81 in $1\frac{1}{2}$ years?

20. The interest on a $680 loan for $2\frac{1}{2}$ years is $153. What rate of interest is charged?

Chapter Review Exercises

1. A baseball park with 20,000 seats had 30% of its seats vacant on a particular day. How many spectators were present?

2. The principal of a school with 2400 students expects a 12% decrease in the number of students next year. How many students does he expect to have in the school next year?

3. Find 32% of 1800.

4. 16 is what percent of 40?

5. Find 125% of 220.

6. 26 is what percent of 20?

7. Find 8.2% of $970.

8. Thirty percent of what number is equal to 72?

9. A turkey loses 15% of its weight in roasting. If a turkey weighs 12 pounds after roasting, how much did it weigh before? (Answer to the nearest pound.)

10. 14 is what percent of 30? (Answer to the nearest percent.)

11. What percent of 80 is 15?

12. A $180 set of books is sold at a discount of $9\frac{1}{2}$%. What is the sale price?

13. a. Find the interest on a loan of $850 at $7\frac{1}{2}$% for 2 years.
    b. What is the total amount returned?

14. A $200 washing machine was purchased for $170. What was the rate of discount?

15. A boy paid $48 for a radio after a $33\frac{1}{3}$% discount. What was the list price of the radio?

16. A salesperson earns 12% of her total sales as commission. If she sells $2500 worth of merchandise one week, how much is her commission?

17. A man deposits some money in a bank at 6% interest and receives $22.50 in interest after $2\frac{1}{2}$ years. How much did he deposit?

18. A woman buys a $42 dress at a 30% discount.
    a. What is the reduced price?
    b. How much does she pay if there is an 8% sales tax on the dress?

19. A girl gets 25 questions right on a 30-question short-answer quiz.
    a. What percent does she have right?
    b. What percent does she have wrong?

20. What principal will yield an interest of $45 at a rate of 6% for $1\frac{1}{2}$ years?

CHECK YOUR SKILLS

1. Subtract: $89.27
          $- 6.98$

2. What is .875 expressed as a common fraction in lowest terms?

3. 6 is what percent of 48?

4. A man buys some shirts for $31.60 and pays 7% tax. How much tax does he pay to the *nearest cent*?

5. 60 is 120% of what number?

6. Solve for x: $2.3x + 4.5 = 16$

7. Find the value of y in the proportion $\dfrac{20}{12} = \dfrac{5}{y}$.

8. A $60 radio is sold at a discount of 20%. What is the sale price?

9. If the yearly income from a $1000 investment is $80, what is the annual interest rate?

10. 126 is $\frac{6}{5}$ of what number?

8

Linear Measure, Area, and Volume

1. MEASUREMENT OF LENGTH

Measurements are always made by comparing the object being measured with a standard unit of measure. In the U.S.A., the customary units for measuring lengths are the inch, the foot, the yard, and the mile. These units are based on the English system of measurement.

The relationships among these units are indicated in the following table:

Units of Linear Measure
12 inches = 1 foot
36 inches = 1 yard
3 feet = 1 yard
5280 feet = 1 mile
1760 yards = 1 mile

As you know, we often abbreviate feet by ft or by a single stroke '. Likewise, we abbreviate inches by in. or by a double stroke ". Thus, 5 feet 3 inches may be written as 5 ft 3 in. or as 5'3".

We measure length with a ruler which is usually marked off in inches; each inch is subdivided into quarters, eighths, and sixteenths. For larger lengths, we use a yardstick or a tape, and indicate the measurements in feet or yards. For greater distances, we indicate measurements in miles.

When we must change from one unit of measure to another, the table of units of linear measure helps us to do this.

ILLUSTRATIVE PROBLEM 1: Change 66 inches to feet.

Solution: As we change from inches to feet to represent a given quantity, we realize that the number of feet will be smaller than the number of inches.

Using the fact that 12 inches = 1 foot, we must reduce the size of 66 by dividing it by 12. Thus,

$$\frac{66}{12} = 5\frac{6}{12} = 5\frac{1}{2}$$

Answer: 66 in. = $5\frac{1}{2}$ ft

ILLUSTRATIVE PROBLEM 2: A racetrack is 440 yards in length. How many laps must a boy run in a 1-mile race?

Solution: Since there are 1760 yards in 1 mile, we must see how many times 440 is contained in 1760. Thus,

$$\begin{array}{r} 4 \\ 440\overline{)1760} \\ \underline{1760} \end{array}$$

Answer: The boy must run 4 laps.

ILLUSTRATIVE PROBLEM 3: How many feet are in: **a.** 5 yd? **b.** *x* yd?

Solution:

a. 5 yards contain 5 × 3 = 15 feet. (*answer*)

b. *x* yards contain $x \cdot 3 = 3x$ feet. (*answer*)

Exercises

1. Change to feet, using mixed numbers or decimals to express your answer when necessary:
a. 7 yd	**b.** $5\frac{1}{3}$ yd	**c.** 48 in.	**d.** *y* in.
e. 2 miles	**f.** 6 yd 2 ft	**g.** $3\frac{1}{2}$ miles	**h.** *x* miles
i. .8 mile	**j.** $1\frac{2}{3}$ yd	**k.** 5 ft 8 in.	

2. Change to inches:
a. 7 ft	**b.** 2 yd 2 ft	**c.** $3\frac{1}{4}$ ft	**d.** *f* ft
e. 1.2 yd	**f.** 4.2 yd	**g.** 5 ft 8 in.	**h.** *y* yards

3. Tell how many yards are in each measurement.

    a. 15 ft   **b.** 90″   **c.** $3\frac{1}{2}$ miles

    d. 2.2 miles   **e.** k ft   **f.** x miles

    g. i inches   **h.** 127 ft   **i.** 8′6″

4. A furlong, used in horse racing, is a distance of $\frac{1}{8}$ mile.

    a. How many miles are in 12 furlongs?

    b. How many yards are in a furlong?

5. Bill is 5 feet 8 inches tall. How tall is he in inches?

6. A jet plane can fly at an altitude of 35,000 feet. How many miles high is this (to the nearest tenth of a mile)?

7. Depth of water is often expressed in fathoms. A fathom is 6 feet. If the depth of a bay at a certain point is 68 feet, what is its depth in fathoms?

8. A board 6 feet long is cut into pieces, each 8 inches long. How many pieces are obtained?

9. What part of a yard is 20 inches?

10. A woman buys $\frac{3}{4}$ yard of ribbon and cuts off 4 pieces, each 5 inches long. How many inches of ribbon are left?

11. A truck 90 inches high enters a tunnel which is 8 feet high. By how many inches does the truck clear the tunnel?

12. Represent the total number of inches in x ft y in.

13. A rod is equivalent to $5\frac{1}{2}$ yards. How many rods are in a mile?

14. A nautical mile is equivalent to 1.15 land miles. How many feet (to the nearest foot) are in a nautical mile?

---

2. METRIC SYSTEM—LINEAR MEASURE

    Like our numeration system, the metric system is based on the number ten. Since each unit is $\frac{1}{10}$ of the next larger unit, the metric system is easy to use. We shall deal here only with the most commonly used metric units of length, as listed in the following table. Note that the symbol $\approx$ means "is approximately equal to."

Metric-English Approximations
1 kilometer (km) = 1000 meters $\approx \frac{5}{8}$ mile
1 meter (m) (basic unit) $\approx$ 39 inches
1 centimeter (cm) = $\frac{1}{100}$ of a meter $\approx \frac{2}{5}$ inch
1 millimeter (mm) = $\frac{1}{1000}$ of a meter $\approx \frac{1}{25}$ inch

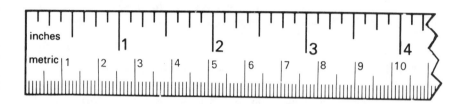

In the figure, each of the numbered markings on the upper edge of the ruler indicates 1 inch. Each of the numbered markings on the lower edge of the ruler indicates 1 centimeter. The small markings within each centimeter indicate millimeters. Since each centimeter is divided into 10 equal parts, 1 millimeter = $\frac{1}{10}$ centimeter. Ten centimeters would be a decimeter, which is $\frac{1}{10}$ of a meter, but this unit is not commonly used. A 1-meter-long ruler (meterstick) is slightly longer than a yardstick and is equal to 100 centimeters.

Let us see how to convert from one system to the other.

ILLUSTRATIVE PROBLEM 1: An artillery shell is 120 mm in diameter.
a. What is its diameter in centimeters?
b. What is its diameter in inches?

Solution:
a. Since 10 mm = 1 cm, we must see how many times 10 mm is contained in 120 mm:

$$120 \div 10 = 12 \text{ cm} \quad (answer)$$

b. Since 1 cm $\approx \frac{2}{5}$ in., 12 cm is approximately equal to:

$$12 \times \frac{2}{5} \approx \frac{24}{5} \approx 4\frac{4}{5} \text{ in.} \quad (answer)$$

ILLUSTRATIVE PROBLEM 2: A carpenter's nail is 1 in. long. What is its length in centimeters?

Solution:

From the preceding table, we see that $\frac{2}{5}$ in. $\approx$ 1 cm. Since $\frac{5}{2} \times \frac{2}{5} = 1$, multiply both sides of the equation by $\frac{5}{2}$.

$$\tfrac{5}{2} \times \tfrac{2}{5} \text{ in.} \approx \tfrac{5}{2} \times 1 \text{ cm}$$

$$1 \text{ in.} \approx \tfrac{5}{2} \text{ cm or 2.5 cm} \quad (answer)$$

Exercises

In the following exercises, use the values given in the table on page 129 to obtain the approximate answers.

1. What is the difference in inches between a yard and a meter?

2. What is the difference in yards between a 100-yard dash and a 100-meter dash?

3. Film comes in widths of 8 mm, 16 mm, and 35 mm. How many inches are in each of these?

4. A one-mile race equals:
   a. __?__ kilometers   b. __?__ meters

5. a. 7 cm = __?__ mm   b. 2.4 cm = __?__ mm   c. x cm = __?__ mm
   d. 73 mm = __?__ cm   e. 87 mm = __?__ cm   f. y mm = __?__ cm

6. a. 6 m = __?__ cm   b. 8.3 m = __?__ cm   c. 700 cm = __?__ m
   d. 340 cm = __?__ m   e. k meters = __?__ cm   f. t cm = __?__ m

7. a. 6 km = __?__ m   b. 4.2 km = __?__ m   c. 3000 m = __?__ km
   d. 4700 m = __?__ km   e. x meters = __?__ km   f. y km = __?__ m

8. The distance between two cities in Europe is 740 kilometers. How many miles is this?

9. The high-jump record a few years ago was 2.1 meters. How high is this in feet and inches? (Answer to the nearest inch.)

10. Which is shorter, a millimeter or $\frac{1}{16}$ of an inch? How much shorter in inches? (Answer to the nearest hundredth.)

11. Tell how many miles are in:
    a. 80 km   b. 400 km   c. 60 km   d. n km

12. Tell how many km are in:

a. 20 mi   **b.** 100 mi   **c.** 72 mi   **d.** *r* mi

13. A plane flies at 450 miles per hour. How many kilometers per hour is this?

14. The circumference of the earth is about 25,000 miles. About how many kilometers is this?

15. A barometer reading one day in Rome is 750 mm. How many inches is this?

16. On a road near Brussels, the speed limit is 80 kilometers per hour. About how many miles per hour is this?

17. About how many centimeters are there in

a. an inch?   **b.** a foot?

---

3. AREA MEASURE—SQUARE AND RECTANGLE

The smaller figure below represents a square one inch on each side. Such a square is called a *square inch.* The larger figure represents a rectangle whose *base* is 4 inches and whose *height* is 3 inches. We count the number of squares in the rectangle and we see that there are 12.

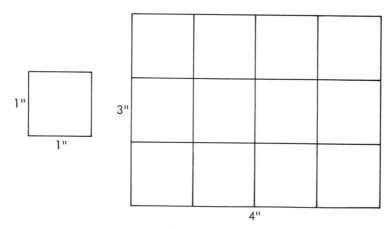

The number of square units that are contained within a figure is called the *area* of the figure. The area of the large rectangle is 12 square inches. We may use other units of area such as square foot, square yard, square mile, square centimeter, or square meter.

Looking at the figure of the rectangle, we see that we could have obtained the area more quickly by multiplying the base (4 in.) by the height (3 in.), which gives us 12 sq in. This example indicates to us that, in general, if the base and height are in the same units, we may obtain the area, A, by the formula

$$A = bh$$

where b is the base of the rectangle and h is its height.

ILLUSTRATIVE PROBLEM 1: A classroom is $30' \times 28'$. How many square feet are in its area?

Solution:
$$A = bh$$
$$A = 30 \times 28$$
$$A = 840$$

Answer: The area is 840 sq ft.

ILLUSTRATIVE PROBLEM 2: Find the area of a rectangular table that is 5 ft 6 in. long and 4 ft wide.

Solution: Convert 5 ft 6 in. to $5\frac{6}{12} = 5\frac{1}{2}$ ft.

$$A = bh$$

$$A = 5\frac{1}{2} \times 4$$

$$A = \frac{11}{2} \times 4$$

$$A = 22$$

Answer: The area is 22 sq ft.

ILLUSTRATIVE PROBLEM 3: The area of a rectangle is 350 sq ft. If its base is 35 ft, what is its height?

Solution:
$$A = bh$$
$$350 = 35h$$
$$\frac{350}{35} = \frac{35h}{35}$$
$$h = 10$$

Answer: The height of the rectangle is 10 ft.

Exercises

In 1 to 13, find the area of the rectangle.

1. 15 ft by 12 ft
2. $18'' \times 11''$
3. $30' \times 10\frac{2}{3}'$
4. 7.2 yd by 5.4 yd
5. $3\frac{1}{2}$ mi by $2\frac{1}{4}$ mi
6. 20 cm by 15 cm
7. 85 cm by 1 m
8. 18 ft by 10 ft 6 in.
9. 5 ft 6 in. by 6 ft 3 in.
10. x ft by y ft
11. k ft by 8 ft 6 in.
12. 7 ft by 3 yd
13. l ft long by w ft wide

14. A rectangular mirror $32'' \times 20''$ is to be resilvered at a cost of 5¢ per square inch. What will be the total cost?

15. What is the cost of a $9' \times 12'$ rug at a cost of $3 per square foot?

16. How many square yards of carpeting are needed to cover a living room 9 yards long and 5 yards 2 feet wide?

17. How many tiles each 1 sq ft in area are needed to cover the floor of a bathhouse 40 ft long and 35 ft wide?

18. A rectangular plot of ground has an area of 3760 sq ft. If its length is 80 ft, find its width in feet.

19. A tennis court has an area of 880 sq yd. If it is 40 yd long, how many yards are in its width?

20. How many ceiling tiles each 1 sq ft in area are needed to cover the ceiling of a room which is 14 ft by 18 ft?

21. A quart of a certain paint will cover 100 sq ft of wall space. How many quarts are needed to cover a wall of a house that is 45 ft long and 20 ft high?

22. How many sq ft of carpeting are needed for the following living room and dining room?

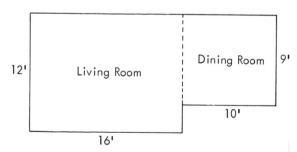

(*Hint:* Divide the figure into two rectangles by drawing the dotted line as shown.)

23. Find the area of the figure shown at the right.

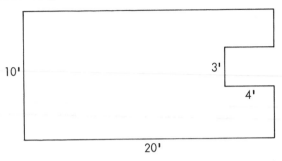

10'

3'

4'

20'

---

4. CHANGING UNITS OF AREA

How many square yards of carpeting are needed to cover a floor that is 72 square feet in area?

To answer this question, we must know how many square feet are in a square yard. A square yard is a square whose base and height are both 1 yard or 3 feet long. Hence, $A = bh$

$$A = 3 \times 3 = 9 \text{ sq ft}$$

Thus, 1 sq yd = 9 sq ft.

In the preceding problem, we see that, since 1 sq yd = 9 sq ft, then 72 sq ft is equivalent to $\frac{72}{9} = 8$ sq yd.

A square foot is a square whose base and height are both 1 foot or 12 inches long. Hence, $A = bh$

$$A = 12 \times 12 = 144 \text{ sq in.}$$

Thus, 1 sq ft = 144 sq in.

ILLUSTRATIVE PROBLEM: How much linoleum is needed to cover a rectangular tabletop 72 inches by 46 inches? Answer in **(a)** square inches, **(b)** square feet, and **(c)** square yards.

Solution: **a.** $A = bh$

$A = 72 \times 46 = 3312$ sq in. (*answer*)

b. $A = \dfrac{3312}{144} = 23$ sq ft (*answer*)

c. $A = \dfrac{23}{9} = 2\dfrac{5}{9}$ sq yd (*answer*)

Exercises

1. 7 sq yd = __?__ sq ft
2. 3 sq ft = __?__ sq in.
3. 63 sq ft = __?__ sq yd
4. $7\frac{2}{3}$ sq yd = __?__ sq ft
5. 72 sq in. = __?__ sq ft
6. 1 sq yd = __?__ sq in.

7. What is the cost of carpeting a floor 15 ft by 21 ft if the carpeting costs $12 per sq yd?

8. What is the total charge for resurfacing a tennis court 78 ft by 36 ft if the surfacing costs $7.50 per sq yd?

9. A man completely panels a closet 9 feet long, 6 feet wide, and 15 feet high. If he must panel the ceiling, the floor, and all four walls, how many square yards of paneling are needed?

10. A room is 18 ft long, 14 ft wide, and 8 ft 6 in. high. How much will it cost to paper the walls and ceiling at $4.50 per sq yd? (Disregard waste due to doors and windows.)

11. Find the number of square yards of linoleum needed to cover a kitchen floor 12 ft wide and 15 ft long.

12. At $7.50 per square yard, find the cost of paving a concrete walk 120 feet long and 6 feet wide.

13. How many tiles one square inch in size are needed to cover a bathroom floor 4 ft by 5 ft?

---

5. AREA OF A TRIANGLE

In right triangle *PQS*, base *b* and height *h* are also the legs of the triangle.

In the figure, we see rectangle *PQRS* with diagonal *QS* drawn. If we cut the rectangle along diagonal *QS*, we find that the figure is divided into two triangles that are equal in

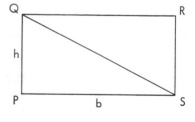

area. Thus, triangle *PQS* is equal in area to one-half the area of the rectangle. Since the area of a rectangle is equal to *bh*, we may write the formula for the area of the triangle as

$$A = \tfrac{1}{2}bh$$

where *b* = base and *h* = height.

Note that the triangle and the rectangle have the same base and height. Although the preceding figure shows how we get the area of a right triangle (a triangle with a 90-degree angle), we can show that the formula holds for any triangle. In the figures below, the area of each triangle is found by using the formula $A = \frac{1}{2}bh$.

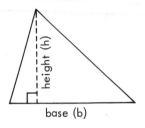
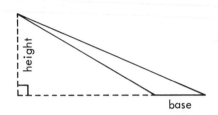

ILLUSTRATIVE PROBLEM 1: Find the area of a triangle whose base is 18″ and height is 7″.

Solution:

$$A = \frac{1}{2}bh$$

$$A = \frac{1}{\overset{}{\underset{1}{2}}} \cdot \overset{9}{\cancel{18}} \cdot 7$$

$$A = 9 \cdot 7 = 63$$

Answer: The area is 63 sq in.

ILLUSTRATIVE PROBLEM 2: The area of a triangle is 192 sq in. If the height is 16 in., find the base.

Solution:

$$A = \frac{1}{2}bh$$

$$192 = \frac{1}{2} \cdot b \cdot 16$$

$$192 = 8b$$

$$\frac{192}{8} = \frac{8b}{8}$$

$$b = 24$$

16″

A = 192

base = ?

Answer: The base is 24 inches.

Exercises_____

1. Find the area of each triangle, given:
   a. $b = 18''$, $h = 5''$ **b.** $b = 12$ ft, $h = 8$ ft
   c. $b = 15$ yd, $h = 7$ yd   **d.** $b = 14$ ft, $h = 7$ ft 6 in.
   e. $b = 18$ ft, $h = 6\frac{1}{2}$ ft   **f.** $b = 10$ in., $h = 2\frac{1}{2}$ ft

2. A triangular sail has a base of 8 ft and a height of 6 ft 3 in. How many square feet are in its area?

3. A triangular piece of plywood weighing $1\frac{1}{2}$ lb per sq ft has a base 2 ft 8 in. and a height 1 ft 6 in. How many pounds does it weigh?

4. A triangle has an area of 46 sq ft. If its base is 16 ft, find its height.

5. Fill in the missing dimensions of the following triangles:

	Base	Height	Area
a.	28 in.	?	140 sq in.
b.	?	26 yd	390 sq yd
c.	60 cm	?	720 sq cm
d.	?	9 in.	3 sq ft

6. A triangular plate has a base of 9″ and a height of 6″. A square slot 2″ on a side is cut out of the plate. Find the area of the remaining metal surface of the plate.

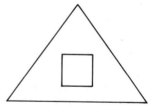

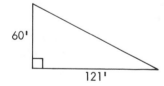

7. A triangular plot of ground has a base of 121 ft and a height of 60 ft. How many acres of land are in this plot? (1 acre = 43,560 sq ft)

8. The figure at the right shows a triangle mounted on a rectangle. Using the dimensions given, find the area of the entire figure.

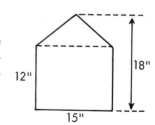

9. A triangular gas station has an area of 2400 sq ft. If the base of the triangle is 80 ft, how many feet are in the height?

10. Find the area of the figure below by breaking it up into triangles and a rectangle.

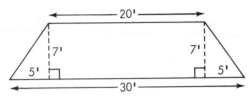

6. USE OF EXPONENTS

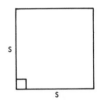

Earlier in the chapter, we saw that the area of a rectangle is given by the formula $A = bh$. In a square, the base and height are equal, so that, if we call the side of the square s, the formula for the area of a square becomes $A = s \times s$.

We usually write this $A = s^2$ (read "s squared") where the small 2 indicates that s appears twice in the product $s \times s$. This small 2 is called an **exponent** and we say that s is raised to the second power. The number s used as a factor is called the **base**.

Exponents are very commonly used in algebra. Thus, y^3 (read "y cubed") would mean $y \times y \times y$, and $aaaa$ would be written a^4 and read "a to the fourth power." An expression such as $ppqqq$ would be written p^2q^3 and is read "p squared q cubed." An expression with numerical base such as $2 \cdot 2 \cdot 2 \cdot 2$ is written 2^4, and is read "2 to the fourth power" or simply "2 to the fourth"; its value is 16.

If we write $2x^3$, only the x appears 3 times in the product. If we wish to indicate that the 2 is also to be cubed, we write $(2x)^3$, which would mean $(2x)(2x)(2x)$.

ILLUSTRATIVE PROBLEM 1: Write, using exponents, $3bbbcc$.

Solution: $3b^3c^2$ *(answer)*

ILLUSTRATIVE PROBLEM 2: Find the value of $2a^3 + 6b^2$ if $a = 4$ and $b = 5$.

Solution:
$$2a^3 + 6b^2 = 2(4)^3 + 6(5)^2$$
$$= 2 \cdot 4 \cdot 4 \cdot 4 + 6 \cdot 5 \cdot 5$$
$$= 128 + 150 = 278 \quad (answer)$$

Exercises———————————————————————————

1. Name the base and exponent in each of the following:
   a. t^2 **b.** d^3 **c.** z^7 **d.** 3^4 **e.** 10^5

2. Find the value of each of the following:
   a. 7^2 **b.** 2^3 **c.** 1^7 **d.** 3^4 **e.** 4^3

3. Write each of the following using exponents:
   a. *rrr*   **b.** *ccccc*   **c.** *tt* + *pppp*
   d. 3*xxx* + 2*yy*   **e.** 7*bbmmm*   **f.** 5*aab* − 7*dddxx*

4. Find the area of a square whose side is:
   a. 8″   **b.** 10 ft   **c.** $\frac{1}{2}$ in.   **d.** 2.5 cm   **e.** $2\frac{1}{3}$ yd

5. If $x = 2$, $y = 3$, and $z = 4$, find the value of each of the following:
   a. $3y^2$ **b.** $4x^3$ **c.** z^2 **d.** x^2y
   e. $y^2 - x^2$ **f.** $3x^2 + y^2$ **g.** $z^3 - xy^2$ **h.** $(5x)^3$
   i. $5x^3$ **j.** x^3y^2z **k.** $(x + y)^2$ **l.** $(z - y)^5$

6. Find the area of a square whose side is 4 ft 6 in.

7. A baseball diamond is a square 30 yd on a side. What is the area enclosed by the base lines in sq yd?

8. The formula $s = 16t^2$ gives the distance, s, in feet, that an object above ground falls in t seconds. How many feet does an object fall in:
   a. 10 seconds?   **b.** 20 seconds?

9. A cube has 6 equal faces, each of which is a square. If each edge of the cube is e, then the total surface area of the cube, S, is given by the formula $S = 6e^2$. Find the total surface area of a cube whose edge is:
   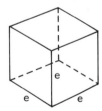
   a. 4 inches   **b.** $2\frac{1}{3}$ yd   **c.** 3 ft 6 in.

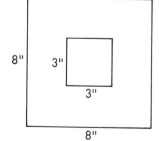

10. A square metal plate 8″ on a side has a smaller square hole punched out of it 3″ on a side. How many square inches of metal remain?

7. AREA OF A CIRCLE

In Chapter 6, we discussed the problem of finding the circumference of a circle when its radius or diameter is given.

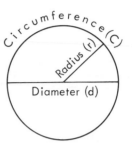

We used the formula

$$C = \pi d$$

where d is the diameter and π is $\frac{22}{7}$ or $3\frac{1}{7}$ or 3.14, approximately. Since $d = 2r$, the formula may be written

$$C = 2\pi r$$

We can find the area of a circle, when its radius is given, by means of the formula

$$A = \pi r^2$$

This formula has been discovered by mathematicians and is in common use. Remember that r is in linear units (inches, feet, etc.) and that A is in square units (sq in., sq ft, etc.).

ILLUSTRATIVE PROBLEM 1: Find the area of a circle whose radius is $3\frac{1}{2}$ inches.

Solution:

$$A = \pi r^2$$

$$A = \frac{22}{7} \left(3\frac{1}{2} \right)^2$$

$$A = \frac{22}{7} \left(\frac{7}{2} \right)^2$$

$$A = \frac{\overset{11}{\cancel{22}}}{\underset{1}{\cancel{7}}} \cdot \frac{\overset{7}{\cancel{49}}}{\underset{2}{\cancel{4}}}$$

$$A = \frac{77}{2} = 38\frac{1}{2}$$

Answer: The area is $38\frac{1}{2}$ sq in.

ILLUSTRATIVE PROBLEM 2: Find, to the nearest square inch, the area of a phonograph record with a 9-inch diameter. (Use $\pi = 3.14$.)

Solution:

$$A = \pi r^2$$

Since $d = 9''$, $r = 4\frac{1}{2}'' = 4.5''$.

$$A = 3.14 \, (4.5)^2$$

$$A = 3.14 \, (20.25)$$

$$A = 63.5850$$

$$A = 64 \text{ sq in.}$$

Answer: The area is 64 sq in., to the nearest sq in.

Exercises

1. Using $\pi = \frac{22}{7}$, find the area of a circle whose radius is:
   a. 7 cm   b. 21 ft   c. 14 yd   d. 35 in.

2. Find the area of the top of a circular piston of diameter 14 inches. (Use $\pi = \frac{22}{7}$.)

3. An artillery shell is 40 mm in diameter. Find the area of the circular base of the shell to the nearest sq mm. (Use $\pi = 3.14$.)

4. A circular lawn is 12 yd in diameter.
   a. Find its area to the nearest sq yd. (Use $\pi = 3.14$.)
   b. About how many square feet of sod are needed to cover this lawn?

5. How many square feet of linoleum are needed to cover a circular tabletop that is 4.2 feet in diameter? (Use $\pi = \frac{22}{7}$.)

6. A circular flower bed has a diameter of 40 feet. What is its area? (Use $\pi = 3.14$.)

7. Using a tape measure, a boy measures the circumference of a half-dollar and finds it to be about 3.96 inches.
   a. Find its diameter to the nearest hundredth of an inch. (Use $\pi = 3.14$.)
   b. Find its area to the nearest hundredth of a square inch.

8. A circular plate 14″ in diameter is cut out of a square piece of tin 14″ on a side, as shown in the figure at the right.

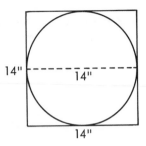

a. What is the area of the square tin plate?

b. What is the area of the circular piece cut out, to the nearest sq in.? (Use $\pi = \frac{22}{7}$.)

c. What is the area of tin that remains?

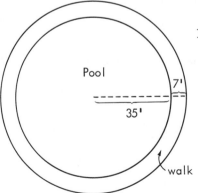

9. a. What is the diameter of the largest circle that can be cut from a square 20 cm on a side?

b. What is the area of the circular cut-out? (Use $\pi = 3.14$.)

c. What is the area of the part of the square that remains?

10. A cow is tied with a 10-yard rope to a post where two fences meet at right angles. She can graze freely over an area allowed by the stretched rope, as shown in the figure.

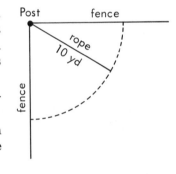

a. Over what fractional part of a circle can she graze?

b. Over how many square yards can she graze, to the nearest square yard? (Use $\pi = 3.14$.)

11. A circular pool 35 ft in radius has a walk 7 ft wide around it.

a. What is the area of the pool? (Use $\pi = \frac{22}{7}$.)

b. What is the area of the circle that includes the pool and the walk?

c. What is the area of the walk?

12. A certain window is made up of a rectangle with a semicircle on top, as shown. How many square feet of stained glass are in this window? (Use π = 3.14.)

5'

2'

13. A football field consists of a rectangle with semicircles at each end, as shown. Using the dimensions given, find, to the nearest square yard, the number of square yards of sod needed to cover the field. (Use π = 3.14.)

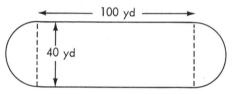

100 yd

40 yd

8. VOLUME OF A RECTANGULAR SOLID

The **volume** of a solid figure is the amount of space it contains. Volume is measured in cubic units. A cubic inch is the space contained in a cube one inch on each edge, as shown in the figure. We may also use units of volume such as the cubic foot, cubic yard, cubic centimeter, and cubic meter.

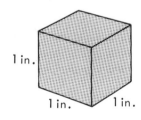

1 in.

1 in.   1 in.

A common solid figure is the **rectangular solid.** Such objects as a box, a board, a book, a room, etc., are in the shape of a rectangular solid, as shown in the figure. In a rectangular solid, the top, bottom, front, back, and both sides

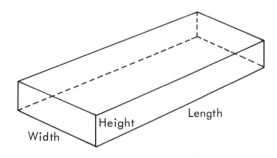

Height   Length

Width

are all rectangles or squares meeting at right angles to each other. These rectangles are called the **faces** of the solid. There are always six faces. The lines in which the faces meet are called the **edges** of the solid, of which there are twelve.

The lengths of these edges are the dimensions of the solid, called *length, width,* and *height,* as shown in the preceding figure.

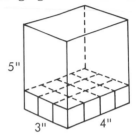

The rectangular solid in the figure has a length of 4″, a width of 3″, and a height of 5″.

In the figure, we see that there are 3 × 4 or 12 cubic inches in the bottom layer. Since the solid is 5″ high, we see that it could contain 5 such layers. Thus, its volume is 3 × 4 × 5 = 60 cubic inches.

This illustrates the general formula for finding the volume of a rectangular solid. If the length (*l*), width (*w*), and height (*h*) are all in the same linear unit, then the volume (*V*) is given by the formula

$$V = lwh$$

ILLUSTRATIVE PROBLEM 1: How many cubic feet are there in a coal bin 12 feet long, 6 feet high, and 5 feet wide?

Solution:
$$V = lwh$$
$$V = 12 \times 6 \times 5$$
$$V = 360 \text{ cu ft } (answer)$$

ILLUSTRATIVE PROBLEM 2: If 1260 cu in. of water are poured into a rectangular aquarium whose base is 12″ × 15″, how high will the water rise?

Solution:
$$V = lwh$$
$$1260 = 15 \times 12 \times h$$
$$1260 = 180h$$
$$\frac{1260}{180} = \frac{180h}{180}$$
$$h = 7″ \ (answer)$$

When the length, width, and height of a rectangular solid are all equal, the figure is a cube. If we represent each edge of the cube by e, then the volume, V, becomes $V = e \times e \times e$ or

$$V = e^3$$

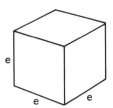

ILLUSTRATIVE PROBLEM 3: How many cubic feet are in the volume of one cubic yard?

Solution: Each edge of a cubic yard has a length of 1 yard, or 3 feet. Therefore, we substitute $e = 3$ in the formula:

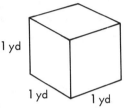

$$V = e^3$$

$$V = 3^3$$

$$V = 27 \text{ cu ft} \quad (answer)$$

In like manner, we can find the volume of a cubic foot in inches:

$$1 \text{ cu ft} = 12^3 = 1728 \text{ cu in.}$$

Exercises

1. Find the volume of a rectangular solid whose dimensions are:
   a. $l = 8''$, $w = 5''$, $h = 4''$
   b. $l = 7'$, $w = 6'$, $h = 5'$
   c. $l = 10$ yd, $w = 6$ yd, $h = 8$ yd
   d. $l = 8\frac{1}{2}$ ft, $w = 6$ ft, $h = 7$ ft
   e. $l = 12$ ft, $w = 8$ ft 9 in., $h = 3$ ft 4 in.
   f. $l = 8$ cm, $w = 7$ cm, $h = 5$ cm

2. Find the missing dimension for each of the following rectangular solids:

	Length	Width	Height	Volume
a.	10″	10″	?	250 cu in.
b.	3′9″	?	2 ft	30 cu ft
c.	?	8 cm	5 cm	420 cu cm
d.	2.5 ft	1.5 ft	?	15 cu ft

3. How many cu ft of air space are there in a room 30′ long, 20′ wide, and 12′ high?

4. A cu ft of ice weighs 57 lb. What is the weight of a block of ice in the shape of a cube 4 ft on each edge?

5. A walk 40 ft long and 5 ft wide is to be paved. It is dug to a depth of 6 in. and filled with crushed rock.

   a. How many cu ft of crushed rock are needed?

   b. Find the cost of the crushed rock at $1.25 per cu ft.

6. An aquarium is 44 in. long, 28 in. wide, and 15 in. high. How many gallons of water will the tank hold? (1 gallon = 231 cu in.)

7. What is the weight, in pounds, of a plate of glass 2 ft long, $1\frac{1}{2}$ ft wide, and $\frac{1}{2}$ in. thick? (1 cu in. of glass weighs 2 oz.)

8. A swimming pool is 30 ft long, 15 ft wide, and has an average depth of $4\frac{1}{2}$ ft. How many gallons of water does it hold if there are $7\frac{1}{2}$ gallons to 1 cu ft?

9. Find how many cu ft there are in:

   a. 5 cu yd   **b.** $\frac{1}{3}$ cu yd   **c.** $\frac{1}{6}$ cu yd
   d. $1\frac{2}{3}$ cu yd   **e.** 864 cu in.

10. A concrete driveway is 60 ft long, 12 ft wide, and 6 in. thick.

    a. How many cu ft of concrete does it contain?

    b. How many cu yd of concrete are there in the driveway?

11. Find the weight of a steel plate that is 20 in. long, 8 in. wide, and $\frac{3}{4}$ in. thick. (A cu in. of steel weighs .28 lb.)

12. What is the volume of a boxcar that is 30 ft long by 10 ft wide by 8 ft high?

13. A liter is 1000 cubic centimeters. How many liters of water will fill a tank 60 cm long, 30 cm wide, and 20 cm deep?

14. The unit of weight in the metric system is the gram, which is the weight of 1 cu cm of water. (cu cm is an abbreviation for "cubic centimeter.")

    a. How many grams of water are in a full tank 25 cm × 40 cm × 10 cm? (*Note:* The symbol × means "by.")

    b. A kilogram is 1000 grams or the weight of a liter of water. How many kilograms of water are in this tank?

15. A rectangular tank has dimensions 80 cm × 50 cm × 20 cm.

    a. How many liters of water does this tank hold?

    b. How many kilograms (kg) does this water weigh?

16. How many cubes, 6 in. on each edge, can be stored in a rectangular box 3 ft wide, 4 ft long, and 2 ft high?

17. How many liters are in the volume of a cube 20 cm on each edge?

18. A coal bin contains 360 cu yd. Find its height if its length is 12 yd and its width is 6 yd.

Chapter Review Exercises

1. A rectangular garden plot is 24 feet long. If its area is 288 square feet, how many feet are in its width?

2. Explain the meaning of each of the following formulas:
  a. $A = s^2$ **b.** $A = bh$ **c.** $V = lwh$
  d. $A = \frac{1}{2}bh$ **e.** $V = e^3$ **f.** $A = \pi r^2$

3. A living room is 18 ft by 14 ft. Find the cost of carpeting the floor of the room if the carpet used costs $17.50 per square yard.

4. A man owns $\frac{1}{2}$ acre of land in the shape of a rectangle. If the length of the plot is 180 ft, how many feet are in its width? (1 acre = 43,560 sq ft)

5. **a.** Find the value of $5t^3$ if $t = 2$.
  b. Find the value of $8s^3 - 3t^4$ if $s = 3$ and $t = 2$.

6. A circular tablecloth has a diameter of 70 in. Find its area:
  a. to the nearest sq in.   **b.** to the nearest sq ft (Use $\pi = \frac{22}{7}$.)

7. A triangular plate is cut out of a rectangular plate 18″ by 12″, as shown in the figure. What is the area of the triangular plate?

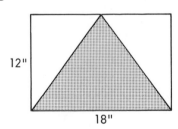

12″

18″

8. A mason is making a sidewalk 4 feet wide and 12 feet long. After digging out the area, he must fill it with concrete to a depth of 4 inches. How many cubic feet of concrete does he need?

9. A cube measures 4 in. on each edge.

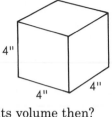

  a. What is the area of each square face of the cube?

  b. What is the total surface area of the cube? (Surface area is the sum of all 6 faces.)

  c. What is the volume of the cube?

  d. If each edge of the cube is doubled, what is its volume then?

  e. If we double the edge of a cube, by what number do we multiply its volume?

10. a. How many cc (cubic centimeters) are in the volume of a rectangular solid 30 cm × 40 cm × 50 cm?

  b. How many liters would it hold? (1 liter = 1000 cc)

11. A coal bin has a capacity of 650 cu ft. Find its height if its length is 13 ft and its width is 10 ft.

12. In the figure at the right, a triangle is drawn in a circle, as shown. The radius of the circle is 10″, the base of the triangle is 15″, and its height is 16″.

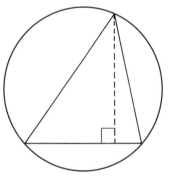

  a. What is the area of the circle? (Use π = 3.14.)

  b. What is the area of the triangle?

  c. What is the area of the region between the circle and the triangle?

CHECK YOUR SKILLS

1. Solve the equation: $\frac{3}{4}y = \frac{9}{20}$

2. What is the value of a^2b if $a = 2$ and $b = 3$?

3. Find the area in square feet of a square whose side is 3 ft 6 in.

4. What is the average of the numbers represented by $x + 2$, $2x - 1$, and $3x + 5$?

5. Find the area, in square feet, of a circular tabletop 7 feet in diameter. (Use $\pi = \frac{22}{7}$.)

6. A television set that usually sells for $400 is to be reduced to $340. What is the percent of discount being offered?

7. If you borrow $900 at 8% interest, for what period of time, in years, would you have the money if the interest is $108?

8. Bernie started a trip at 9:45 A.M. and arrived at his destination at 1:30 P.M. How long, in hours, did the trip take?

9. A rectangular aquarium tank has dimensions that are 90 cm × 60 cm × 40 cm. How many liters of water does it hold? (1 liter = 1000 cubic centimeters)

10. What is the circumference of a circle whose diameter is 20 cm? (Use $\pi = 3.14$.)

9

Signed Numbers

1. MEANING OF SIGNED NUMBERS

If we look at a thermometer, such as the one in the figure, we see that there is a zero reading (0°), readings above zero, and readings below zero. Those above zero are called *positive* readings and those below are called *negative* readings. These readings, which are numbers with plus and minus signs, are called **signed numbers.**

For example, the thermometer in the figure shows a reading of 5° below zero. We indicate this temperature as −5°, using the signed number −5.

In reading a thermometer, we used signed numbers (also called *directed numbers*) to represent temperatures above and below zero. We may use signed numbers to represent any quantities that are *opposite* in direction from a starting point, a *zero point*, or an *origin*. Such numbers are called *opposite numbers.*

One use of signed numbers is in representing profit and loss. If we talk about a *profit* as a *positive* number (+), we would treat a *loss* as a *negative* number (−). If we treat a $10 deposit in the bank as +$10, we would treat a $10 withdrawal as −$10.

If no sign appears before a number, it is understood to be positive. Thus the number 4 means +4.

We find it convenient to represent signed numbers on a number line as shown below:

We represent the positive numbers to the right of the zero point, or origin, and the negative numbers to the left of the origin. Signed whole numbers along with zero are also called **integers.**

Suppose that a football team gains 10 yards (+10) and then loses 7 yards (−7). We can show the result of both events on the number line by first moving 10 units to the right of zero, and then moving seven units to the left from +10. We thus arrive at +3, which indicates a 3-yard gain as the combined effect of both plays.

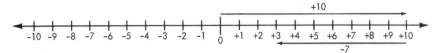

ILLUSTRATIVE PROBLEM: If the temperature rises from −7° to +5°, what is the increase in temperature?

Solution: In going from −7° to zero, the temperature rises 7°. In going from zero to +5°, the temperature rises 5°. Thus the increase is

$$7° + 5° = 12° \quad (answer)$$

Exercises

In 1 to 14: **a.** Represent the expression by a signed number.
　　　　　　 b. State the opposite number.
　　　　　　 c. Indicate what the opposite number represents.

EXAMPLE: a saving of $14
Answer: **a.** +14 **b.** −14 **c.** spending $14

1. 200 feet above sea level
2. a profit of $30
3. 150 miles north of the equator
4. an 8-yard loss in football
5. a $20 deposit in the bank
6. a $3 drop in price
7. 9 pounds overweight
8. 18 miles east
9. 30° Celsius (above zero)
10. 250 B.C. (a date)
11. 50 kilometers south of Paris
12. 8 points below average
13. a saving of $14
14. a loss of $\frac{1}{2}$ of a point

15. **a.** If the temperature goes from −8° to +12°, what is the change?
　　 b. Represent the change as a signed number.
16. MBC stock went from $19\frac{3}{4}$ on Thursday to $20\frac{3}{8}$ on Friday.
　　 a. What was the change in value of the stock?
　　 b. Represent the change as a signed number.
17. An auto driver in the area of the Dead Sea went from an altitude of 1500 feet above sea level to one of 1200 feet below sea level.
　　 a. What was the change in altitude?
　　 b. Represent the change as a signed number.

18. The average of a class on a math test was 78. For each mark, indicate by a signed number how much each student's mark was above (+) or below (−) the average.

	Name	Mark
a.	Carlos	83
b.	Terry	81
c.	Beverly	72
d.	Joan	90
e.	Dale	69

19. Represent each change in temperature as a signed number.
   a. from +49° to +52°   **b.** from +12° to −7°
   c. from −8° to +24°   **d.** from −15° to −6°
   e. from −18° to zero   **f.** from −8° to −19°
   g. from +17° to zero   **h.** from +13° to −12°

20. By using a number line in each problem below, represent by a signed number the change in going from:
   a. +3 to +12   **b.** −4 to +14   **c.** −6 to −15
   d. −7 to −2   **e.** +20 to −12   **f.** 0 to −13
   g. −32 to −20   **h.** −11 to 0

---

2. IDENTITY AND INVERSE ELEMENTS

Note that

$$3 + 0 = 3 \quad -5 + 0 = -5 \quad 4\tfrac{1}{2} + 0 = 4\tfrac{1}{2}$$

We observe from these examples that when zero is added to a number, the sum is the number itself. Thus, the number to which zero is added "preserves its *identity*" (does not change its value) under this addition. Therefore, zero is called the identity element of addition, or the **additive identity**. We may state this in equation form:

$$n + 0 = n \quad \text{or} \quad 0 + n = n$$

where n may assume any value.

In like manner, note that

$$7 \times 1 = 7 \quad 1 \times (-5) = -5 \quad \tfrac{2}{3} \times 1 = \tfrac{2}{3}$$

We observe that the product of 1 and any number is the number itself. Thus, the number by which 1 is multiplied "preserves its *identity*" (does not change its value) under this multiplication. Therefore, 1 is called the identity element of multiplication, or the **multiplicative identity.**

We may state this in equation form:

$$n \cdot 1 = n \quad \text{or} \quad 1 \cdot n = n$$

where n may assume any value.

Now note that

$$(+3) + (-3) = 0 \quad (+\tfrac{7}{8}) + (-\tfrac{7}{8}) = 0 \quad (-.6) + (+.6) = 0$$

We observe from these examples that the sum of a number and its opposite is zero. We may state this in equation form:

$$n + (-n) = 0$$

where n may assume any value.

If the sum of two numbers is zero (additive identity), we say that one number is the **additive inverse** of the other. Thus, (-3) is the additive inverse (or opposite) of $(+3)$, and $(+\tfrac{1}{2})$ is the additive inverse of $(-\tfrac{1}{2})$.

Now note that

$$5 \times \tfrac{1}{5} = 1 \quad (-\tfrac{1}{3}) \times (-3) = 1 \quad \tfrac{2}{3} \times \tfrac{3}{2} = 1$$

When the product of two numbers is 1, we say that one number is the **multiplicative inverse** (or **reciprocal**) of the other. Thus, $\tfrac{1}{5}$ is the multiplicative inverse of 5, and (-3) is the multiplicative inverse of $(-\tfrac{1}{3})$. What is the multiplicative inverse of $\tfrac{5}{7}$? of $-\tfrac{5}{6}$?

We may state this in equation form:

$$n \cdot \frac{1}{n} = 1 \quad \text{or} \quad \frac{1}{n} \cdot n = 1$$

where n may assume any value except 0, since division by zero is undefined.

Thus, 0 does not have a multiplicative inverse. It is the only number that does not have a multiplicative inverse.

ILLUSTRATIVE PROBLEM 1: For which value of y is the fraction $\dfrac{2}{y-9}$ undefined?

Solution: The fraction is undefined when its denominator is equal to 0.

$$\text{Let } y - 9 = 0$$

$$y = 9 \quad (answer)$$

ILLUSTRATIVE PROBLEM 2: What is:

a. The additive inverse of $\frac{3}{4}$?

b. The multiplicative inverse of $\frac{3}{4}$?

c. The multiplicative inverse of k? Can $k = 0$?

d. The additive inverse of k? Can $k = 0$?

Solution:

a. The additive inverse of $\frac{3}{4}$ is $-\frac{3}{4}$, since $\frac{3}{4} + (-\frac{3}{4}) = 0$.

  Answer: $-\frac{3}{4}$

b. The multiplicative inverse of $\frac{3}{4}$ is $\frac{4}{3}$, since $\frac{3}{4} \cdot \frac{4}{3} = 1$.

  Answer: $\frac{4}{3}$

c. The multiplicative inverse of k is $\dfrac{1}{k}$, since $k \cdot \dfrac{1}{k} = 1$. k cannot equal 0, since division by 0 is not allowed.

  Answer: $\dfrac{1}{k}$; No

d. The additive inverse of k is $-k$, since $k + (-k) = 0$. k can equal 0. Note, 0 is its own additive inverse.

  Answer: $-k$; Yes

The following table summarizes the special terms studied in this section.

Term	Example	Equation Form
additive identity (0)	$7 + 0 = 7$	$n + 0 = n$
multiplicative identity (1)	$7 \times 1 = 7$	$n \times 1 = n$
additive inverse (opposite)	$7 + (-7) = 0$	$n + (-n) = 0$
multiplicative inverse (reciprocal)	$7 \cdot \dfrac{1}{7} = 1$	$n \cdot \dfrac{1}{n} = 1$ $n \neq 0$

Exercises

1. What is the *additive inverse* of each of the following?
   a. -7 **b.** $\frac{3}{4}$ **c.** $.8$ **d.** 1 **e.** 0 **f.** $-\frac{4}{5}$

2. Complete each of the following:
   a. $r + \underline{\ ?\ } = 0$.
   b. The additive inverse of $-t$ is $\underline{\ ?\ }$.
   c. The additive identity is $\underline{\ ?\ }$.
   d. The multiplicative inverse of $-p$, $(p \neq 0)$, is $\underline{\ ?\ }$.
   e. The multiplicative identity is $\underline{\ ?\ }$.

3. What is the *multiplicative inverse* of each of the following?
   a. 10 **b.** $\frac{4}{7}$ **c.** -8 **d.** 1 **e.** $-\frac{5}{9}$ **f.** 0

4. **a.** When a number is added to its additive inverse, what is the result?
   b. When a number is multiplied by its multiplicative inverse, what is the result?

5. In each of the following expressions, for what value of x is the expression undefined?
   a. $\dfrac{12}{x - 5}$ **b.** $\dfrac{9}{x}$ **c.** $\dfrac{8}{6 - x}$
   d. $\dfrac{3}{2x - 7}$ **e.** $\dfrac{x - 2}{3}$

3. THE ABSOLUTE VALUE OF A NUMBER

In order to develop rules for computation with signed numbers, we shall refer to the **absolute value** of a signed number, which is simply the numerical value of a number regardless of its sign. Thus, the absolute value of -3 is 3, which is the same as the absolute value of $+3$.

The absolute value of a number is designated by a pair of vertical bars, $|\ |$. Thus, $|+7| = 7$ and $|-7| = 7$. Therefore, $|+7| = |-7| = 7$.

Likewise, $|-3\frac{1}{2}| = |+3\frac{1}{2}| = 3\frac{1}{2}$ and $|-5| = |+5| = 5$.

Note that the absolute value of 0 is 0, written $|0| = 0$. In general, the absolute value of a positive number is the number itself; the absolute value of a negative number is the opposite of the number.

ILLUSTRATIVE PROBLEM 1: Find the value of $|-9| + |3|$.

Solution:
$$|-9| = 9$$
$$|3| = 3$$
$$|-9| + |3| = 9 + 3 = 12 \quad (answer)$$

ILLUSTRATIVE PROBLEM 2: What value(s) of y satisfy $|y| = 3$?

Solution:

We know that $|+3| = 3$ and that $|-3| = 3$. There are no other numbers whose absolute value is equal to 3. Therefore, y may have the value $+3$ or -3. (*answer*)

Exercises

In 1 to 9, find the value of each expression.

1. $|-18|$ **2.** $|+7\frac{1}{4}|$ **3.** $|-15|$ **4.** $|-8| + |-7|$

5. $|-9| + |+4|$ **6.** $|-12| \times |-4|$ **7.** $|10 - 6|$ **8.** $|-15| \times |+3|$

9. $|12 - 7| + |-8|$

In 10 to 15, for what values of x are the equations true?

10. $|x| = 4$ **11.** $|x| = 9$ **12.** $|x| = 0$

13. $|-x| = 3$ **14.** $|x| = -3$ **15.** $|x| + 2 = 7$

4. ADDING SIGNED NUMBERS

If the temperature rises 5° (+5) and then rises 4° (+4), the combined effect is a rise of 9° (+9). We have added +5 and +4 and obtained +9. This process of finding the combined effect is called the **combination** or **addition** of signed numbers.

We may apply this process to both positive and negative numbers. If a storekeeper loses $3 (−3) on one sale and then loses $5 (−5) on the next sale, the combined effect is a loss of $8 (−8). Thus, we see that the sum of −3 and −5 is −8. We refer to the sum of signed numbers as their **algebraic sum.**

In the above examples, we added two numbers with like signs. In the first case, (+5) + (+4), both numbers were positive. In the second case, (−3) + (−5), both were negative. Note that, in both of these cases, the absolute value of the sum was the sum of the absolute values of the two signed numbers being added. We may state this idea in the following rule:

RULE: To add two numbers having the same sign, add their absolute values and place the common sign before the sum.

Let us examine the case where the signs of the two numbers are unlike. If a football team gains 7 yards $(+7)$ in its first play and then loses 9 yards (-9) in its second play, its total loss is 2 yards (-2). We may write the sum vertically:

$$\text{Add: } +7 \\ \underline{-9} \\ -2$$

Or we may write the sum horizontally:

$$(+7) + (-9) = -2$$

If a team gains 9 yards $(+9)$ and then loses 4 yards (-4), the result would be a gain of 5 yards $(+5)$. This may be written as

$$\text{Add: } +9 \quad \text{or} \quad (+9) + (-4) = +5 \\ \underline{-4} \\ +5$$

Note that, in both of these examples, the absolute value of the sum is the difference of the absolute values of the two signed numbers being added. In the first case, $(+7) + (-9)$, the sum has the same sign as -9, which is the number with the larger absolute value. In the second case, $(+9) + (-4)$, the sum has the same sign as $+9$, which is the number with the larger absolute value.

We may summarize this idea in the following rule:

RULE: To add two signed numbers having unlike signs, find the difference of their absolute values and place before it the sign of the number having the larger absolute value.

ILLUSTRATIVE PROBLEM 1:

$$\text{Add: } -15 \\ \underline{+7}$$

Solution: Signs are unlike. Difference of absolute values is 8. Sign of larger absolute value (-15) is negative. Sum is (-8). (*answer*)

ILLUSTRATIVE PROBLEM 2:

$$\text{Add:} \quad \begin{array}{r} -14 \\ -8 \\ \hline \end{array}$$

Solution: Signs are alike. Sum of absolute values is 22. Common sign is negative. Sum is (-22). (*answer*)

It is easy to show that, as with the counting numbers in arithmetic, the *order* of addition of signed numbers does not matter (the commutative principle). Thus,

$$(+10) + (-3) = (-3) + (+10) = +7$$

In adding three or more signed numbers, the way we *group* the numbers does not matter (the associative principle). Thus,

$$[(+5) + (-2)] + (-4) = +5 + [(-2) + (-4)] = -1$$

ILLUSTRATIVE PROBLEM 3:

$$\text{Add: } (+12) + (-7) + (-2)$$

Solution: Since the arrangement does not matter, we may take the numbers in any order. The sum of $(+12)$ and (-7) is $(+5)$, and the sum of $(+5)$ and (-2) is $(+3)$. (*answer*)

Exercises _____

In 1 to 28, add.

1. $\begin{array}{r}+4\\-1\\\hline\end{array}$	**2.** $\begin{array}{r}+3\\-7\\\hline\end{array}$	**3.** $\begin{array}{r}-2\\-6\\\hline\end{array}$	**4.** $\begin{array}{r}+12\\-9\\\hline\end{array}$
5. $\begin{array}{r}-18\\+12\\\hline\end{array}$	**6.** $\begin{array}{r}+12\frac{1}{8}\\+7\frac{3}{8}\\\hline\end{array}$	**7.** $\begin{array}{r}-4.7\\-8.2\\\hline\end{array}$	**8.** $\begin{array}{r}-9\frac{1}{2}\\-3\frac{1}{4}\\\hline\end{array}$
9. $\begin{array}{r}+7\\-7\\\hline\end{array}$	**10.** $\begin{array}{r}+14\frac{1}{3}\\-8\frac{2}{3}\\\hline\end{array}$	**11.** $\begin{array}{r}+11.4\\-11.4\\\hline\end{array}$	**12.** $\begin{array}{r}+52\\-75\\\hline\end{array}$
13. $\begin{array}{r}-8.7\\+3.9\\\hline\end{array}$	**14.** $\begin{array}{r}-93\\+47\\\hline\end{array}$	**15.** $\begin{array}{r}-27\frac{1}{6}\\+14\frac{1}{3}\\\hline\end{array}$	

16. a. $(+17) + (+11)$ **b.** $(-45) + (+20)$ **c.** $(-16) + (-15)$

     d. $(-13) + (+13)$ **e.** $(-12) + (0)$ **f.** $(-10) + (-19)$

17. $\begin{array}{r}-7\frac{1}{2}\\+12\\\hline\end{array}$	**18.** $\begin{array}{r}-9.4\\-11.7\\\hline\end{array}$	**19.** $\begin{array}{r}-12\\+15\frac{3}{4}\\\hline\end{array}$	**20.** $\begin{array}{r}+9\\-4\\-7\\\hline\end{array}$

21. -23	22. -21	23. $+11$	24. $+12$
$+11$	-17	-17	-27
$+6$	$+24$	-23	-19
		$+19$	-8

25. $+4.2$	26. $-5\frac{3}{4}$	27. -18.4	28. $+5\frac{2}{3}$
-5.8	$+2\frac{1}{2}$	-17.8	$-3\frac{1}{6}$
-9.7	$-1\frac{7}{8}$	$+23.5$	$+11\frac{1}{3}$
-12.3			

In 29 to 35, represent the quantities by signed numbers and find their algebraic sum.

29. A team gained 12 yards and then lost 9 yards.

30. A dealer made a profit of $250 and then had a loss of $175.

31. The temperature rose 14° and then dropped 20°.

32. Nancy lost 7 pounds and then gained 12 pounds.

33. In one month, Mr. Jones showed deposits of $18 and $42 and a withdrawal of $36.

34. An auto went 180 miles south and then traveled 115 miles north.

35. A plane rose 850 feet into the air and then dropped 260 feet.

---

5. SUBTRACTING SIGNED NUMBERS

If the temperature rises from 15° above zero ($+15°$) to 20° above zero ($+20°$), how many degrees is the rise? Here we simply *subtract* 15 from 20 and get $20 - 15 = 5$. In vertical form:

$$\text{Subtract:} \quad \begin{array}{r} +20 \\ +15 \\ \hline +5 \end{array} \quad \begin{array}{l} \text{(minuend)} \\ \text{(subtrahend)} \\ \text{(difference)} \end{array}$$

In general, when we are subtracting x from y, x is called the **subtrahend** and y the **minuend.** The result is called the **difference** of y and x.

$$\begin{array}{r} y \\ x \\ \hline y - x \end{array} \quad \begin{array}{l} \text{(minuend)} \\ \text{(subtrahend)} \\ \text{(difference)} \end{array}$$

We can show the subtraction of signed numbers very conveniently on a number line. If we wish to subtract $+6$ from $+10$, we first locate $+6$ (subtrahend) on the number line, then locate $+10$ (minuend), and then count the spaces from subtrahend to minuend (4). Since we moved to the *right*, the difference is positive and is equal to +4.

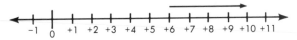

Let us subtract −3 from +6 in a similar manner. Locate −3 (subtrahend) and then +6 (minuend). We move 9 units to the *right*, so the result is +9. We are saying, in effect, that, to get from −3 to +6, we must add +9 to −3.

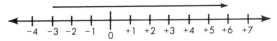

Now subtract −8 from −5. Locate −8 on the number line and then −5. To get from −8 to −5, we move 3 units to the *right* so that the difference is +3. If we subtract −5 from −8, we move 3 units to the *left* and the difference is −3.

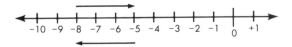

We always move *from the subtrahend to the minuend*. If the movement is to the right, the difference is positive. If the movement is to the left, the difference is negative.

In order to avoid the constant use of a number line, we can develop a simple rule for subtracting signed numbers. We write the preceding four examples in vertical form:

Subtract:

1. +10	2. +6	3. −5	4. −8
+6	−3	−8	−5

Then we *change the sign of the subtrahend* in each case and proceed as in algebraic addition:

1. +10	2. +6	3. −5	4. −8
⊕6	⊖3	⊖8	⊖5
+4	+9	+3	−3

In each case, we have circled the sign, written the opposite sign above it, and then found the *algebraic sum*. Note that the correct difference is obtained in each case.

RULE: To subtract one signed number from another, change the sign of the *subtrahend* and proceed as in algebraic addition.

By means of this rule, all problems in subtraction of signed numbers can be done by addition. We may indicate the subtraction vertically, as shown, or horizontally. Thus, example 1 becomes $(+10) - (+6)$. Note that the sign between parentheses is a sign of operation (subtraction), while the other signs are number signs. Then, using the rule, change $(+10) - (+6)$ to $(+10) + (-6)$, which gives $+4$.

Exercises

In 1 to 16, subtract.

1. $+5$
$\underline{+9}$

2. $+8$
$\underline{-5}$

3. -9
$\underline{+3}$

4. -11
$\underline{-6}$

5. $+17$
$\underline{-10}$

6. -12
$\underline{+8}$

7. -9
$\underline{-13}$

8. $+7$
$\underline{+16}$

9. -23
$\underline{-17}$

10. $+7\frac{5}{8}$
$\underline{-3}$

11. $-4\frac{5}{6}$
$\underline{-1\frac{2}{3}}$

12. -7.4
$\underline{+3.8}$

13. 0
$\underline{+9\frac{3}{4}}$

14. $+6\frac{1}{3}$
$\underline{-6\frac{1}{3}}$

15. -8.2
$\underline{+12.9}$

16. $-3\frac{3}{4}$
$\underline{+7\frac{1}{8}}$

In 17 to 28, perform the indicated operation.

17. $(+9) - (+7)$

18. $(+1) - (-5)$

19. $(-9) - (-2)$

20. $(-6) - (-8)$

21. $(+15) - (-4)$

22. $(-28) - (+23)$

23. From $+23$ subtract -11.

24. From -18 subtract -10.

25. From $+11\frac{3}{4}$ subtract $-4\frac{5}{8}$.

26. From -8.7 subtract -12.

27. From 0 subtract 15.

28. From -12 subtract 0.

29. How much is 24 decreased by -15?

30. What number is 11 less than -10?

31. From the sum of $+32$ and -16, subtract -7.

32. How much greater is $+12$ than -4?

In 33 to 36, find the value of the given expression.

33. $(+8) + (-11) - (-3)$

34. $(-19) + (+12) - (-20)$

35. $(-7.4) - (-2.3) + (-8.5)$

36. $(+7\frac{1}{2}) + (-5\frac{1}{4}) - (-3\frac{1}{8})$

In 37 to 46, set up each exercise as a subtraction problem and solve.

37. The temperature went from $-8°$ to $+14°$. What was the change in temperature?

38. A motorist drove from 200 feet above sea level to 100 feet below sea level. What was his change in altitude?

39. A Roman emperor was born in the year 40 B.C. and died in the year 35 A.D. How old was he at the time of death?

40. Bill had $15 on Friday and owed $8 on Monday. How much did he spend over the weekend?

41. The temperature in Stockholm was $-20°$ when it was $-32°$ in Oslo? How much colder was it in Oslo?

42. Tom weighed 6 pounds less than his sister in January and 7 pounds more than his sister in December of the same year. Assuming his sister's weight did not change, how many pounds did he gain over the year?

43. Subtract a from b.

44. To the difference of a and b, add c.

45. Subtract p from the sum of q and r.

46. From the difference of c and d, subtract x.

---

6. MULTIPLYING SIGNED NUMBERS

To illustrate the multiplication of two signed numbers, we will consider a man playing a game for money. We may think of *winning* a certain amount as a *positive* number and *losing* a certain amount as a *negative* number. We may also think of hours in the *future* as *positive* numbers and hours in the *past* as *negative* numbers.

CASE 1. The Product of Two Positive Numbers.

If the man wins $4 every hour, then in 3 hours he will have won $12. This may be represented by:

$$(+4) \times (+3) = +12$$

Thus, the product of two positive numbers is a positive number.

CASE 2. The Product of a Negative Number and a Positive Number.

If the man loses $4 every hour, then in 3 hours he will have lost $12. This may be represented by:

$$(-4) \times (+3) = -12$$

Thus, the product of a negative number and a positive number is a negative number.

CASE 3. The Product of a Positive Number and a Negative Number.

If the man wins $4 every hour, then 3 hours ago he had $12 less than now. This may be represented by:

$$(+4) \times (-3) = -12$$

Thus, the product of a positive number and a negative number is a negative number.

CASE 4. The Product of Two Negative Numbers.

If the man is losing $4 every hour, then 3 hours ago he had $12 more than now. This may be represented by:

$$(-4) \times (-3) = +12$$

Thus, the product of two negative numbers is a positive number.

Note that, in each case, the absolute value of the product 12 is the product of the absolute values of the signed numbers 4 and 3.

From the four cases, we get the following rules for multiplying signed numbers:

RULE 1: The product of two numbers having the *same* sign is a *positive* number whose absolute value is the product of the absolute values of the two numbers.

RULE 2: The product of two numbers having *different* signs is a *negative* number whose absolute value is the product of the absolute values of the two numbers.

EXAMPLES: Multiply.

1. $+8$
   $+5$
   $\overline{+40}$

2. -12
   $+4$
   $\overline{-48}$

3. -15
   -6
   $\overline{+90}$

The *order* of multiplication does not matter (the commutative principle), so that

$$(+10)(-7) = (-7)(+10) = -70.$$

Also, in multiplying three or more signed numbers, the *grouping* does not matter (the associative principle). In multiplying three numbers, find the product of *any two*, and multiply this product by the third number.

ILLUSTRATIVE PROBLEM 1: Multiply: $(-5)(+2)(-7)$

Solution: $[(-5)(+2)](-7) = (-10)(-7) = +70$ *(answer)*

We may also use exponents to show repeated multiplication of signed numbers.

ILLUSTRATIVE PROBLEM 2: Find the value of $(-2)^3$.

Solution: $(-2)^3 = (-2)(-2)(-2) = (+4)(-2) = -8$ *(answer)*

Exercises _____

In 1 to 27, multiply.

1. $+7$
    $+8$

2. -8
    $+4$

3. $+12$
    -3

4. $+4$
    $+6$

5. -7
    -9

6. -12
    $+5$

7. 0
    -14

8. $+32$
    -7

9. $+8$
    0

10. -18
     $+\frac{2}{3}$

11. -3.4
     $+2.7$

12. -24
     $-3\frac{1}{2}$

13. $(+7)(-12)$

14. $(-38)(-10)$

15. $(+14)(-28)$

16. $(+9)(-4\frac{1}{3})$

17. $(-24)(+\frac{5}{6})$

18. $(-\frac{5}{7})(-\frac{7}{5})$

19. $(-3\frac{1}{2})(+3\frac{1}{7})$

20. $(-.45)(-60)$

21. $(-25)(4\frac{1}{5})$

22. $(+7)(-5)(-2)$

23. $(+8)(-3)(-1)$

24. $(-6)(-4)(-3)$

25. $(+8)(-7)(0)$

26. $(-4)(-5)(-6)$

27. $(27)(0)(-15)$

In 28 to 43, find the value of the given expression.

28. $(-3)^2$

29. $(-4)^2$

30. $(-1)^4$

31. $(+\frac{1}{3})^2$

32. $(-3)^3$

33. $(+5)^3$

34. $(-6)^2$

35. $(-\frac{1}{2})^3$

36. $(-4)^3$

37. $(+3)^4$

38. $(-12)^2$

39. $(-10)^3$

40. $(-.5)^2$

41. $(-\frac{2}{3})^3$

42. $(-1)^{11}$

43. $(-10)^6$

7. DIVIDING SIGNED NUMBERS

When we first learned to divide, we learned that $6 \div 2$ is defined as the number which, when multiplied by 2, gives a product of 6. Thus $\frac{6}{2} = 3$. The 6 is called the **dividend**, 2 is the **divisor**, and 3 is the **quotient**.

We may thus determine the quotient in all four cases of division of signed numbers.

CASE 1. $\dfrac{+12}{+4} = +3$ since $(+3)(+4) = +12$.

CASE 2. $\dfrac{-12}{-4} = +3$ since $(+3)(-4) = -12$.

CASE 3. $\dfrac{+12}{-4} = -3$ since $(-3)(-4) = +12$.

CASE 4. $\dfrac{-12}{+4} = -3$ since $(-3)(+4) = -12$.

Note that, in cases 1 and 2, the signs of the dividend and the divisor are *alike*, and the quotient in both cases is *positive*. In cases 3 and 4, the signs of the dividend and the divisor are *different*, and the quotient in both cases is *negative*. In all cases, the absolute value of the quotient is the quotient of the absolute values of the dividend and the divisor.

From the four cases, we get the following rules for the division of signed numbers:

RULE 1:   The quotient of two numbers having the *same* sign is a *positive* number which is the quotient of their absolute values.

RULE 2:   The quotient of two numbers having *different* signs is a *negative* number whose absolute value is the quotient of the absolute values of the two numbers.

EXAMPLES: Perform the indicated divisions.

1. $\dfrac{+18}{+6} = +3$
2. $\dfrac{-15}{+5} = -3$

3. $\dfrac{-36}{-9} = +4$
4. $\dfrac{+63}{-7} = -9$

Exercises

In 1 to 20, perform the indicated divisions.

1. $\dfrac{+21}{+3}$
2. $\dfrac{-22}{-2}$
3. $\dfrac{-26}{+13}$
4. $\dfrac{45}{-9}$

5. $\dfrac{-7}{7}$
6. $\dfrac{+27}{-9}$
7. $\dfrac{+81}{-9}$
8. $\dfrac{-121}{-11}$

9. $\dfrac{0}{-11}$ **10.** $\dfrac{20}{-5}$ **11.** $\dfrac{+8.8}{-4}$ **12.** $\dfrac{-75}{+15}$

13. $\dfrac{+36}{-7}$ **14.** $\dfrac{-22}{-5}$ **15.** $\dfrac{-100}{+12}$ **16.** $\dfrac{-4.8}{1.2}$

17. $(+63) \div (-9)$ **18.** $(-75) \div (-5)$

19. $(+15) \div (-\frac{5}{9})$ **20.** $(-7.2) \div (+.9)$

21. Divide each of the following numbers by -7:

    a. $+21$ **b.** -14 **c.** -56 **d.** $+49$ **e.** -50

    In 22 to 27, divide as indicated.

22. $14 \div \frac{7}{9}$ **23.** $(-\frac{5}{4}) \div (-20)$ **24.** $(-3\frac{3}{4}) \div (+6\frac{2}{3})$

25. $(-7250) \div (-10)$ **26.** $(-3.14) \div (100)$ **27.** $(+25.25) \div (-5)$

---

8. EVALUATING EXPRESSIONS WITH SIGNED NUMBERS

In Chapter 1 we found the numerical value of algebraic expressions by substituting the numbers of arithmetic (the non-negative numbers) into the expressions. We follow the same procedure when we substitute any given signed numbers—positive or negative—into algebraic expressions.

ILLUSTRATIVE PROBLEM 1: Find the value of $3xy^2$ when $x = -2$ and $y = -5$.

Solution: Substitute the given values in the expression.

$$\begin{aligned}
3xy^2 &= 3(-2)(-5)^2 \\
&= 3(-2)(+25) \\
&= (-6)(+25) \\
&= -150 \;\; (answer)
\end{aligned}$$

ILLUSTRATIVE PROBLEM 2: Find the value of $3p - 4q$ when $p = 5$ and $q = -6$.

Solution: Think of $3p - 4q$ as $3p + (-4q)$; then substitute the given values.

$$\begin{aligned}
3p + (-4q) &= (3)(5) + (-4)(-6) \\
&= 15 + 24 \\
&= 39 \;\; (answer)
\end{aligned}$$

ILLUSTRATIVE PROBLEM 3: Find the value of $r^2 - s^2$ when $r = -2$ and $s = -3$.

Solution:

$$r^2 - s^2 = (-2)^2 - (-3)^2$$
$$= 4 - 9$$
$$= (+4) + (-9) = -5 \quad (answer)$$

Exercises

In 1 to 36, find the value of the following expressions when $c = -2$, $d = +3$, $r = -4$, and $s = +5$.

1. $7c$

2. $-5d$

3. $6cr$

4. $-8rs$

5. $2r + 7s$

6. $c^2 - 2cd$

7. $cr - ds$

8. $c^3 - d^2$

9. $3c - 4d$

10. $5r - 4s$

11. $r^2 - s^2$

12. $4c^2d^3$

13. $-r^2s$

14. $\frac{1}{2}c^3$

15. $\frac{2}{3}dr^2$

16. $d^2 - c^2$

17. $cd - rs$

18. $(c + d)^2$

19. $3s^2 - r^2$

20. $r^2 + 2r$

21. $3c + 2d - r$

22. $s^2 - 2r$

23. $(r - s)^2$

24. $-2rs$

25. $c^2 - 2cd + d^3$

26. c^2dr

27. $3(d - r)$

28. $2r^2 + c^2$

29. $s^2 + 2cs$

30. $2c + d + r$

31. $(r + s)(r - s)$

32. $c(s - r)$

33. $2(c - d) + 8$

34. $(r + 5)(s - 1)$

35. $r^2 - 2r + 1$

36. $(r - 1)^2$

In 37 to 45, find the value of the expression when $x = -10$, $y = +6$, and $z = -1$.

37. $\dfrac{x}{5}$

38. $\dfrac{y}{-3}$

39. $\dfrac{yz}{2}$

40. $\dfrac{xy}{20}$

41. $\dfrac{xz}{y}$

42. $\dfrac{x^2z}{5y}$

43. $\dfrac{3x + 10z}{5y}$

44. $\dfrac{x^2 + y^2}{8z}$

45. $\dfrac{x^2 + y^2}{x^2 - y^2}$

Chapter Review Exercises

1. Represent each quantity below with a signed number.
   a. 100 ft below sea level   **b.** a profit of $28
   c. 5 miles south   **d.** 12° below zero
   e. the year 60 B.C.   **f.** 10 pounds overweight
   g. $18 in debt   **h.** 9 miles west

2. What number added to $+9$ gives zero?

3. What number added to $-\frac{2}{3}$ gives zero?

4. Represent the change in each statement below by a signed number.
   a. a loss of 7 pounds   **b.** a 3% rise in cost
   c. a deposit of $20   **d.** a withdrawal of $10
   e. a loss of $400   **f.** a growth of 2 inches

5. A football team makes a gain of 8 yards on one play and a loss of 3 yards on the next play. What was its total yardage for both plays?

6. On a number line, start at 0, move -8 units, $+5$ units, then -3 units. At what signed number do you stop?

   In 7 to 20, perform the indicated operations.

7. $(+8) + (+11)$ 8. $(-5) + (-7)$

9. $(-7.5) + (+.7)$ 10. $(-2.8) + (+3.5) + (-7.4)$

11. $(-16) - (+9)$ 12. $(-24) - (-30)$

13. $(-12)(+10)$ 14. $(-3\frac{1}{2})(-18)$

15. $(-5)(-7)(-1)$ 16. $(+11)(-3)(-2)$

17. $(-27) \div (+3)$ 18. $(-45) \div (-5)$

19. $\dfrac{+48}{-12}$ 20. $\dfrac{-7.5}{+.5}$

   In 21 and 22, subtract.

21. $\begin{array}{r} -23 \\ +15 \\ \hline \end{array}$ 22. $\begin{array}{r} -12\frac{1}{2} \\ -2\frac{3}{4} \\ \hline \end{array}$

23. Find the value of:
   a. $(-4)^2$ **b.** $(-1)^7$ **c.** $(-3)^4$

   In 24 to 47, find the value of the expressions when $p = 3$, $q = -1$, and $r = -4$.

24. $6pq$ 25. pqr 26. $5pr$ 27. $4p + 5q$

28. $3q - 2p$ 29. $4q - 5r$ 30. p^2q 31. $q^2 + r^2$

32. $pq - qr$ **33.** $pq + r$ **34.** $(p + q)^2$ **35.** p^3q^2r

36. $2p^2 - 3r^2$ **37.** $2pq - 2pr$ **38.** $r^2 - r - 1$ **39.** $7(p - q)$

40. $\dfrac{pq}{r}$ **41.** $\dfrac{3qr}{p}$ **42.** $\dfrac{p - q}{r}$ **43.** $|p + r|$

44. $|pq|$ **45.** $|r - q|$ **46.** $|pr|$ **47.** $|p^2q|$

48. What is the additive inverse of each of the following?
    a. -2 **b.** 8 **c.** $-\frac{2}{3}$ **d.** y

49. What is the multiplicative inverse of each of the following?
    a. $\frac{4}{5}$ **b.** $-2\frac{1}{3}$ **c.** $-.8$ **d.** 0

CHECK YOUR SKILLS

1. Solve for x: $.01x = 3$

2. What is the absolute value of -5?

3. What is the greatest common factor of the numbers 12, 30, and 60?

4. Find the area of a triangle whose base is 20 cm and whose height is 15 cm.

5. If p and q represent positive integers, which expression must represent another positive integer?

    (a) $\dfrac{q}{p}$ **(b)** $p + q$ **(c)** $p - q$ **(d)** $p \div q$

6. If $x = -3$, what is the value of $2x^2$?

7. Find the value of the expression $\dfrac{15 - 50}{-5}$.

8. Using the formula $P = I^2R$, find P if $I = 2$ and $R = 15$.

9. Solve for y: $\dfrac{5}{y} = \dfrac{30}{18}$

10. A man has $1.35 in his pocket, all in nickels and dimes. He has 6 more nickels than dimes. How many nickels does he have?

10

Operations with Algebraic Expressions

1. ADDING MONOMIALS

In Chapter 1, we learned that we can add terms such as $5x$ and $3x$ and obtain the sum $8x$. We will now do this when the numerical coefficients are signed numbers.

First, let us define some of the words we will be using. A **term** is a number or letter or an expression written as a product of numbers and letters. Examples of terms are 7, x, $5x^2y$, $-3ab$, $\frac{2}{3}pq^3$, and $\frac{rst}{7}$. An expression such as $a + b$ is not considered a one-term expression since the letters are not combined by multiplication, but by addition. We will learn about these expressions later in this chapter.

A **monomial** is an algebraic expression that has only one term. Thus, all the expressions just mentioned as terms are monomials. **Like terms** are terms that have the same letters with the same exponents. For example, $3x^2y$ and $7x^2y$ are like terms. We call x^2y the **common literal factor**. Their sum would be $10x^2y$.

RULE: To add like terms (or monomials), add the numerical coefficients and multiply this sum by the common literal factor.

This same procedure applies to like terms when the coefficients are signed numbers. Thus the sum of $-3ab^2$ and $+8ab^2$ is $5ab^2$.

We cannot add or combine unlike terms, such as $+7xy^2$ and $-2x^2y$.

Here are some examples of adding monomials. Notice that the last two examples require two steps to do the addition.

EXAMPLES

1. $+11x$	**2.** $+10b^2$	**3.** $-9ab$	**4.** $+12b^2c$
$\underline{+5x}$	$\underline{-3b^2}$	$\underline{-4ab}$	$\underline{-12b^2c}$
$+16x$	$+7b^2$	$-13ab$	0

5. $(-7ab^2c) + (-3ab^2c) = -10ab^2c$

6. $(-17st) + (+9st) + (-8st) = (-8st) + (-8st) = -16st$

7. $-8rs - 4rs + 15rs = -12rs + 15rs = +3rs$

The principle that we are using here is an application of what we call the **distributive property of multiplication over addition**.

Note that $3(4 + 5) = 3(9) = 27$; also $3(4) + 3(5) = 12 + 15 = 27$. It thus appears that

$$3(4 + 5) = 3(4) + 3(5)$$

In general, if x, a, and b are any three numbers, we may write the distributive property as follows:

$$x(a + b) = xa + xb$$

We may also write it in the form

$$(a + b)x = ax + bx$$

In adding two like terms or monomials, we are using the distributive property in reverse. Thus,

$$3x + 4x = (3 + 4)x = 7x$$

Exercises

In 1 to 12, add the monomials.

1. $-3x$	**2.** $-7b$	**3.** $+12ab$	**4.** $+11b^2c$
$\underline{+5x}$	$\underline{-4b}$	$\underline{-5ab}$	$\underline{+4b^2c}$
5. $+6pq$	**6.** $-7t^2$	**7.** $+8m$	**8.** $-12k$
$\underline{-10pq}$	$\underline{-5t^2}$	$+5m$	$+7k$
		$\underline{+4m}$	$\underline{-6k}$
9. $-7rs^2$	**10.** $-15dx^2y$	**11.** $+23rst$	**12.** $+1.5ay^2$
$-8rs^2$	$-dx^2y$	$-17rst$	$+7.8ay^2$
$\underline{+5rs^2}$	$\underline{-8dx^2y}$	$\underline{-6rst}$	$\underline{-19.3ay^2}$

In 13 to 17, add like terms.

13. $(-7k) + (-2k) + (+4k)$

14. $(+13xy) + (-8xy) + (-7xy)$

15. $(-27rs^3) + (+18rs^3) + (+12rs^3)$

16. $(+12c^2d) + (-17c^2d) + (-30c^2d)$

17. $(+32p^2q^2) + (-40p^2q^2) + (-8p^2q^2)$

In 18 to 26, think of the signs as signs of the numerical coefficients of the terms, and add.

18. $+12x - 8x - 9x$

19. $-13ab - 15ab + 10ab - 7ab$

20. $s^2t - 5s^2t - 8s^2t + 3s^2t$

21. $\frac{3}{4}xy - \frac{1}{2}xy + \frac{3}{8}xy$

22. $9.4s^2 - 7.2s^2 - 8.7s^2 + s^2$

23. $-23b^2c^2 + 15b^2c^2 - 8b^2c^2 + 5b^2c^2$

24. $+18m^2n - 13m^2n - 15m^2n + 3m^2n$

25. $-32f^2g + 12f^2g + 20f^2g$

26. $-5r^2 + 8r^2 - 21r^2 - 7r^2$

---

2. SUBTRACTING MONOMIALS

Recall that, in subtracting signed numbers, we changed the sign of the subtrahend and then proceeded as in addition of signed numbers. We may do the same in subtracting one monomial from another. Thus,

$$(+8a) - (-5a) = (+8a) + (+5a) = +13a$$

In this case, the subtrahend $(-5a)$ is changed to $(+5a)$ and the subtraction operation is changed to addition.

RULE: To subtract one monomial from another, change the sign of the subtrahend and proceed as in addition.

Here are some examples of subtracting monomials.

EXAMPLES: In 1 to 4, subtract.

1. $+7k$
  $\ominus 2k$
  $+5k$

2. $-8r^2$
  $\ominus 3r^2$
  $-5r^2$

3. $+8p^2q$
  $\ominus 9p^2q$
  $+17p^2q$

4. 0
  $\ominus 8xy^2$
  $+8xy^2$

5. $(-12m^2n) - (-12m^2n) = 0$

6. $(-15z^2) - (0) = -15z^2$

Exercises

In 1 to 6, subtract like terms.

1. $(+8p) - (-2p)$ **2.** $(-17rs) - (-9rs)$ **3.** $(-11.4k) - (+4.7k)$

4. $(+43x^2) - (-22x^2)$ **5.** $(-17s^2t) - (-17s^2t)$ **6.** $0 - (-14mn^2)$

In 7 to 18, subtract.

7. $\begin{array}{r} +34x^2y^3 \\ -12x^2y^3 \\ \hline \end{array}$ **8.** $\begin{array}{r} -9pq^2 \\ -9pq^2 \\ \hline \end{array}$ **9.** $\begin{array}{r} -7rst \\ -7rst \\ \hline \end{array}$

10. $\begin{array}{r} 0 \\ 3y^2z^2 \\ \hline \end{array}$ **11.** $\begin{array}{r} -17x \\ 0 \\ \hline \end{array}$ **12.** $\begin{array}{r} +.72a^2b \\ -.35a^2b \\ \hline \end{array}$

13. $\begin{array}{r} +2\frac{3}{4}m^2n \\ -5\frac{1}{8}m^2n \\ \hline \end{array}$ **14.** $\begin{array}{r} +\frac{3}{7}b^2 \\ +\frac{4}{7}b^2 \\ \hline \end{array}$ **15.** $\begin{array}{r} -7\frac{1}{2}ad^2 \\ +2\frac{3}{4}ad^2 \\ \hline \end{array}$

16. $\begin{array}{r} +9x^2y^3 \\ -x^2y^3 \\ \hline \end{array}$ **17.** $\begin{array}{r} -11(p + q) \\ -8(p + q) \\ \hline \end{array}$ **18.** $\begin{array}{r} -16s^2t \\ +14s^2t \\ \hline \end{array}$

19. Subtract $5r^2h$ from $13r^2h$.   **20.** Subtract $13r^2h$ from $5r^2h$.

21. How much greater is $18x$ than $12x$?

22. How much greater is $17x^2y^3$ than $-3x^2y^3$?

23. What is the difference of $19y^2$ and $10y^2$?

24. What is the difference of $-12ab$ and $-8ab$?

25. From the sum of $-7xy$ and $+13xy$, subtract $-3xy$.

26. From the sum of $-19p^2q$ and $-11p^2q$, subtract $-7p^2q$.

27. What must be added to $7t$ to get $10t$?

28. What must be added to $-11h$ to get $+5h$?

29. By how much does $14rh$ exceed $9rh$?

30. By how much does $-5z^2$ exceed $-20z^2$?

In 31 to 33, write an algebraic expression for each statement.

31. From a subtract b.

32. To the difference between r and s, add t.

33. Subtract y from the difference between p and q.

3. MULTIPLYING POWERS OF THE SAME BASE

We have already learned that $a^3 = a \cdot a \cdot a$ and $a^4 = a \cdot a \cdot a \cdot a$, so that

$$a^3 a^4 = \overbrace{(a \cdot a \cdot a)}^{3} \cdot \overbrace{(a \cdot a \cdot a \cdot a)}^{4}$$

$$= a \cdot a \cdot a \cdot a \cdot a \cdot a \cdot a = a^7$$

Likewise, $b^2 \cdot b^3 = \overbrace{(b \cdot b)}^{2} \cdot \overbrace{(b \cdot b \cdot b)}^{3}$

$$= b \cdot b \cdot b \cdot b \cdot b = b^5$$

Also, $p \cdot p^2 = \overbrace{(p)}^{1} \cdot \overbrace{(p \cdot p)}^{2}$

$$= p \cdot p \cdot p = p^3$$

Note that the exponent in each product above is the sum of the exponents of the factors. We may state, in general, that

$$\mathbf{x}^m \cdot \mathbf{x}^n = \mathbf{x}^{m+n}$$

where x is any signed number and m and n are positive integers.

RULE: To multiply powers of the same base, add the exponents of the factors and make this sum the power of the common base in the product.

This rule may apply to two or more powers of the same base. Thus, $y^2 \cdot y^3 \cdot y^5 = y^{2+3+5} = y^{10}$. However, it does not apply to the product of powers of different bases. For example, $x^3 y^2$ cannot be written with a single base and a single exponent.

Consider $(a^4)^3$:

$$(a^4)^3 = a^4 \cdot a^4 \cdot a^4 = a^{4+4+4} = a^{12}$$

We say that $(a^4)^3 = a^{12}$. Thus, if we are finding the power of a power, we must multiply the two exponents to find the power of a product. In general,

$$(\mathbf{x}^r)^s = \mathbf{x}^{rs}$$

where x is any signed number and r and s are positive integers.

Note also that $(a^2 b^3)^3 = (a^2)^3 (b^3)^3 = a^6 b^9$. That is, we apply the power to each factor in the parentheses.

Here are some examples of multiplying powers of the same base.

EXAMPLES: Multiply:

1. $a^6 \cdot a^3 = a^{6+3} = a^9$ 2. $p^2 \cdot p^5 = p^{2+5} = p^7$

3. $7^3 \cdot 7^8 = 7^{3+8} = 7^{11}$ 4. $r \cdot r^3 = r^{1+3} = r^4$

5. $(t^3)^5 = t^{3 \cdot 5} = t^{15}$ 6. $(c^2 d^3)^4 = c^8 d^{12}$

Exercises

In 1 to 33, multiply.

1. $x^3 \cdot x^2$ 2. $m^6 \cdot m$ 3. $p^5 \cdot p^4$

4. $y^7 \cdot y$ 5. $n^8 \cdot n^3$ 6. $r^{10} \cdot r^5$

7. $3^5 \cdot 3^2$ 8. $2^3 \cdot 2^4$ 9. $10^3 \cdot 10^5$

10. $s^2 \cdot s^3 \cdot s^4$ 11. $c^5 \cdot c \cdot c^6$ 12. $d^3 \cdot d^7 \cdot d^3$

13. $2^3 \cdot 2^2 \cdot 2$ 14. $1^3 \cdot 1^2 \cdot 1^4$ 15. $4^5 \cdot 4^2 \cdot 4^3$

16. $g^8 \cdot g^2 \cdot g^5$ 17. $x^5 \cdot x^a$ 18. $y^3 \cdot y^n$

19. $(b^3)^2$ 20. $(c^2)^3$ 21. $(d^4)^5$

22. $(e^2)^5$ 23. $(2^3)^4$ 24. $(1^5)^3$

25. $(a^3)^2 \cdot (a^2)^3$ 26. $(c^2 d^3)^2$ 27. $(pq^2)^3$

28. $(mn)^4$ 29. $(2^3 \cdot 3^2)^4$ 30. $(3 \cdot 4^2)^3$

31. $b^2 \cdot b^4 \cdot c^3$ 32. $x^5 \cdot y^3 \cdot y^7$ 33. $r^7 \cdot s^2 \cdot r$

4. MULTIPLYING MONOMIALS

When we multiply monomials, we make use of the fact that we may arrange the factors in any order (the commutative principle) and that we may group them as we wish (the associative principle). Thus,

$$(3a)(4b) = 3 \cdot 4 \cdot a \cdot b = 12ab$$

$$(-6y^2)(+3y^3) = (-6)(+3) \cdot y^2 y^3$$

$$= -18y^5$$

$$(-7c^2 d)(-5cd^3) = (-7)(-5)(c^2 \cdot c)(d \cdot d^3)$$

$$= 35c^3 d^4$$

RULE: To multiply monomials:
  1. Multiply the numerical coefficients.
  2. Multiply powers of the same base.

Here are some examples of multiplying monomials.

EXAMPLES: Multiply:

1. $(-3p)(+7q) = (-3)(+7)pq = -21pq$
2. $(-4x^3)(-2x^6) = (-4)(-2)x^3 \cdot x^6 = 8x^9$
3. $(9m^2)(-3m^3) = 9(-3)m^2 \cdot m^3 = -27m^5$
4. $(-3cd^2)(+5c^2d^3)(-2c) = (-3)(+5)(-2)c \cdot c^2 \cdot c \cdot d^2 \cdot d^3$
$$= 30c^4d^5$$

Exercises

In 1 to 20, multiply.

1. $(+4a^2)(-5a^7)$
2. $(-15)(+4x)$
3. $(-6m)(-3n)$
4. $(-7p^2)(-5p^3q)$
5. $(+3a)(-2b)(-4c)$
6. $(5c^2)(-8c^2d^3)$
7. $(\frac{2}{3}r)(-9a)$
8. $(-6p)(+\frac{1}{2}q)(-\frac{2}{3}r)$
9. $(2x)^3$
10. $(3ab)^2$
11. $(-3x^2y)(+6xyz)$
12. $(-9r^2s)(+7r^3s^4)$
13. $(-2c)(+3cd)(-4d^2)$
14. $(-.4b^2)^3$
15. $(-3ak^2)(+5a^2k^3)$
16. $(-5p)^2(-q)^3$
17. $(+5c^2)(-5d^2)$
18. $(-3ab^2)^3$
19. $(-4d^2)(+3e^3)(-5de)$
20. $(+\frac{1}{3}g)^2(9h)$

21. Express the area of a rectangle whose length is $3t^2$ and whose width is $8t$.

22. Express the area of a rectangle whose width is w and whose length is $3w$.

23. Express the area of a square each of whose sides is $7k$ inches.

24. Express the volume of a cube each of whose edges is $3y$ feet.

25. Express the volume of a rectangular solid of height h, width $2h$, and length $3h$.

5. DIVIDING POWERS OF THE SAME BASE

When we divide x^7 by x^3, we are in effect asking, "What quantity multiplied by x^3 results in x^7?" We have learned that $x^4 \cdot x^3 = x^7$. Therefore, $x^7 \div x^3 = x^4$.

Likewise, $a^9 \div a^6 = a^3$, since $a^6 \cdot a^3 = a^9$.

Note that, in both these examples, the exponent of the quotient is the difference between the exponents of the dividend and the divisor. That is,

$$x^7 \div x^3 = x^{7-3} = x^4 \text{ and } a^9 \div a^6 = a^{9-6} = a^3$$

In general, then, we may write

$$\mathbf{x^m \div x^n = x^{m-n}}$$

where $x \neq 0$, m and n are positive integers, and m is greater than n.

RULE: To divide two powers of the same base:
  1. Subtract the exponent in the divisor from that in the dividend.
  2. Make this difference the exponent of the common base in the quotient.

Here are some examples of dividing powers of the same base.

EXAMPLES: Divide:

1. $y^{10} \div y^3 = y^{10-3} = y^7$ **2.** $b^9 \div b^5 = b^{9-5} = b^4$

3. $k^6 \div k = k^{6-1} = k^5$ **4.** $7^5 - 7^2 = 7^{5-2} = 7^3$

5. $\dfrac{r^7}{r^2} = r^{7-2} = r^5$ **6.** $\dfrac{x^a}{x^b} = x^{a-b}$

We know that any nonzero number divided by itself is 1, so that $x^5 \div x^5 = 1$. In general, $x^p \div x^p = 1$, where $x \neq 0$ and p is a positive integer.

Exercises _____

In 1 to 27, divide.

1. $a^5 \div a^2$ **2.** $p^{11} \div p^4$ **3.** $c^8 \div c$ **4.** $y^9 \div y^3$

5. $x^{10} \div x^5$ **6.** $d^8 \div d^7$ **7.** $\dfrac{n^6}{n^4}$ **8.** $\dfrac{t^5}{t^3}$

9. $\dfrac{m^{12}}{m^8}$ **10.** $2^5 \div 2^3$ **11.** $10^6 \div 10^2$ **12.** $3^6 \div 3^3$

13. $a^8 \div a^5$ **14.** $z^{11} \div z^{10}$ **15.** $b^8 \div b^5$ **16.** $k^2 \div k^2$

17. $c^9 \div c^9$ **18.** $s \div s$ **19.** $x^{2a} \div x^a$ **20.** $y^t \div y^3$

21. $b^x \div b^x$ **22.** $\dfrac{r^3 \cdot r^5}{r^2}$ **23.** $\dfrac{t^4 \cdot t^6}{t^5}$ **24.** $\dfrac{a^9 \cdot a^3}{a^3}$

25. $\dfrac{2^4 \cdot 2^7}{2^5}$ **26.** $\dfrac{10^4 \cdot 10^5}{10^3}$ **27.** $\dfrac{3^5 \cdot 3^2}{3^4}$

6. DIVIDING MONOMIALS

If we wish to divide $12x^5$ by $3x^2$, we are in effect asking, "What quantity multiplied by $3x^2$ results in a product of $12x^5$?" We already know that $(4x^3)(3x^2) = 12x^5$. Hence, $(12x^5) \div (3x^2) = 4x^3$.

Likewise, $(-10y^8) \div (5y^4) = -2y^4$, since $(-2y^4)(5y^4) = -10y^8$.

RULE: When dividing monomials:
1. The coefficient of the quotient is the quotient of the coefficients of dividend and divisor.
2. The exponents in the quotient are the result of dividing powers of the same base.

Here are some examples of dividing monomials.

EXAMPLES: Divide:

1. $(-10b^7) \div (-5b^4) = +2b^3$

2. $\dfrac{-15x^8y^5}{+3x^6y^2} = \dfrac{-15}{+3} \cdot \dfrac{x^8}{x^6} \cdot \dfrac{y^5}{y^2} = -5x^2y^3$

3. $\dfrac{+21a^7b^5c^4}{-3a^4b^4c^4} = \dfrac{+21}{-3} \cdot \dfrac{a^7}{a^4} \cdot \dfrac{b^5}{b^4} \cdot \dfrac{c^4}{c^4} = -7a^3b(1) = -7a^3b$

Exercises

In 1 to 21, divide.

1. $30b^7$ by $10b^2$ **2.** $-21m^5n^6$ by $7m^3$ **3.** $18ab$ by $6a$

4. $-42x$ by 6 **5.** $-20y^5$ by $4y^2$ **6.** $-35p^7$ by $-7p^3$

7. $(14a^7k) \div (-2a^5k)$ **8.** $(6ab) \div (-3a)$ **9.** $(a^9b^4) \div (a^6b^3)$

10. $(a^2b^4) \div (ab^2)$ **11.** $(10p^4q^3) \div (-5p^2q^2)$

12. $(15x^2y^3z^4) \div (5xy^2z^3)$

13. $\dfrac{21x^7}{-3x^4}$ **14.** $\dfrac{150r^3s^4}{-50r^3s^4}$ **15.** $\dfrac{-40c^7d^5}{+8c^6d^4}$

16. $\dfrac{-10a^3b^4}{2a^2b}$ **17.** $\dfrac{12x^5y^5}{-4x^4y}$ **18.** $\dfrac{16a^6}{-4a^3}$

19. $\dfrac{-45r^7s^5t^4}{-9r^5s^2t^4}$ **20.** $\dfrac{-31b^3c^4}{-31b^3c}$ **21.** $\dfrac{-56x^7y^3z^3}{8x^5y}$

22. If the area of a rectangle is $48y^5$ and the width is $8y^3$, represent the length.

23. If a dozen eggs cost $36n$ cents, represent the cost of each egg.

24. If $3x$ books cost $21x^4$ dollars, represent the cost of each book.

25. Perform the operations indicated:

   a. $\dfrac{(7rs)(-4r^3s^4)}{14r^2s^2}$ **b.** $\dfrac{(-5p^2q^3)(-8p^4q^5)}{10p^3q^3}$

---

7. ADDING POLYNOMIALS

Expressions such as $3p$, $4x + 5$, and $y^2 + 2y + 7$ are called **polynomials.** A polynomial is merely a sum of terms. The polynomial $3p$ is called a monomial since it has only one term. The polynomial $4x + 5$, which has two terms, is called a **binomial;** and $y^2 + 2y + 7$, a polynomial of three terms, is called a **trinomial.**

An expression such as $2x^2 + (-5x) + (-6)$ is called a *polynomial in* x. The addition signs are generally omitted, and the expression is written $2x^2 - 5x - 6$.

The addition of polynomials is simply the addition of like terms. This is similar to adding units of measure. For example, add 4 lb 5 oz to 6 lb 4 oz:

$$4 \text{ lb} + 5 \text{ oz}$$
$$\underline{6 \text{ lb} + 4 \text{ oz}}$$
$$10 \text{ lb} + 9 \text{ oz} = 10 \text{ lb } 9 \text{ oz}$$

Now add $3a + 4b$ to $6a + 2b$:

$$3a + 4b$$
$$\underline{6a + 2b}$$
$$9a + 6b$$

This method of adding polynomials is to place like terms in columns and to find the algebraic sum of the like terms. It may be convenient to arrange the terms alphabetically or in descending powers of one of the letters (decreasing exponents).

ILLUSTRATIVE PROBLEM 1: Add $2x - 3z + 4y$ to $5y - 7z + 2x$.

Solution: Arrange terms in alphabetical order and add like terms.

$$2x + 4y - 3z$$
$$2x + 5y - 7z$$
$$\overline{4x + 9y - 10z} \quad (answer)$$

In rearranging terms, remember that the sign to the left of any term is part of the term.

ILLUSTRATIVE PROBLEM 2: Add: $3a^2 - 2ab + b^2$, $4b^2 + 7ab - 13a^2$, $-2b^2 + 5a^2 - 3ab$

Solution: Rearrange terms in descending powers of a (decreasing exponents):

$$3a^2 - 2ab + b^2$$
$$-13a^2 + 7ab + 4b^2$$
$$\underline{5a^2 - 3ab - 2b^2}$$
$$-5a^2 + 2ab + 3b^2 \quad (answer)$$

ILLUSTRATIVE PROBLEM 3: Simplify: $4x - 3y + 2x + 5y - 3x$

Solution: Combine like terms in the polynomial:

$$4x + 2x - 3x - 3y + 5y$$

$$= (4x + 2x - 3x) + (-3y + 5y)$$

$$= 3x + 2y \quad (answer)$$

Exercises _____

In 1 to 12, add.

1. $3a + b$
$2a + 5b$

2. $5x - 2y$
$3x + y$

3. $2r - 3s + t$
$5r + 9s - 5t$

4. $6b - 4c$
$-3b + 2c$
$8b - 5c$

5. $4y + 5z$
$-3y - 2z$
$y + z$

6. $7x^2y - 5y^2z$
$-2x^2y + 3y^2z$
$-11x^2y + 8y^2z$

7. $6x^2 - 3x + 8$
$-2x^2 + 5x + 3$

8. $3a^2 + 2ab + b^2$
$a^2 + 4ab - 3b^2$

9. $7r^2 - 8r + 9$
$-2r^2 - 3r - 12$

10. $4x + 2y$
$3x - y + z$
$x \quad\quad - z$

11. $4c^2 - 2c + d$
$-3c^2 + 7c + 5d$
$9c^2 - 8c + 4d$

12. $7a - 9b$
$-3a + 2b - 4$
$6a - 3b + 11$

In 13 to 23, add.

13. $3a - 2c + 4b$ and $2a - 7b + 9c$

14. $5x - 2y, y - x,$ and $3y + 4x$

15. $4a + b + c, a + c - b,$ and $3a + 2b + 5c$

16. $-3x + 2y, 7x - 9y,$ and $-4x - 5y$

17. $8a - 2b, 10a + b,$ and $-2a - 3b$

18. $7x^2 - 2x + 9$ and $5 + 4x - 3x^2$

19. $a^2 - b^2 + c^2, 3a^2 - 2c^2,$ and $4b^2 - 5c^2 + 3a^2$

20. $5r^2 - 2rs + 4s^2, 3r^2 + s^2 - 6rs,$ and $-s^2 + r^2$

21. $3c^2 - c^3 + 2c - 7, 5 + 2c^2 + 3c^3,$ and $5c^2 - 9$

22. $7a^2 - 5ab + 3b^2, 8b^2 + 6a^2 - 9ab,$ and $-4ab + 3a^2 - 11b^2$

23. $9p - 3q + 2r, 4r - 3p + 8q,$ and $p - r$

In 24 to 37, simplify by combining like terms.

24. $6x - 4y + 3x + 9y$

25. $7a - 3b - 5b + 6a$

26. $3r - 5s + 6r - 9t + 7s$

27. $-5a^2 - 3a + 8 + 2a + 12a^2 + 10$

28. $7pq - 8qr - 9pq + 12qr$

29. $10y^2 - 7 + 3y - 8 + 2y^2 - 7y$

30. $3k + (2k - 7)$

31. $(-8x + y) + 3y$

32. $4t^2 + (-6t^2 - 9)$

33. $8m^2 + (5m - 3m^2)$

34. $(9s^2 - 3) + (4s^2 + 7)$

35. $(12a + 7b) + (2b - 5a)$

36. $(y^2 + 3 + 4y) + (3y^2 - 9y)$

37. $(2p^3 - 7 + 9p^2) + (3 - 12p + 7p^2)$

In 38 to 42, represent in simplest form the perimeters of the figures whose sides are given.

38. $2a - 9, 3a + 4, 5a - 2$

39. $4x + 2y, 3x - 5y, x + 4y$

40. $2r - 3s, 4r + 2s, 3r + 7s, 5r$

41. $3a - 2, 2b + 4, 3a - 2, 2b + 4$

42. $7s + 2t, 3s - 5t, 2s + 6t, 3t$

43. Find the perimeter of a square each of whose sides is $2x - 3$.

8.   SUBTRACTING POLYNOMIALS

We can subtract polynomials in the same way we subtracted monomials; that is, we change the sign of the subtrahend and proceed as in addition.

For convenience, we may arrange like terms under one another as we did in the addition of polynomials.

ILLUSTRATIVE PROBLEM 1:   Subtract: $(8a - 2b) - (10a + b)$

Solution:
$$8a - 2b$$
$$\overline{\oplus 10a \oplus b}$$
$$-2a - 3b \quad (answer)$$

ILLUSTRATIVE PROBLEM 2:   Subtract $3x^2 - 2x + 5$ from $7 + 8x^2 - 4x$.

Solution 1: Arrange in descending order in columns:
$$8x^2 - 4x + 7$$
$$\overline{\oplus\, 3x^2 \ominus 2x \oplus 5}$$
$$5x^2 - 2x + 2 \quad (answer)$$

Solution 2: We may use a horizontal arrangement:
$$(8x^2 - 4x + 7) - (3x^2 - 2x + 5)$$

Now change the signs of all terms in the subtrahend and change the operation of subtraction to addition ($-$ to $+$). Thus,
$$(8x^2 - 4x + 7) + (-3x^2 + 2x - 5)$$
$$(8x^2 - 3x^2) + (-4x + 2x) + (7 - 5)$$
$$5x^2 - 2x + 2 \quad (answer)$$

Exercises

In 1 to 12, subtract the lower polynomial from the upper one.

1. $7x + 5y$
   $2x + 3y$

2. $8m - 5n$
   $2m + n$

3. $12p - 5q$
   $-3p - 7q$

4. $12x^2 - 9y^2$
   $x^2 + y^2$

5. $5a$
   $2a + 3b$

6. $x^2 + 2x - 5$
   $3x^2 - 4x + 9$

7. 0
$\underline{4r - 5s}$

8. $9y^2 + 3y$
$\underline{7y^2 - 2y + 4}$

9. $2a^2 - 11$
$\underline{a^3 \quad - 5}$

10. $8p - 9q + 3r$
$\underline{-3p \qquad + 5r}$

11. $10c - 3d + 7$
$\underline{-2c - 3d}$

12. $3rs - 7st$
$\underline{-6rs - 12st}$

13. From $8x - 4y + 5z$ take $3z - 2y - 7x$.

14. Subtract $a^2 + 12 - 7a$ from $19 + 3a^2 - 4a$.

15. From $5c^2 - 7cd + d^2$ take $3d^2 - c^2 + 2cd$.

16. From $r - s - t$ take $s + t - r$.

17. From $2b^2 - 3bc + 4c^2$ take $2b^2 - 4c^2 - 3bc$.

18. Subtract $5k^2 - 12$ from 0.

19. How much greater is $9p - 7q$ than $3p + 2q$?

20. From the sum of $2x - 3y + 4z$ and $8x + 2y - 7z$, subtract $x + y + z$.

21. By how much does $3c^2 - 5cd + 9d^2$ exceed $2c^2 - 7d^2 - 2cd$?

22. How much less than $7x + 9$ is $3x - 2$?

23. Subtract $a^2 + b^2 - c^2$ from the sum of $3a^2 - 2b^2 + 5c^2$ and $-2a^2 + b^2 - 3c^2$.

In 24 to 33, simplify.

24. $(2a + b) - (3a + 5b)$

25. $(3x + 7) - (x + 6)$

26. $(5x^3y + 3x^2y) - x^3y$

27. $(4a^2 - b) - (2a^2 + b)$

28. $(8p + 2q) - (10p + q)$

29. $5r - (3r + 7)$

30. $(9y^2 - 8y + 1) - (3y^2 + 2y + 4)$

31. $(5a^2 + 3ab - 8b^2) - (a^2 - b^2)$

32. $(7x - 2) + (8x - 3) - (3x + 5)$

33. $(12k^2 - 4k + 7) - (3 - 7k^2)$

---

9. MULTIPLYING A POLYNOMIAL BY A MONOMIAL

The distributive property of multiplication over addition tells us that

$$x(2x + 3) = x(2x) + x(3)$$

$$= 2x^2 + 3x$$

We can illustrate this geometrically by considering the area of a rectangle whose length is $2x + 3$ and whose width is x. See the figure on the following page.

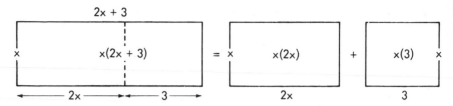

The area of the original rectangle is the sum of the areas of the two smaller rectangles, so that

$$x(2x + 3) = x(2x) + x(3) = 2x^2 + 3x$$

RULE: To multiply a polynomial by a monomial, multiply each term of the polynomial by the monomial.

Here are some examples of multiplying a polynomial by a monomial.

EXAMPLES: Multiply:

1. $3(a + b) = 3a + 3b$

2. $2x(x - y) = 2x(x) - 2x(y) = 2x^2 - 2xy$

3. $4r^2 (r^2 + 5r + 2) = 4r^2 (r^2) + 4r^2 (5r) + 4r^2 (2)$
$$= 4r^4 + 20r^3 + 8r^2$$

4. $5p - 3(2p - 1) = 5p - 3(2p) - 3(-1)$
$$= 5p - 6p + 3 = -p + 3$$

Exercises

In 1 to 28, multiply.

1. $5(2x + 3y)$	**2.** $7(r - s)$	**3.** $-4(3 + 7a)$
4. $4x(y - 3z)$	**5.** $3a(a + b)$	**6.** $2a^3(a^3 - ab)$
7. $c(c^2 - 2d^2)$	**8.** $6(2e + 3f)$	**9.** $-k(2k - 1)$
10. $n(a + 1)$	**11.** $p^2 (2p - 7)$	**12.** $-r(r^2 - s)$

13. $5m^2n(2m - 3n)$ **14.** $r^2t(r^2 - s^2)$

15. $-5xy^2(y^3 - x^2)$ **16.** $5(3p^2 - 2p - 7)$

17. $2k(k^2 - 2k + 1)$ **18.** $6c(2c^2 - c - 1)$

19. $x^2 (3 - 5x + 7x^2 - x^3)$ **20.** $-1(q^2 - 3q + 2)$

21. $\frac{1}{2}p(4p^2 - 6p + 2)$ **22.** $.5r(r^2 - 10r)$

23. $-3n(n^2 - n - 4)$ **24.** $2ab(a^2 - 2ab + b^2)$

25. $\frac{3}{4}s(4s^2 - 8s + 12)$ **26.** $-30(\frac{2}{3}x^2 - \frac{1}{5}x + \frac{5}{6})$

27. $5xy(y^2 - yx - x^2)$ **28.** $-2t(a^2t^2 - 3at + a^3)$

In 29 to 38, simplify by multiplying as indicated and combining like terms.

29. $5(n - 2) + 7n$ **30.** $3t + 4(t - 3)$

31. $7p + 1(p - 9)$ **32.** $2r - 5(3 - r)$

33. $-6(2k + 1) + 9k$ **34.** $-1(5t - 4) - 6$

35. $-3(4x + 5) + 20$ **36.** $-4(2y - 3) + 5y - 1$

37. $-1(c + d) + 3(c - d)$ **38.** $5q(q^2 - 10) + 18q$

In 39 to 44, represent the answer to each question in simplest form.

39. Express the area of a rectangle of length $3x - 4$ and width $5x$.

40. Express the area of a rectangle of length $\frac{1}{2}p$ and width $(4p - 2)$.

41. Express the perimeter of a square each of whose sides is $(3x - 5)$ inches.

42. An egg costs $(3c - 2)$ cents. What is the cost of a dozen eggs?

43. A can of milk weighs $(2k - 3)$ pounds. How many ounces does it weigh?

44. The length of a board is $(2k - 1)$ meters. How many centimeters is this?

---

10. SIMPLIFYING EXPRESSIONS THAT HAVE GROUPING SYMBOLS

In some of the exercises in the preceding section, we removed parentheses by multiplying and then simplified by combining like terms. We can do this in a number of cases by using the distributive property.

Occasionally, we find a parenthetic expression within brackets [ ]. In cases of this kind, we begin by removing the parentheses first. For example,

$$5y - 3[2y - 7(y + 1)]$$

$$= 5y - 3[2y - 7y - 7] = 5y - 3[-5y - 7]$$

$$= 5y + 15y + 21 = 20y + 21$$

Note that first we remove the parentheses and combine, and then we remove the brackets and combine like terms.

EXAMPLES: Rewrite without parentheses and combine like terms.

1. $4a + (5 - 2a) = 4a + 1(5 - 2a)$
$$= 4a + 5 - 2a = 2a + 5$$

2. $4a - (5 - 2a) = 4a - 1(5 - 2a)$
$$= 4a - 5 + 2a = 6a - 5$$

Exercises

In 1 to 18, simplify.

1. $5 - (2b - 7)$ **2.** $10 - (3a + 7)$

3. $8 + (6k - 5)$ **4.** $12 - (-4c + 9)$

5. $6p + (5 - 2p)$ **6.** $12r - (3r - 6)$

7. $7c - 3c(2c - 1)$ **8.** $(9 + 5y) - (9 - 5y)$

9. $6a - 4b - (a + b)$ **10.** $3(x - y) - (y - x)$

11. $5d - (d - 3) + 4$ **12.** $m(m - 5) - m(m + 2)$

13. $3s + 4[s - 2(s - 7)]$ **14.** $9t - 3[t - 4(t - 6)]$

15. $5[p - 6(p - 1)] - 3p$ **16.** $7[8z - 3(z - 4)]$

17. $9[3e - (e - 4) + 1]$ **18.** $p^2q - p[pq - p(q - r)]$

11. DIVIDING A POLYNOMIAL BY A MONOMIAL

We already learned that $3(a + b) = 3a + 3b$.
It then follows that

$$\frac{3a + 3b}{3} = a + b$$

Note that the quotient $a + b$ can be obtained by dividing each term of the dividend $3a + 3b$ by 3 as follows:

$$\frac{3a + 3b}{3} = \frac{3a}{3} + \frac{3b}{3} = a + b$$

It thus appears that division by a monomial is also distributive over addition.

RULE: To divide a polynomial by a monomial, divide each term of the polynomial by the monomial.

EXAMPLES

1. $\dfrac{4p^2 + 8p}{4p} = \dfrac{4p^2}{4p} + \dfrac{8p}{4p} = p + 2$

2. $\dfrac{3r^3 - 6r^2 + 12r}{3r} = \dfrac{3r^3}{3r} - \dfrac{6r^2}{3r} + \dfrac{12r}{3r}$

$$= r^2 - 2r + 4$$

Exercises

In 1 to 31, divide.

1. $\dfrac{6m - 9n}{3}$

2. $\dfrac{15c - 20d}{-5}$

3. $\dfrac{21p - 14}{7}$

4. $\dfrac{cr + cs}{c}$

5. $\dfrac{k - tk}{k}$

6. $\dfrac{2m + 2}{-2}$

7. $\dfrac{6x^2 - 12y^2}{-3}$

8. $\dfrac{8p^2 - 20q^2}{4}$

9. $\dfrac{12d^2 - 6}{-6}$

10. $\dfrac{n^2 - 5n}{n}$

11. $\dfrac{15d^2 - 2d}{d}$

12. $\dfrac{9k^2 - 5k}{k}$

13. $(7x^2 - 3x) \div x$

14. $(18t^2 - 12t) \div (-t)$

15. $(a^2 - a) \div (-a)$

16. $\dfrac{7r^3 - 2r^2}{r^2}$

17. $\dfrac{9t^4 - 8t^2}{t^2}$

18. $\dfrac{15y^3 - 10y^2}{5y^2}$

19. $\dfrac{3p - 6q + 9r}{3}$

20. $\dfrac{7a + 14b - 28}{-7}$

21. $\dfrac{2x^2 - 4x - 8}{2}$

22. $\dfrac{\pi a^2 - \pi b^2}{\pi}$

23. $\dfrac{r^3s^2 - r^2s^2 + rs}{rs}$

24. $\dfrac{x^4 - x^2 + x}{-x}$

25. $\dfrac{-x^2 - xy + y^2}{-1}$

26. $\dfrac{21a^2b - 14ab^2 + 7a^3b}{7ab}$

27. $\dfrac{p^3t - pt^3}{-pt}$

28. $\dfrac{22x^4 - 11x^3 + 33x^2}{11x^2}$

29. $\dfrac{18r^2s^3 - 9r^2s^2 + 27r^3s^2}{9r^2s^2}$

30. $\dfrac{3.2x^2y - 2.4xy^2}{.8xy}$

31. $\dfrac{1.5p^3q^3 - 2.5p^2q^2 + 2p^3q^2}{.5p^2q^2}$

32. The area of a rectangle is $15w^2 + 10w$. If the width is $5w$, represent the length in simplest form.

33. Represent in simplest form the number of feet in $(24a + 36b)$ inches.

34. Represent in simplest form the number of quarters in $(50n + 75)$ cents.

35. A train travels $(26h^2 + 39h)$ miles in $13h$ hours. Express this rate in miles per hour in simplest form.

36. The perimeter of a square is $(8t^2 + 12)$ inches. Express the length in inches of each side in simplest form.

37. A building is $(6x^2 + 9x)$ feet high. Express, in simplest form, the height of the building in yards.

Chapter Review Exercises

1. Add: $(-3x^2y) + (-5x^2y)$

2. Combine like terms: $6.4x + 8.2x - 5.8x$

3. Subtract $7a^2b$ from $12a^2b$.

4. From the sum of $-3y^2$ and $8y^2$, take $-4y^2$.

5. Multiply: $m^2 \cdot m^3 \cdot m^4$

6. Multiply: $(-7r^2s^3)(+4rs^2)$

7. Divide $24p^3q^2r$ by $-6pqr$.

8. Represent in simplest form the number of days in
    a. n weeks   **b.** $(n + 3)$ weeks   **c.** $(2n - 4)$ weeks

9. Add: $-3x^2 + 4xy - 7y^2,\ 5xy + 6x^2,\ x^2 - y^2$

10. Subtract $5a - 4c + 6b$ from $3b - 2c + 4a$.

11. Perform the indicated operations and simplify:
    a. $6(5p - 4q)$ **b.** $y - (3y + 7)$
    c. $8m^2n(m^2 - n^2)$ **d.** $7n - 4(n - 8)$

12. Divide and simplify:
    a. $\dfrac{6x^2 - 9xy - 3y^2}{3}$ **b.** $\dfrac{15r^2s - 10rs^2}{-5rs}$

13. Simplify: $(5p^2 - 3p + 7) - (2p^2 + 4p - 10)$

14. What number must be added to $5m^2 - 3$ to obtain $6m^2 + 7$?

15. Divide:
    a. $\dfrac{25n^2 - 15n}{-5n}$ **b.** $\dfrac{8x^3 - 6x^2 + 4x}{-2x}$

CUMULATIVE REVIEW

1. From $3x - 5y + z$, subtract $2x - 6y + z$.

2. Solve for a: $7a + 16 = -5$

3. If the length of a rectangle is represented by x and its width is represented by y, express its perimeter in terms of x and y.

4. Solve for r: $4(r + 3) = 16$

5. If s represents the side of a regular hexagon (six-sided polygon), a formula for its area is $A = 2.6s^2$. Using this formula, find the area of a regular hexagon whose side is 10.

6. What is the multiplicative inverse of $-\frac{2}{3}$?

7. If $a = 2$ and $b = 3$, evaluate $\dfrac{1}{a} + \dfrac{1}{b}$.

8. Three partners in a business are to divide the profits in the ratio $3:4:5$. If the profits for the first year are $7200, how much does each partner receive?

9. Divide $12x^3y$ by $-4xy$.

10. 32% of a certain number is 24. Find the number.

Answers to Cumulative Review

If you get an incorrect answer, refer to the chapter and section shown in brackets for review. For example, [5-2] means chapter 5, section 2.

1. $x + y$ [10-8]
2. -3 [9-5]
3. $2(x + y)$ [3-1]

4. 1 [10-9]
5. 260 [8-6]
6. $-\frac{3}{2}$ [9-2]

7. $\frac{5}{6}$ [4-5]
8. $1800
   $2400 [6-2]
   $3000

9. $-3x^2$ [10-6]

10. 75 [7-6]

11

Equations and Problem Solving

1. SOLVING EQUATIONS BY ADDING SIGNED NUMBERS

In Chapter 2, we learned how to solve simple equations by adding or subtracting the numbers of arithmetic. For example, in the solution of $x + 3 = 8$, we subtracted 3 from both sides and arrived at the solution $x = 5$. We learned in our discussion of signed numbers that subtracting 3 is the same as adding -3. Now, therefore, we need to use only the addition principle of equality. Here is an example:

ILLUSTRATIVE PROBLEM 1: Solve and check: $x + 11 = 4$

Solution: Add -11 to both members: *Check:*

$$
\begin{array}{rl}
x + & 11 = 4 \\
& \underline{-11 \quad -11} \\
x & = -7 \ (answer)
\end{array}
$$

Check:
$$x + 11 = 4$$
$$-7 + 11 = 4 \ (x = -7)$$
$$4 = 4 \ \checkmark$$

Recall that our aim is always to get the unknown term by itself. In the above problem, we wanted to get the x by itself. We therefore added -11 to both sides of the equation. Since $+11$ and -11 add up to zero, the left member became $x + 0$, or x.

ILLUSTRATIVE PROBLEM 2: Solve and check: $y - 5 = 7$

Solution: Add $+5$ to both members: *Check:*

$$
\begin{array}{rl}
y - & 5 = 7 \\
& \underline{+5 \quad +5} \\
y & = 12 \ (answer)
\end{array}
$$

Check:
$$y - 5 = 7$$
$$12 - 5 = 7 \ (y = 12)$$
$$7 = 7 \ \checkmark$$

Exercises_____

In 1 to 16, solve and check.

1. $y + 15 = 9$ 2. $x - 10 = -4$ 3. $18 = n - 6$
4. $7 + t = -15$ 5. $-12 = -20 + r$ 6. $-15 = s + 22$
7. $p + 2.4 = 8.7$ 8. $7.6 + k = -15.4$ 9. $n - 5\frac{1}{2} = 8$
10. $f - 6\frac{1}{3} = -7$ 11. $g - 3\frac{1}{4} = -5\frac{1}{2}$ 12. $3\frac{1}{8} + y = 7\frac{1}{2}$
13. $\frac{1}{2} + m = -\frac{3}{4}$ 14. $3.14 = w + 6.28$ 15. $-7 + x = -7$
16. $13 = y + 13$

---

2. SOLVING EQUATIONS BY MULTIPLICATION AND DIVISION OF SIGNED NUMBERS

In Chapter 2, we solved equations by multiplying or dividing both members by the numbers of arithmetic. We may now extend these principles of equality to the same operations with signed numbers.

ILLUSTRATIVE PROBLEM 1: Solve and check: $6x = -24$

Solution: Divide both members by 6:   *Check:*

$$6x = -24$$

$$\frac{6x}{6} = \frac{-24}{6}$$

$$x = -4 \quad (answer)$$

Check:

$$6x = -24$$

$$6(-4) = -24$$

$$-24 = -24 \ \checkmark$$

ILLUSTRATIVE PROBLEM 2: Solve and check: $-\frac{3}{4}y = 15$

Solution: We may divide both sides by $-\frac{3}{4}$. As we learned in our study of fractions, this is the same as multiplying both sides by $-\frac{4}{3}$ (the reciprocal of $-\frac{3}{4}$). Thus,

$$-\frac{3}{4}y = 15$$

$$\left(-\frac{4}{3}\right)\left(-\frac{3}{4}y\right) = \left(-\frac{4}{3}\right)(15)$$

$$1 \cdot y = -20$$

$$y = -20 \quad (answer)$$

Check:

$$-\frac{3}{4}y = 15$$

$$\left(-\frac{3}{4}\right)(-20) = 15$$

$$15 = 15 \ \checkmark$$

ILLUSTRATIVE PROBLEM 3: Solve and check: $\dfrac{p}{8} = -3$

Solution: $\dfrac{p}{8}$ is the same as $\dfrac{1}{8}p$. In order to obtain $1 \cdot p$ or p in the left member of the equation, we must multiply both members of the equation by 8. Thus,

$$\frac{p}{8} = -3$$

$$8\left(\frac{p}{8}\right) = 8(-3)$$

$$1 \cdot p = -24$$

$$p = -24 \quad (answer)$$

Check:

$$\frac{p}{8} = -3$$

$$\frac{-24}{8} = -3$$

$$-3 = -3 \ \checkmark$$

Exercises

In 1 to 27, solve and check.

1. $3t = -18$

2. $4y = -7$

3. $7k = -4$

4. $-6x = 42$

5. $-11r = 99$

6. $-8 = 3p$

7. $-5m = 15.5$

8. $-z = 17$

9. $-9a = -36$

10. $-12b = -6$

11. $-13c = -39$

12. $-8d = -44$

13. $-\frac{1}{5}p = 8$

14. $\frac{1}{6}m = -7$

15. $\frac{2}{3}k = -10$

16. $\frac{4}{5}r = -1.6$

17. $-\frac{5}{3}x = \frac{3}{5}$

18. $2\frac{2}{3}y = -48$

19. $-\frac{5}{2}t = -7\frac{1}{2}$

20. $\dfrac{a}{3} = -10$

21. $-\dfrac{f}{5} = 13$

22. $\dfrac{m}{-7} = -5$

23. $-\frac{3}{8}p = -15$

24. $\dfrac{b}{12} = -\dfrac{5}{6}$

25. $\dfrac{k}{8} = -\dfrac{5}{4}$

26. $-\dfrac{3s}{4} = \dfrac{5}{12}$

27. $\dfrac{r}{-7} = -\dfrac{3}{28}$

3. SOLVING EQUATIONS USING TWO OPERATIONS

If the solution of an equation requires two operations, it is usually simpler to use the addition principle first, and then the multiplication or division principle. Before either principle is used, however, like terms (if any) should be combined.

ILLUSTRATIVE PROBLEM 1: Solve and check: $2x - 6x + 9 = -7$

Solution:

$$(2x - 6x) + 9 = -7 \quad \text{(combine like terms)}$$
$$-4x + 9 = -7$$
$$\underline{ -9 -9} \quad \text{(add } -9 \text{ to both members)}$$
$$-4x = -16$$
$$\frac{-4x}{-4} = \frac{-16}{-4} \quad \text{(divide both members by } -4)$$
$$x = 4$$

Check:
$$2x - 6x + 9 = -7$$
$$2(4) - 6(4) + 9 = -7$$
$$8 - 24 + 9 = -7$$
$$-7 = -7 \checkmark$$

Answer: $x = 4$

ILLUSTRATIVE PROBLEM 2: Solve and check: $\dfrac{y}{5} - 13 = 17$

Solution:

$$\frac{y}{5} - 13 = 17$$
$$\underline{\phantom{\frac{y}{5}} +13 \quad +13} \quad \text{(add 13 to both members)}$$
$$\frac{y}{5} = 30$$
$$5\left(\frac{y}{5}\right) = 30 \cdot 5 \quad \text{(multiply both members by 5)}$$
$$y = 150 \quad (\textit{answer})$$

Check:
$$\frac{y}{5} - 13 = 17$$
$$\frac{150}{5} - 13 = 17$$
$$30 - 13 = 17$$
$$17 = 17 \checkmark$$

Exercises

In 1 to 26, solve and check.

1. $4x + 3 = -21$ **2.** $9y - 5 = 22$ **3.** $6 = 7k - 15$

4. $-5 + 6t = -23$ **5.** $5p - 8 = 12$ **6.** $9 = 4a - 7$

7. $7r - 3r = -10$ **8.** $2s - 5s = 18$ **9.** $8y - 2y - 11 = 19$

10. $\dfrac{c}{8} + 1 = -3$ **11.** $3x - 6x + 2 = -7$ **12.** $\dfrac{b}{3} - 7 = 1$

13. $-\dfrac{3k}{4} + 7 = 13$ **14.** $\dfrac{5n}{8} - 8 = 2$ **15.** $-11 + \dfrac{2c}{3} = -15$

16. $-\dfrac{2a}{3} + 4 = 9$ **17.** $2x + 4 = 16$ **18.** $\dfrac{3p}{2} - 4 = 11$

19. $8a + 3 = 11$ **20.** $\dfrac{t}{3} + 4 = 15$ **21.** $\dfrac{d}{8} + 45 = 64$

22. $2x - 6 = 0$ **23.** $2y - y + 7 = -9$ **24.** $8x + 4 - 7x = 9$

25. $6m - 8 - 3m = -17$ **26.** $-7 + 11y + y = 17$

27. If 3 times a number is added to 7, the result is 40. Find the number.

28. If 6 times a number is subtracted from 40, the result is -14. Find the number.

4. SOLVING EQUATIONS WITH THE UNKNOWN ON BOTH SIDES

If terms involving the unknown are on both sides of an equation, we try to bring them to the same side by adding a suitable monomial to both sides of the equation.

ILLUSTRATIVE PROBLEM 1: Solve and check: $7t = 24 - 5t$

Solution: Add $5t$ to both sides. Thus,

$$7t = 24 - 5t$$
$$\underline{+5t \cdot \qquad +5t}$$
$$12t = 24$$

Divide both members by 12:

$$\frac{12t}{12} = \frac{24}{12}$$

$$t = 2 \quad (answer)$$

Check:

$$7t = 24 - 5t$$

$$7(2) = 24 - 5(2)$$

$$14 = 24 - 10$$

$$14 = 14 \ \checkmark$$

ILLUSTRATIVE PROBLEM 2: Solve and check: $4y + 12 = 48 - 2y$

Solution:

$$4y + 12 = 48 - 2y$$
$$\underline{+2y \qquad\qquad\quad +2y}$$
$$6y + 12 = 48$$
$$\underline{\quad -12 \quad -12}$$
$$6y \qquad\quad = 36$$
$$\frac{6y}{6} = \frac{36}{6}$$
$$y = 6 \quad (answer)$$

Check:

$$4y + 12 = 48 - 2y$$
$$4(6) + 12 = 48 - 2(6)$$
$$24 + 12 = 48 - 12$$
$$36 = 36 \;\checkmark$$

Exercises

In 1 to 28, solve and check.

1. $8p = 10 + 3p$ **2.** $7p = 55 - 4p$ **3.** $15t = 6t + 36$

4. $17a = 28a + 44$ **5.** $9k = 70 - k$ **6.** $7d = 6 - 2d$

7. $9x = 36 + 5x$ **8.** $12r = 19r - 35$ **9.** $4y = 39y + 70$

10. $-3m = 4m - 49$ **11.** $.9s = .3s - 30$

12. $3\frac{3}{4}n = 60 - 1\frac{1}{4}n$ **13.** $15t + 19 = -3t - 17$

14. $8k - 17 = -29 - 4k$ **15.** $6p + 17 = -2p - 15$

16. $3x + 11 = 6x + 2$ **17.** $7b + 23 = 7 - b$

18. $t - 7 = -6t + 21$ **19.** $5r - 12 = 12r - 5$

20. $-3m + 4 = -7m - 14$ **21.** $10n - 2 = 3n + 47$

22. $-a + 18 = 53 - 6a$ **23.** $-7x = -8 + x - 14$

24. $4y + 33 = 15y - 11$ **25.** $10p - 6 - 5p = -5p + 4$

26. $4m + 1 = 8m + 3$ **27.** $15a = 30a + 33 - 4a$

28. $14n + 3 = 15n + 8 - 6n$

In 29 to 36, write and solve an equation to find the number.

29. Seven times a number equals 45 more than twice the number.

30. Three times a number is equal to 20 less than five times the number.

31. If 3 times a number is increased by 12, the result is equal to 7 times the number.

32. If 5 times a number is decreased by 8, the result is the same as 3 times the number increased by 30.

33. If 4 times a number is increased by 7, the result is the same as 31 more than the number.

34. If 3 times a number is increased by 7, the result is the same as when 72 is decreased by twice the number.

35. If 7 times a number is decreased by 8, the result is the same as when 80 is decreased by 4 times the number.

36. Six times a number exceeds 25 by the same amount that twice the number exceeds 15.

---

5. SOLVING EQUATIONS CONTAINING PARENTHESES

To solve an equation containing parentheses, first remove the parentheses by use of the distributive property or by performing the operation indicated. Then proceed to solve the resulting equation.

ILLUSTRATIVE PROBLEM 1: Solve and check: $6y - 3(y - 4) = -6$

Solution:

$$6y - 3(y - 4) = -6$$
$$6y - 3y + 12 = -6 \quad \text{(remove parentheses)}$$
$$3y + 12 = -6 \quad \text{(combine terms)}$$
$$\underline{ -12 \quad -12} \quad \text{(add } -12 \text{ to both sides)}$$
$$3y = -18$$
$$\frac{3y}{3} = \frac{-18}{3} \quad \text{(divide both sides by 3)}$$
$$y = -6$$

Check:

$$6(-6) - 3(-6 - 4) = -6$$
$$-36 - 3(-10) = -6$$
$$-36 + 30 = -6$$
$$-6 = -6 \; \checkmark$$

Answer: $y = -6$

ILLUSTRATIVE PROBLEM 2: Solve and check: $9p - (p + 3) = 21$

Solution:

$$9p - 1(p + 3) = 21$$

$9p - p - 3 =$	21	(remove parentheses)
$8p \qquad -3 =$	21	(combine like terms)
$+3$	$+3$	(add +3 to both sides)
$8p \qquad\qquad =$	24	
$p =$	3	(divide both sides by 8)

Check:

$$9p - (p + 3) = 21$$
$$9(3) - (3 + 3) = 21$$
$$27 - 6 = 21$$
$$21 = 21 \;\checkmark$$

Answer: p = 3

ILLUSTRATIVE PROBLEM 3: Solve and check: $2x + 3 = 3(x - 2)$

Solution:

$2x + 3 =$	$3x - 6$	(remove parentheses)
$-2x$	$-2x$	(add $-2x$ to both sides)
$3 =$	$x - 6$	
$+6$	$+6$	(add +6 to both sides)
$9 =$	x	
or $x =$	9	

Check:

$$2x + 3 = 3(x - 2)$$
$$2(9) + 3 = 3(9 - 2)$$
$$18 + 3 = 3(7)$$
$$21 = 21 \;\checkmark$$

Answer: x = 9

Exercises

In 1 to 18, solve and check.

1. $2t + 3(t - 5) = -20$

2. $3p - 2(p + 1) = 13$

3. $2n + (3n - 8) = -28$

4. $7k + (3k - 5) = 55$

5. $7n - (4n + 9) = -15$

6. $y - (8y - 11) = 46$

7. $(2k + 7) - 18 = +3$

8. $(4x - 7) - 12 = 13$

9. $a - (15 - a) = 45$

10. $5(7n - 3) = 20$

11. $6(2b - 3) = 54$

12. $8(y + 2) = 5(2y - 4)$

13. $8p - (4p + 6) = 46$

14. $8t - (5t + 3) = 12$

15. $4(2a - 9) - 6(10 - a) = 2$

16. $2(c - 3) - 7 = 12 - 5(c - 2)$

17. $10 + 3(5m + 2) = -m$

18. $3(p - 2) - 19 = 13 - 5(p + 2)$

19. The larger of two numbers is 7 more than the smaller. The larger number plus 5 times the smaller equals 67. Find the numbers. (*Hint:* Let x = the smaller number and $x + 7$ = the larger number.)

20. One number is 5 less than another. If 5 times the larger is subtracted from 8 times the smaller, the result is 11. Find the numbers. (*Hint:* Let x = the larger number and $x - 5$ = the smaller number.)

21. A shirt costs $7 more than a tie. Three shirts and 5 ties cost $53. Find the cost of each.

22. The length of a rectangle is 7 inches shorter than 4 times its width. Find the width if the perimeter is 66 feet.

23. Solve and check: $15y = 13 + 4(3y + 2)$

6. SIMPLIFYING EQUATIONS BY COLLECTING LIKE TERMS

First-degree equations often become more complicated as more terms appear. Consider the equation $3x - 2 + 4x = 12$.

We must first simplify such an equation by collecting like terms. In this case, we are dealing only with like terms that are already on the same side of the equal sign. The solution of the equation is shown as follows:

$$3x - 2 + 4x = 12$$

$7x - 2$	$= 12$	(combine like terms)
$+ 2$	$= +2$	(add $+2$ to both members)
$\dfrac{7x}{7}$	$= \dfrac{14}{7}$	(divide both members by 7)
x	$= 2$	(*answer*)

Now consider an equation still more complicated, such as

$$3x - 2 + 4x = 13 + 9x - 3$$

In this case, we can combine like terms on each side of the equation before we use inverse operations.

$3x - 2 + 4x = 13 + 9x - 3$		
$7x - 2$	$= 10 + 9x$	(combine like terms)
$-9x$	$-9x$	(add $-9x$ to both members)
$-2x - 2$	$= 10$	
$+ 2$	$= +2$	(add $+2$ to both members)
$-2x$	$= 12$	
$(-\tfrac{1}{2})(-2x)$	$= 12(-\tfrac{1}{2})$	(multiply both members by $-\tfrac{1}{2}$)
x	$= -6$	

Thus, the solution is -6. Check by substituting -6 for x in the original equation.

Check:

$$3x - 2 + 4x = 13 + 9x - 3$$

$$3(-6) - 2 + 4(-6) = 13 + 9(-6) - 3$$

$$-18 - 2 - 24 = 13 - 54 - 3$$

$$-20 - 24 = -41 - 3$$

$$-44 = -44 ✔$$

Answer: x = -6

7. USING EQUATIONS TO SOLVE VERBAL PROBLEMS

In Chapter 3, we learned to solve some simple verbal problems by the use of equations. Among others, we dealt with number problems, coin problems, perimeter problems, and problems arising from the use of various formulas.

Now that we have learned more about solving equations, we are in a position to solve more difficult verbal problems. Review the steps in

Section 2, Chapter 3 for solving verbal problems. The following illustrative problems indicate how this method is extended to some slightly more difficult problems.

ILLUSTRATIVE PROBLEM 1: Find two consecutive integers whose sum is -31.

Solution:

$$\text{Let } n = \text{the first integer}$$
$$\text{And } n + 1 = \text{the second integer.}$$

$$
\begin{aligned}
n + (n + 1) &= -31 \quad \text{(write equation)} \\
n + n + 1 &= -31 \quad \text{(remove parentheses)} \\
2n + 1 &= -31 \quad \text{(combine like terms)} \\
\underline{-1 \qquad} &\underline{\;-1} \quad \text{(add } -1 \text{ to both sides)} \\
2n &= -32 \\
n &= -16 \quad \text{(divide both sides by 2)} \\
n + 1 &= -16 + 1 = -15
\end{aligned}
$$

Check: Do -16 and -15 add up to -31?

$$(-16) + (-15) = -31 \;\checkmark$$

Answer: The numbers are -16 and -15.

ILLUSTRATIVE PROBLEM 2: Mary has 30 coins in her coin bank, consisting only of nickels and dimes. If the total value of the coins is $2.40, how many of each kind does she have?

Solution:

$$
\begin{aligned}
\text{Let } x &= \text{number of nickels.} \\
\text{Then } (30 - x) &= \text{number of dimes.} \\
\text{And } 5x &= \text{value of the nickels in cents.} \\
\text{And } 10(30 - x) &= \text{value of the dimes in cents.}
\end{aligned}
$$

$$
\begin{aligned}
5x + 10(30 - x) &= 240 \quad \text{(write equation)} \\
5x + 300 - 10x &= 240 \quad \text{(remove parentheses)} \\
-5x + 300 &= 240 \quad \text{(combine like terms)} \\
\underline{-300} &= \underline{-300} \quad \text{(add } -300 \text{ to both sides)} \\
-5x &= -60 \\
x &= 12 \quad \text{(divide both sides by } -5) \\
30 - x &= 30 - 12 = 18
\end{aligned}
$$

Check: Is the total value of 12 nickels and 18 dimes equal to $2.40?

$$12(5) + 18(10) = 60 + 180 = 240 \text{ cents} = \$2.40 \;\checkmark$$

Answer: Mary has 12 nickels and 18 dimes.

ILLUSTRATIVE PROBLEM 3: Tom's father is now 3 times as old as Tom. Four years ago, he was 4 times as old as Tom. Find the age of each now.

Solution:

$$\text{Let } t = \text{Tom's age now (in years).}$$
$$\text{Then } 3t = \text{father's age now.}$$
$$\text{And } t - 4 = \text{Tom's age 4 years ago.}$$
$$\text{And } 3t - 4 = \text{father's age 4 years ago.}$$

$3t - 4 =$	$4(t - 4)$	(write equation)
$3t - 4 =$	$4t - 16$	(remove parentheses)
$+4 =$	$+4$	(add 4 to both sides)
$3t \quad =$	$4t - 12$	
$-4t$	$-4t$	(add $-4t$ to both sides)
$-t \quad =$	-12	
$t \quad = 12$		(divide both sides by -1)
$3t \quad = 3(12) = 36$		

Check: Four years ago, Tom was 8 and his father was 32.

$$32 = 4(8)$$
$$32 = 32 \ \checkmark$$

Answer: Tom's age is now 12 and his father's age is 36.

ILLUSTRATIVE PROBLEM 4: The length of a rectangle is 8 inches more than its width. If the perimeter of the rectangle is 68 inches, find its dimensions.

Solution:

$$\text{Let } w = \text{width in inches.}$$
$$\text{Then } w + 8 = \text{length in inches.}$$
$$\text{And perimeter} = 2 \times \text{length} + 2 \times \text{width.}$$

$68 = 2(w + 8) + 2w$	(write equation)
$68 = 2w + 16 + 2w$	(remove parentheses)
$68 = 4w + 16$	(combine like terms)
$-16 \qquad -16$	(add -16 to both sides)
$52 = 4w$	
$w = 13$	(divide both sides by 4)
$w + 8 = 13 + 8 = 21$	

Check: perimeter $= 2l + 2w$
$$68 = 2(21) + 2(13)$$
$$68 = 42 + 26$$
$$68 = 68 \ \checkmark$$

Answer: The dimensions are 21 in. and 13 in.

Exercises _____

In 1 to 25, write and solve an equation.

1. Felicity has 20 coins in her purse consisting of dimes and quarters. If the total value of the coins is $3.50, find the number of each kind of coin.

2. The length of a rectangular room is 5 feet less than twice the width. If the perimeter is 110 feet, find the dimensions of the room.

3. In a math class of 34 students, there are 8 more boys than girls. How many boys and how many girls are in the class?

4. A fourth-grade class contributed $3.50 in nickels and dimes to the Red Cross. In all there were 45 coins. How many of each kind were there?

5. The larger of two numbers is 5 more than twice the smaller. The larger number exceeds the smaller by 14. Find the numbers.

6. The length of a rectangle is twice the width. If the length is increased by 4 and the width diminished by one, the new perimeter will be 198. Find the dimensions of the original rectangle.

7. One weekend Bill earned 3 times as much as Jim. Tom earned $5 more than Jim. In all, they earned $60. How much did each earn?

8. Nancy bought 50 gift-wrap bows for which she paid $7.20. Some were 12-cent bows and some were 20-cent bows. How many of each bow did she buy?

9. Kim and Joe bought the same number of savings bonds. Kim still has all of his. Joe has two-thirds of his. Together they still have 75 bonds. How many did each purchase?

10. The perimeter of a rectangle is 40. The length is 2 more than 5 times the width. Find the dimensions.

11. Paula has 5 times as many marbles as Dolores. Tania has 3 fewer than 4 times as many as Dolores. In all they have 167 marbles. How many does each have?

12. At a school game, student tickets were 50¢ each and adult tickets were $1.00 each. If the total receipts from 900 tickets were $500, how many tickets of each kind were sold?

13. Three partners in a business are to divide the profits in the ratio of 2:3:4. If the profits for the first year are $7200, how much does each partner receive?

14. Using the formula $C = \frac{5}{9}(F - 32)$, find F when $C = 20$.

15. At an afternoon movie, tickets cost $2.10 for children, and $4.50 for adults. If a total of 550 tickets were sold and the receipts were $1395, how many adults attended?

16. The sum of the measures of the angles of any triangle is 180°. If the three angles of a triangle are in the ratio of 2:5:8, find the measure of each angle of the triangle.

17. A man is now 28 years of age and his daughter is 4 years old. In how many years will he be 4 times as old as his daughter will be then?

18. Separate 84 into two parts such that one part will be 12 less than twice the other.

19. One number is 3 times another number. If 17 is added to each, the first resulting number is twice the second resulting number. Find the two numbers.

20. Find two consecutive integers such that twice the smaller diminished by the larger is 71.

21. A man is 35 years old, and his son is 7 years old. In how many years will the father be 3 times as old as his son will be then?

22. Mary is twice as old as Lynn. Five years ago she was three times as old as Lynn. How old is Lynn now?

23. Mr. Newton is now 56 years old, and his son is 30 years old. How many years ago was Mr. Newton twice the son's age then?

24. Peter is now 3 times as old as Fred. Four years ago, Peter was 4 times as old as Fred was then. How old is Peter now?

25. In 10 years, Merv will be three times as old as he was 10 years ago. How old is Merv now?

---

8. MOTION PROBLEMS: INTRODUCTION

If an auto travels at a rate (or speed) of 55 miles per hour, then, in 5 hours, it will travel 55(5) = 275 miles. This relationship may be expressed by the formula

$$d = rt$$

where d is the distance (in miles), r is the rate or average speed (in miles per hour, mph), and t is the time (in hours).

By dividing both sides of this formula, first by r and then by t, we obtain the equivalent formulas

$$t = \frac{d}{r} \text{ and } r = \frac{d}{t}$$

which are also frequently used in solving uniform motion problems (constant rate of motion).

If a car travels 100 miles at the rate of 40 miles per hour, the time it takes is given by the formula $t = \frac{d}{r}$:

$$t = \frac{d}{r} = \frac{100}{40} = 2\frac{1}{2} \text{ hours}$$

If a plane flies 600 miles in 3 hours, then its average rate is

$$r = \frac{d}{t} = \frac{600}{3} = 200 \text{ mph}$$

Problems may be literal as well as numerical. How many miles does a boy ride his bike if his rate is x miles per hour for 4 hours?

$$d = rt$$
$$d = x(4) = 4x \text{ miles}$$

Exercises

1. How far will a plane fly if it goes 450 mph for
   a. 3 hours?   b. $4\frac{1}{2}$ hours?   c. n hours?
2. How far will a plane fly if it goes y miles an hour for
   a. 3 hours?   b. h hours?   c. $(h + 2)$ hours?
3. a. If a man walks 9 miles in 3 hours, what is his rate?
   b. If a man walks m miles in h hours, what is his rate?
4. How long does it take a boy to ride his bike 15 miles
   a. at 5 mph?   b. at 3 mph?   c. at k mph?
5. How fast must a train travel to cover 350 miles
   a. in 7 hours?   b. in 5 hours?   c. in $(n + 3)$ hours?
6. How far will an auto travel in 4 hours at an average speed of
   a. 45 miles per hour?   b. 54 miles per hour?
   c. $(r - 10)$ miles per hour?

7. Two cars start from the same place and travel for h hours in opposite directions at rates of 40 mph and 50 mph, respectively.

  a. Represent in terms of h the distance traveled by the slower car.

  b. Represent in terms of h the distance traveled by the faster car.

  c. Represent how far apart the two cars are at the end of h hours.

  d. Write an equation that would indicate that the cars are 270 miles apart at the end of h hours.

8. Two planes started from different points and flew toward each other. The slower plane flew at 200 mph and the faster at 250 mph. They met in n hours.

  a. Represent in terms of n the distance traveled by the slower plane.

  b. Represent in terms of n the distance traveled by the faster plane.

  c. Represent in terms of n the total distance they traveled.

  d. Write an equation that would indicate that the planes were originally 1000 miles apart.

---

9. SOLVING MOTION PROBLEMS

ILLUSTRATIVE PROBLEM 1:   Two trains are 700 miles apart and begin traveling toward each other. If one is moving at 40 mph and the other at 30 mph, in how many hours will they meet?

Solution: Draw a diagram of the situation, as shown below.

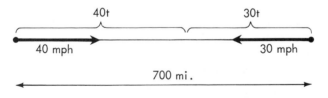

Let t = number of hours until they meet.
Then $40t$ = distance traveled by faster train.
And $30t$ = distance traveled by slower train.

$$40t + 30t = 700$$
$$70t = 700$$
$$t = 10$$

Check: $40t = 40(10) = 400$ miles
$\ 30t = 30(10) = 300$ miles
$400 + 300 = 700$ miles ✔

Answer: The trains will meet in 10 hours.

ILLUSTRATIVE PROBLEM 2: Two planes start from the same point and travel in opposite directions, one at 60 mph faster than the other. In 5 hours, they are 1500 miles apart. Find the rate of each plane.

Solution: Draw a diagram, as shown below.

$$\underleftarrow{\qquad 5(r + 60) \text{ mi.} \qquad}\bullet\underrightarrow{\quad 5r \text{ mi}. \quad}$$

$$\underleftarrow{\qquad\qquad 1500 \text{ mi.} \qquad\qquad}\rightarrow$$

Let r = rate of slower plane.
Then $r + 60$ = rate of faster plane.
And $5r$ = distance traveled by slower plane.
And $5(r + 60)$ = distance traveled by faster plane.

$$5r + 5(r + 60) = 1500$$
$$5r + 5r + 300 = 1500$$
$$10r + 300 = 1500$$
$$\underline{\quad -300 \qquad -300\quad}$$
$$10r \qquad = 1200$$
$$r \qquad = 120 \text{ mph} \qquad \text{(rate of slower plane)}$$
$$r + 60 \ = 180 \text{ mph} \qquad \text{(rate of faster plane)}$$

Check: $5(120) = \ \ 600$ mi   (distance, slower plane)
$5(180) = \underline{\ \ 900 \text{ mi}}$ (distance, faster plane)
$\qquad\qquad\quad 1500$ mi ✓

Answer: The rates are 120 mph and 180 mph.

ILLUSTRATIVE PROBLEM 3:   A freight train leaves Buffalo for St. Louis and travels at the rate of 45 mph. Two hours later, a passenger train leaves Buffalo for St. Louis and travels at 60 mph. In how many hours will the passenger train overtake the freight train?

Solution: Since the freight train left 2 hours earlier, it travels 2 hours longer than the passenger train.

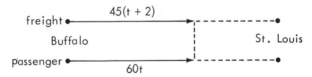

Let t = number of hours passenger train travels.
Then $t + 2$ = number of hours freight train travels.
And $60t$ = distance traveled by passenger train.
And $45(t + 2)$ = distance traveled by freight train.

The distance traveled by the passenger train is the same as the distance traveled by the freight train. Thus,

$$
\begin{aligned}
60t &= 45(t + 2) \\
60t &= 45t + 90 \\
-45t &\quad -45t \\
\hline
15t &= 90 \\
t &= 6
\end{aligned}
$$

Check: 60(6) = 360 miles (distance, passenger train)

45(6 + 2) = 45(8) = 360 miles (distance, freight train)

The distances are equal. ✓

Answer: The passenger train will overtake the freight train in 6 hours.

Note from the preceding illustrative problems that it is helpful to draw distance diagrams to clarify the relationship needed to form an equation.

Exercises _____

In 1 to 12, write and solve an equation.

1. Two cars leave the same place at the same time and travel in opposite directions. At the end of 7 hours, they are 455 miles apart. Find the rate of each car if one travels 5 mph faster than the other.

2. Two trains starting from the same place and traveling in opposite directions are 432 miles apart in 6 hours. One rate is 12 mph faster than the other. Find the rate of each.

3. Two trains start toward each other at the same time from stations that are 570 miles apart. One is a passenger train that averages 55 mph, while the other is a freight train that averages 40 mph. In how many hours will they meet?

4. Two airplanes start from Chicago at the same time and fly in opposite directions. One flies 20 mph faster than the other. After 3 hours, they are 1140 miles apart. Find the rate of each plane.

5. Two trains are 276 miles apart, and are traveling toward each other at rates of 42 mph and 50 mph, respectively. In how many hours will they meet?

6. Two planes leave the airport at the same time, and travel in opposite directions. The rate of one plane is 30 mph faster than that of the other. If they are 860 miles apart at the end of 2 hours, what is the rate of each plane?

7. An eastbound freight train left the station at 8 A.M. and traveled at the rate of 40 mph. At 10 A.M. an eastbound passenger train left the same station and traveled 60 mph. At what time will the passenger train overtake the freight train?

8. A truck and a car leave the city at the same time and travel along the same road. The car is traveling twice as fast as the truck. If, at the end of 3 hours, the car is 72 miles ahead of the truck, find the rate of each.

9. A boy walked a certain distance at the rate of 3 mph. One and one-half hours after he left, his father followed him at the rate of 4 mph. How many miles had the boy gone when his father overtook him?

10. Ted left home on his bicycle traveling at the rate of 6 mph. One hour later, Jack set out to overtake him traveling at the rate of 8 mph. In how many hours will Jack overtake Ted?

11. A jet and a propeller plane leave the same airport at the same time and travel in opposite directions. The rate of the jet is 550 mph and the rate of the propeller plane is 300 mph.

   a. In how many hours will they be 2550 miles apart?

   b. How many miles from the airport will the jet be at this time?

12. At 10 A.M., two women start driving from two different towns that are 270 miles apart. They travel toward each other and meet at 1 P.M. If one woman's rate is 10 mph faster than the rate of the other, find the rate of each.

Chapter Review Exercises

In 1 to 16, solve and check.

1. $n - 9 = -3$
2. $-k + 11 = -4$
3. $y + 12 = -5$
4. $4x = 3$
5. $-3p = 10$
6. $4t + 7 = 6$
7. $3a + 7 = a - 5$
8. $3r + 7 + 2r = 2$
9. $3y + 1 = 8y + 16$
10. $10n - 2(3n + 1) = 26$
11. $7b + 5 = 3b + 17$
12. $2(3r - 1) = 5r + 4$
13. $6t + 13 = 4t + 15$
14. $4(d + 6) + d = 8d - 3$
15. $5a - 4 = 3a + 6$
16. $6(y + 3) = 2y - 2$

In 17 to 24, write an equation and solve.

17. Three children each contributed toward a birthday present for their mother. The oldest contributed three times as much as the youngest, while the second oldest contributed 50 cents more than the youngest. If the present cost $10.50, how much did each contribute?

18. A rectangular playground is enclosed by 440 feet of fencing. If the length of the playground is 20 feet less than 3 times the width, find the dimensions of the playground.

19. Two trains, traveling in opposite directions, left the station at the same time. The rate of one train was 12 mph faster than that of the other train. At the end of 3 hours, the trains were 276 miles apart. What was the rate of each train?

20. There are 153 students in the senior class. If the ratio of boys to girls is 5:4, how many boys are in the class?

21. A boy saved nickels and dimes until he had $1.45. If he had 8 more nickels than dimes, how many dimes did he have?

22. During one month, Jack, Bob, and Jerry earned a total of $115. If Bob earned twice as much as Jack, and Jerry earned $15 more than Jack, how much did each earn?

23. Irving is now twice as old as his sister. Four years ago the sum of their ages was 10. What are their ages now?

24. A salesperson sold 200 pairs of slippers. Some were sold at $6 per pair and the remainder were sold at $11 per pair. Total receipts from this sale were $1600. How many pairs of slippers were sold at $6 each?

CHECK YOUR SKILLS

1. How much larger is $25\frac{1}{8}$ than $17\frac{3}{4}$?

2. Subtract $2x^2 - 3x + 2$ from $x^2 + 4x - 1$.

3. If a car can be driven 25 miles on a gallon of gasoline, how many gallons will be needed to drive this car m miles?

4. Solve for y: $3y + 1 = 8y + 16$

5. Solve for x: $10x - 2(3x + 1) = 26$

6. Divide $15x^3y^4$ by $5xy^4$.

7. In a furniture store the list price of a chair is $85. If a 10% discount is given, what is the reduced price of the chair?

8. On a scale drawing of the floor plans for a house, the scale is $\frac{1}{8}$ inch = 1 foot. A room is 14 feet long. How long a line segment will represent this length?

9. At 8:00 A.M., two planes leave an airport and travel in opposite directions. At 10:00 A.M., the planes are 840 miles apart. If one plane travels 20 miles per hour faster than the other plane, what is the rate of the slower plane?

10. The sum of the measures of the angles of any triangle is 180°. If the three angles of a triangle are in the ratio 2:5:8, find the measure of the smallest angle of the triangle.

12

Graphing Linear Equations

1. GRAPHING A FORMULA

In some of your work in previous grades, you learned to read and make graphs of a statistical nature, such as bar and line graphs. It is frequently desirable to graph a formula or an equation.

For example, if a boy is riding his bicycle at 4 miles per hour, the distance (d), in miles, which he travels is given by the formula $d = 4t$, where t is the number of hours he travels.

The relation between d and t may also be shown by a table:

When t is:	0	1	2	3	4	5
Then d is:	0	4	8	12	16	20

The table can show only a few of the possible pairs of values for t and d, since t can have any positive or zero value and d will change as t changes. Thus, we call t and d **variables** of the formula.

We can also picture the relations between variables by graphing the relations. We shall illustrate this by using the table we established above for the formula $d = 4t$.

Draw a vertical line and a horizontal line meeting at a point. These lines are called the **vertical axis** and the **horizontal axis,** respectively. The point of intersection of the axes is called the **origin**. See Fig. 1 on the following page.

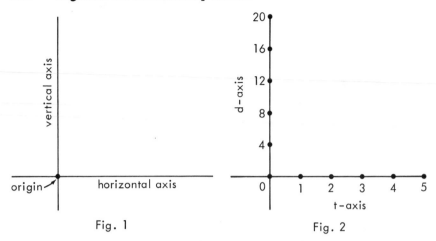

Fig. 1   Fig. 2

We now choose appropriate scales for each axis so that we can represent all the values in the table on the graph. In Fig. 2, each space on the horizontal axis is 1 and each space on the vertical axis is 4.

We now represent each pair of numbers on the table as a point. The point for $t = 3$ and $d = 12$ is located by moving 3 units to the right on the horizontal or t-axis and then moving vertically up until we reach the line representing $d = 12$ on the vertical or d-axis. For the point representing $t = 5$ and $d = 20$, we move 5 to the right from the origin and then up to 20.

We thus obtain five points from the pairs of numbers in the table. When we locate these points, we say that we are "plotting" them. In Fig. 3, note that the five points so plotted lie in a straight line, which is then drawn on the graph. The line starts at the origin and extends infinitely to the right and up. Any pair of values that satisfies the formula $d = 4t$ would be represented on the graph as a point on this line, and any point on this line would give us values of t and d that would satisfy the relation $d = 4t$.

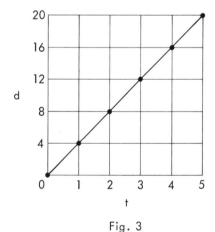

Fig. 3

The relation $d = 4t$ is a first-degree equation with two variables. It is first degree because the exponents of d and t are both understood to be 1. Since the graph of a first-degree equation is always a straight line, such an equation is often called a **linear equation.**

Once we have drawn the graph, we may take readings from it which were not in the original table. For example, when $t = 7$, find d.

Using Fig. 4, locate $t = 7$ on the horizontal axis. Draw a vertical line up to the graph and, at that point, draw a horizontal line across to the d-axis. We see that $d = 28$.

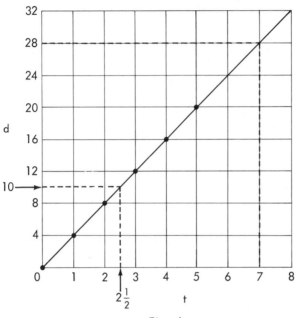

Fig. 4

Likewise, to find t when $d = 10$: locate 10 on the d-axis, draw a horizontal line across to the graph and, at that point, draw a vertical line down to the t-axis. We see that $t = 2\frac{1}{2}$.

Exercises

Use separate paper to make the tables of values; use graph paper to make the graphs.

1. The perimeter (p) of a square is given by the formula $p = 4s$ where s is the length of a side of the square.
   a. Make a table of values for this formula, using whole number values of s from 1 to 5 inclusive.
   b. Make a graph from the table.
   c. From the graph, find (1) the value of p when $s = 7$; (2) the value of s when $p = 14$.

2. The velocity V (speed in feet per second) of an object falling to the ground is given by the formula $V = 32t$ where t is the time in seconds that the object is falling.

   a. Make a table of values for the formula, using $t = 0, 2, 4, 6, 8, 10$.

   b. Make a graph from the table of values.

   c. From the graph, find (1) the value of V when $t = 7$; (2) the value of t when $V = 176$.

3. The distance in miles a car travels at 30 miles per hour, d, is given by the formula $d = 30t$ where t is the number of hours it travels.

   a. Make a table of values for the formula, using $t = 0, 2, 4, 6$, and 8 hours.

   b. Make a graph from the table of values.

   c. From the graph, find (1) the value of d when t is 9; (2) the value of t when $d = 195$ miles.

4. The approximate number of miles m that is equivalent to a given number of kilometers k is given by the formula $m = .6k$. Make a table of values and a graph for this formula for values of $k = 0, 10, 20, 30, 40, 50$.

5. The equation $p = 2l + 5$ shows a relation between the length l and the perimeter p of a rectangle whose width is $2\frac{1}{2}$ inches.

   a. Make a table of values for this formula, using values of $l = 2, 5, 9, 12$.

   b. Make a graph from the table of values.

   c. From the graph, find (1) the value of p when l is 7 inches; (2) the value of l when p is 28 inches.

6. Draw graphs of the relations shown by the following equations:

   a. $r = \frac{2}{3}n$ $(n = 0, 3, 6, 9, 12)$

   b. $s = 3t + 2$ $(t = 0, 2, 4, 6, 8)$

2. GRAPHS USING SIGNED NUMBERS

Thus far, we have discussed only graphs that use positive numbers or zero as values of the variables. However, many formulas and relations may also use negative numbers as values of the variables. For example, a formula involving temperatures must also allow for negative values of temperature. How do we represent negative numbers on a graph?

We extend the horizontal axis to the left of the origin; we extend the vertical axis below the origin. We call the horizontal axis the x-axis and the vertical axis the y-axis, as shown in the following diagram:

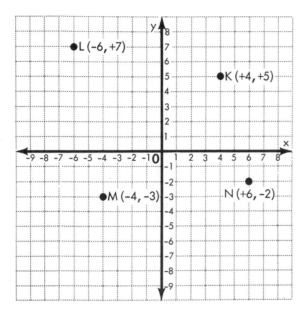

On the x-axis, negative numbers are indicated to the left of the origin (or 0); positive numbers are indicated to the right of the origin, as before. Likewise, on the y-axis, negative numbers are indicated below the origin; positive numbers are above the origin, as before.

To locate the point K on the graph, we move 4 units to the right of the origin along the x-axis and then 5 units up. The point K is designated by the number pair (4, 5) or (+4, +5). Note that the order of the numbers in a pair is important; the point (5, 4) would give us a point other than K. Therefore, we refer to the numbers associated with any point on the graph as an ***ordered number pair***.

The ordered pair of numbers for any point represents the ***coordinates*** of the point. The first number of the number pair is called the ***x-coordinate*** or the ***abscissa*** of the point. The second number is called the ***y-coordinate*** or the ***ordinate*** of the point.

The x-coordinate of a point indicates how many units to the left or right of the y-axis it is. The y-coordinate of a point indicates how many units above or below the x-axis it is. We must be careful not to interchange the numbers in an ordered pair because the reversed pair usually represents a point different from that represented by the original pair.

The point L is located by moving 6 units to the left of the origin and then 7 units up. Its coordinates are $(-6, +7)$.

The point M is located by moving 4 units to the left and 3 units down. Its coordinates are $(-4, -3)$.

The point $N(+6, -2)$ is located by moving 6 units to the right of the origin and then 2 units down.

Exercises

1. Write as ordered number pairs the coordinates of points A, B, C, D, E, F, G, H, and O in the following graph:

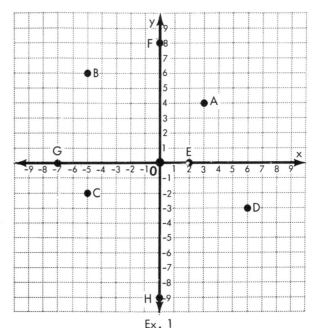

Ex. 1

2. Draw a pair of coordinate axes on a sheet of graph paper. Number the axes, and plot the following points:

   $A(4, 7)$, $B(-2, 5)$, $C(-1, -6)$, $D(4, -5)$, $E(0, -3)$, $F(-4, 0)$, $G(-2\frac{1}{2}, -3)$, $H(0, 0)$, $J(6.5, -4.5)$

3. Draw axes on graph paper, number the axes, and locate the points $P(4, -3)$ and $Q(-3, 4)$. Join them with a straight line.

   a. At what point does the line cut the y-axis?

   b. At what point does the line cut the x-axis?

4. Draw axes on graph paper, number the axes, and locate the points $R(11, 4)$ and $S(3, -4)$. Draw the line RS and extend it in both directions.

   a. At what point does the line cut the x-axis?

   b. At what point does the line cut the y-axis?

5. Draw axes on graph paper, number the axes, and locate these points: $A(4, 2)$, $B(0, 2)$, $C(-1, -1)$, and $D(3, -1)$. Join them in succession. Describe the figure revealed by the graph.

6. Draw axes on graph paper, number the axes, and locate these points: $(3, 4)$, $(3, 6)$, $(3, 0)$, $(3, -2)$, $(3, -5)$.

   a. What is true of the abscissas of all five points?

   b. If you join all the points in succession, what geometric figure do you get?

7. On the same axes as in exercise **6,** locate these points: $(6, 2)$, $(3, 2)$, $(0, 2)$, $(-2, 2)$, $(-5, 2)$

   a. What is true of the ordinates of all five points?

   b. If you join all the points in succession, what geometric figure do you get?

8. Plot the points $C(-2, 1)$ and $D(2, -3)$. Draw the straight line CD and extend it in both directions. At what point does this line cut

   a. the x-axis?   **b.** the y-axis?

---

3. GRAPHS OF LINEAR EQUATIONS

The equation $y = x - 2$ shows us how the two variables x and y are related. It tells us that any value of y is 2 less than the corresponding value of x. For example, if $x = 3$, then $y = 3 - 2 = 1$. Thus, the ordered pair $(3, 1)$ satisfies the equation $y = x - 2$. Or we may say that $(3, 1)$ is a solution of the equation $y = x - 2$, where it is understood that 3 is the x-value and 1 the y-value.

By making a table of values that satisfy the equation, we can determine several solutions of the equation.

When x is	-3	-2	-1	0	1	2	3
Then y is	-5	-4	-3	-2	-1	0	1

We may now plot these points on a pair of axes and obtain the graph of $y = x - 2$. Note that there is an infinite number of ordered pairs that satisfy the equation, but we are using only a few. Remember that we may use fractional or decimal values of the variables as well as whole numbers.

The number pairs in the table are plotted as points on the graph below. Note that all the points lie in a straight line, which is extended indefinitely in both directions.

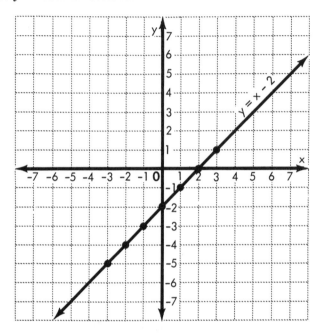

This line is the graph of the equation $y = x - 2$. Every ordered pair of numbers that satisfies this equation represents a point on the line, and every point on the line has coordinates that satisfy the equation.

Since we need only two points to draw a straight line, we need only two ordered pairs of numbers. However, we usually take a third pair of numbers as a check. If all three do not lie in a straight line, then we know we have made an error in calculating our pairs of values.

ILLUSTRATIVE PROBLEM: Draw the graph of $y = 2x - 3$.

Solution: First make a table showing at least three pairs of numbers satisfying the equation. Choose a value for x and calculate the corresponding value of y. Thus, if $x = 2$, then $y = 2(2) - 3 = 4 - 3 = 1$.

When x is	0	2	4
Then y is	−3	1	5

Now, plot a point for each number pair.

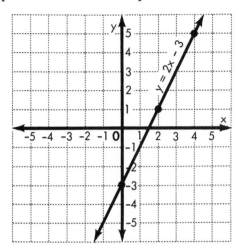

Note that the three points lie in a straight line. Draw the line and extend it indefinitely in both directions. Label the line with the equation $y = 2x - 3$.

Exercises

On separate paper, make a table with at least three pairs of values for each equation below. Use graph paper to draw the graph of each equation.

1. $y = x + 3$ **2.** $y = 3x - 4$ **3.** $y = 4x + 1$

4. $y = 2x$ **5.** $y = 5x - 4$ **6.** $y = -3x + 2$

7. $y = \frac{1}{2}x + 2$ **8.** $y = 4 - 3x$ **9.** $y = 5 + \frac{1}{3}x$

10. $y = -\frac{2}{5}x + 4$

4. GRAPHING MORE DIFFICULT LINEAR EQUATIONS

ILLUSTRATIVE PROBLEM 1: Graph the equation $3x + 2y = 6$.

Solution: Here y is not expressed directly in terms of x. However, we may form a table of values by substituting values of x in the equation and solving for y; or by substituting values of y and solving for x.

For example, if $x = 4$, then $y = $?

$$3(4) + 2y = \quad 6$$
$$12 + 2y = \quad 6$$
$$\underline{-12 \qquad\quad = -12}$$
$$2y = \quad -6$$
$$y = \quad -3$$

Substituting $x = 0$ and $y = 0$ usually gives us easy calculations for the table. Thus,

x	0	4	2
y	3	-3	0

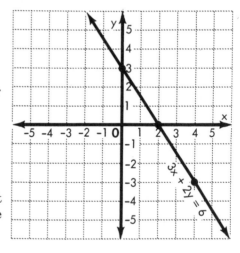

The calculation of the pair $(2, 0)$ is as follows:

$$3x + 2(0) = 6$$
$$3x + \quad 0 = 6$$
$$3x = 6$$
$$x = 2$$

We may then proceed to plot the three points and draw the graph as before.

ILLUSTRATIVE PROBLEM 2: Which of the following points lies on the graph of $2x + 3y = 4$?   **(a)** $(3, 0)$ **(b)** $(-1, 2)$ **(c)** $(-2, 2)$

Solution: Here we need only substitute each number pair in the equation to see if the pair satisfies the equation.

(a) $2(3) + 3(0) = 4$
      $6 + 0 = 4$

Not true. Point $(3, 0)$ is not on the line.

(b) $2(-1) + 3(2) = 4$
$\quad\quad -2 \;+\; 6 \;= 4$

True. Point $(-1, 2)$ is on the line.

(c) $2(-2) + 3(2) = 4$
$\quad\quad -4 \;+\; 6 \;= 4$

Not true. Point $(-2, 2)$ is not on the line.

Answer: **(b)**

ILLUSTRATIVE PROBLEM 3: If the point whose abscissa (x-value) is 3 lies on the graph of $2x + y = 1$, what is the ordinate (y-value) of the point?

Solution: Substitute $x = 3$ in $2x + y = 1$. Thus,

$$2(3) + y = 1$$

$$6 + y = 1$$

$$\underline{-6 \quad\quad\; = -6}$$

$$y = -5 \quad (answer)$$

Exercises

In 1 to 4, choose your answers from the given choices.

1. Which point lies on the graph of $2x + y = 10$?
   (a) $(10, 0)$ **(b)** $(3, 4)$ **(c)** $(0, 8)$ **(d)** $(4, 3)$

2. Which number pair corresponds to a point which lies on the graph of $3x + 2y = 4$?
   (a) $(-1, 2)$ **(b)** $(-4, 4)$ **(c)** $(2, -1)$ **(d)** $(2, 0)$

3. The graph of $x + 3y = 6$ intersects the y-axis in which point?
   (a) $(0, 2)$ **(b)** $(0, 18)$ **(c)** $(6, 0)$ **(d)** $(3, 6)$

4. Which is an equation whose graph does *not* pass through the point whose coordinates are $(2, 3)$?
   (a) $2x - y = 1$ **(b)** $3x - 2y = 0$ **(c)** $x + y = 5$ **(d)** $x - y = 5$

5. If a point with ordinate (y-value) 3 lies on the graph of $x + y = 2$, what is the abscissa (x-value) of the point?

6. Point P lies on the graph of $2x - y = 1$. If the abscissa of P is 2, what is the ordinate of P?

In 7 to 16, graph the equations. (Use graph paper.)

7. $2x + y = 8$ **8.** $x + 3y = 10$ **9.** $y - 2x = 0$

10. $3x + y = 6$ **11.** $2x - y = 5$ **12.** $4x + 3y = 12$

13. $2x - 3y = 6$ **14.** $3x + 4y = -12$ **15.** $3x + 6y = 0$

16. $5x + 3y = 15$

---

5. GRAPHING LINES PARALLEL TO THE AXES

ILLUSTRATIVE PROBLEM 1: Graph the equation $y = 3$.

Solution: We may think of this equation as $y = 0 \cdot x + 3$. Now make a table:

$y = 0 \cdot x + 3$

x	-2	0	$+4$
y	3	3	3

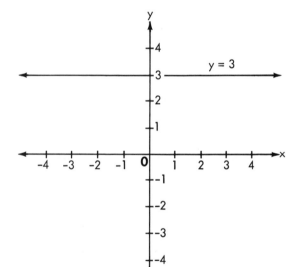

Note that, regardless of the value of x chosen, y is always 3. The graph is a straight line parallel to the x-axis and passing through the point (0, 3).

The **_y-intercept_** of a line is the y-coordinate of the point where the line crosses the y-axis. In this problem, the y-intercept is 3.

ILLUSTRATIVE PROBLEM 2: Graph the equation $x = 4$.

Solution: We may think of this equation as $x = 0 \cdot y + 4$. Now make a table, choosing values of y and calculating values of x:

$x = 0 \cdot y + 4$

x	4	4	4
y	-2	0	3

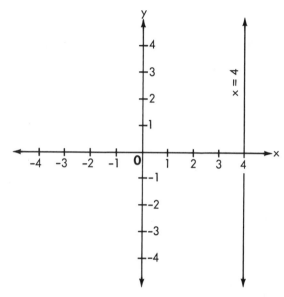

The graph is a straight line parallel to the y-axis and passing through the point (4, 0).

The **x-intercept** of a line is the x-coordinate of the point where the line crosses the x-axis. In this problem, the x-intercept is 4.

Summary:

1. The graph of the equation $y = k$, where k is a constant (a number), is a line parallel to the x-axis and $|k|$ units away from it.

2. The graph of the equation $x = c$, where c is a constant, is a line parallel to the y-axis and $|c|$ units away from it.

Exercises _____

In 1 to 6, draw the graph of each equation. (Use graph paper.)

1. $x = 5$ 2. $x = -2$ 3. $x = 2\frac{1}{2}$

4. $y = 4$ 5. $y = -3$ 6. $y = 3.5$

7. What line is the graph of $x = 0$?

8. What line is the graph of $y = 0$?

9. Write an equation of a line that is parallel to the y-axis and whose x-intercept is:

   a. 7   **b.** -7 **c.** $4\frac{1}{2}$

10. Write an equation of a line that is parallel to the x-axis and whose y-intercept is:

    a. 6   **b.** -5 **c.** $3\frac{1}{4}$

---

6. SLOPE OF A LINE

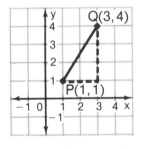

Most straight lines drawn on a set of coordinate axes are oblique (slanted). One way of measuring the amount of slant of a line is to determine the ratio of the vertical change to the horizontal change as we go from one point to another on the line. In the accompanying figure, as we go from P to Q, our vertical position changes by 3 units and our horizontal position changes by 2 units. How do we compute these changes? The vertical change is the ordinate (y-value) of Q minus the ordinate of P, or $4 - 1 = 3$. The horizontal change is the abscissa of Q (x-value) minus the abscissa of P, or $3 - 1 = 2$. The ratio of the vertical change to the horizontal change is defined as the **slope** of the line.

Thus, the slope of this particular line is $\frac{3}{2}$.

In general, to find the slope of a straight line, we choose any two points on the line and find the ratio of the vertical change to the horizontal change as we go from one point to the other.

Many times we are asked to choose any two general points on coordinate axes. Since we do not know the exact coordinates, we often refer to the coordinates of the first point as (x_1, y_1), which is read "x sub 1, y sub 1." Similarly, we can refer to the coordinates of the second point as (x_2, y_2), which is read "x sub 2, y sub 2."

In the figure,

$$\text{Slope of line} = m = \frac{\text{Vertical change}}{\text{Horizontal change}} = \frac{y_2 - y_1}{x_2 - x_1}$$

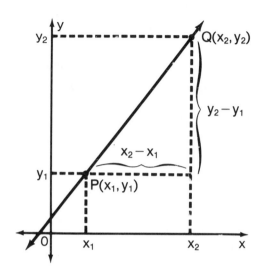

It does not matter which two points we choose on the line. *Any* two points on the line will give us the same slope.

Also, it does not matter whether we use the expression $\dfrac{y_2 - y_1}{x_2 - x_1}$ or $\dfrac{y_1 - y_2}{x_1 - x_2}$ to determine the slope. Both expressions will give the same slope. The important thing to remember is that the *order* must be the same in the numerator and in the denominator. If y_2 comes first in the numerator, then x_2 must come first in the denominator. Similarly, if y_1 comes first in the numerator, then x_1 must come first in the denominator. For example, in the case of points $P(1, 1)$ and $Q(3, 4)$, both of the following expressions give the same slope:

$$m = \frac{y_2 - y_1}{x_2 - x_1} = \frac{4 - 1}{3 - 1} = \frac{3}{2}$$

$$m = \frac{y_1 - y_2}{x_1 - x_2} = \frac{1 - 4}{1 - 3} = \frac{-3}{-2} = \frac{3}{2}$$

Some observations about the slopes of oblique lines are listed below:

Oblique lines slant in different ways.

Looking at an oblique line from *left* to *right,* we can say that it is either *rising* or *falling.*

When we compute the slope of an oblique line, we may arrive at either a positive or a negative result.

A line with a *positive slope* is said to be *rising.*

A line with a *negative slope* is said to be *falling.*

When we compare two lines that slant in the same direction, one may appear to be *steeper* than the other. If both lines have positive slopes, the line with the greater slope will be the steeper line.

ILLUSTRATIVE PROBLEM 1: Find the slope of the line that joins $P_1(-2, 6)$ and $P_2(4, -3)$. Draw the line. Label it "Line 1." Look at Line 1 from left to right. Is it rising or falling?

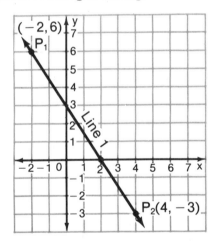

Solution:

$$\text{Slope} = m = \frac{\text{Vertical change}}{\text{Horizontal change}} = \frac{y_2 - y_1}{x_2 - x_1}$$

$$m = \frac{(-3) - (6)}{(4) - (-2)} = \frac{-3 - 6}{4 + 2} = \frac{-9}{6} = \frac{-3}{2} \quad (answer)$$

Note that the slope is negative and, looking from left to right, we say that the line is *falling.* (*answer*)

ILLUSTRATIVE PROBLEM 2: Find the slope of the line that joins the points $Q_1(-8, -4)$ and $Q_2(13, 10)$. Draw the line. Label it "Line 2." Look at Line 2 from left to right. Is it rising or falling?

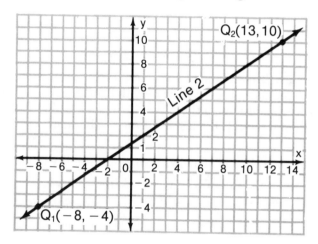

Solution:

$$\text{Slope} = \frac{y_2 - y_1}{x_2 - x_1}$$

$$= \frac{10 - (-4)}{13 - (-8)} = \frac{10 + 4}{13 + 8} = \frac{14}{21} = \frac{2}{3} \quad (answer)$$

The slope is positive and, looking from left to right, we say that the line is *rising*. (*answer*)

ILLUSTRATIVE PROBLEM 3: Find the slope of the line joining $R_1(-1, 4)$ and $R_2(7, 6)$. Draw the line. Label it "Line 3." Is Line 3 rising or falling?

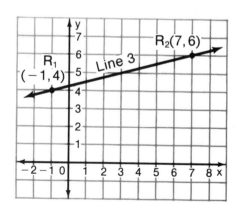

Solution:

$$\text{Slope} = m = \frac{y_2 - y_1}{x_2 - x_1}$$

$$= \frac{(6) - (4)}{(7) - (-1)} = \frac{6 - 4}{7 + 1} = \frac{2}{8} = \frac{1}{4} \quad (answer)$$

The slope is positive and the line is *rising*. (*answer*)

Now compare Lines 2 and 3. They both have positive slopes and they are both rising. Can you see, however, that Line 2 is steeper (rising faster or more "slanted up") than Line 3? Note also that $\frac{2}{3}$, the slope of Line 2, is greater than $\frac{1}{4}$, the slope of Line 3. The line with the greater slope is steeper.

Using the definition of slope, we can now state the following facts:

The slope of a horizontal line is *zero*.

The slope of a vertical line is *undefined*.

We arrive at both of these facts by examining what happens as we trace our way on these lines:

On a horizontal line as x changes, the change in y is zero. Since the fraction that gives the slope would have zero in the numerator, the resulting value of the slope is *zero*.

On a vertical line as y changes, the change in x is zero. Thus, the fraction that gives the slope would have zero in the denominator. Since we can not divide by zero, we say that the value of the slope is *undefined*.

Exercises

1. In **a** to **f**, indicate whether the line has positive slope, negative slope, zero slope, or an undefined slope.

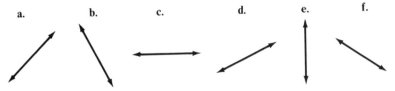

a.   b.   c.   d.   e.   f.

2. Find the slopes of the lines joining the following pairs of points and sketch their graphs:
   a. $(0, 0)$ and $(5, 5)$ **b.** $(1, 2)$ and $(8, 4)$ **c.** $(-3, -2)$ and $(10, 6)$
   d. $(2, 5)$ and $(7, 5)$ **e.** $(-3, -5)$ and $(5, 7)$ **f.** $(-4, 3)$ and $(6, -2)$
   g. $(-2, 1)$ and $(3, -1)$ **h.** $(-4, 3)$ and $(8, 3)$

3. The line joining the points $P(2, 2)$ and $Q(6, y)$ has a slope of $\frac{3}{4}$. What is the value of y?

4. The line joining the points $P(x, 4)$ and $Q(4, 2)$ has a slope of $-\frac{1}{3}$. What is the value of x?

5. a. What is the slope of the line joining $(2, 2)$ and $(5, 5)$?

  b. What is the slope of a line joining any two points, each of which has its ordinate equal to its abscissa?

6. a. What is the slope of the line joining $(-2, 2)$ and $(-5, 5)$?

  b. What is the slope of a line joining any two points, each of which has its abscissa equal to the negative of its ordinate?

---

7. SLOPE-INTERCEPT FORM OF A LINEAR EQUATION

Recall the following:

The ***y-intercept*** of a straight line is the y-coordinate (ordinate) of the point where the line crosses the y-axis. This value can be determined from the linear equation by substituting 0 for x in the equation and *solving for y*.

The ***x-intercept*** of a straight line is the x-coordinate (abscissa) of the point where the line crosses the x-axis. This value can be determined from the linear equation by substituting 0 for y in the equation and *solving for x*.

ILLUSTRATIVE PROBLEM 1: Find the x-intercept and the y-intercept of the equation $y = 2x - 4$.

Solution: Substitute 0 for x in the equation and solve for y:

$$y = 2x - 4$$
$$y = 2(0) - 4$$
$$y = 0 - 4$$
$$y = -4 \quad (y\text{-intercept})$$

Substitute 0 for y in the equation and solve for x:

$$y = 2x - 4$$
$$0 = 2x - 4 \quad (\text{add 4 to both members})$$
$$4 = 2x, \text{ or } 2x = 4$$
$$x = 2 \quad (x\text{-intercept})$$

Answer: The x-intercept is 2 and the y-intercept is -4.

In Illustrative Problem 1, we may draw the line that is the graph of $y = 2x - 4$ by joining the two points where the line intersects the axes as shown in the diagram. Going from the point $(0, -4)$ where the line crosses the y-axis to the point $(2, 0)$ where the line crosses the x-axis, we see that the slope of the line is given by

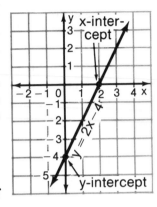

$$m = \frac{0 - (-4)}{2 - 0} = \frac{4}{2} = 2$$

Note that the slope, 2, is the coefficient of x in the equation $y = 2x - 4$ and the y-intercept, -4, is the constant term in this equation.

In general, if a linear equation is written in a form that expresses y in terms of x, the coefficient of x is equal to m (the slope of the line), and the constant term is equal to b (the y-intercept). We refer to this form of the linear equation as the **slope-intercept** form.

$$y = mx + b$$
where m = the slope of the line
and b = the y-intercept of the line

ILLUSTRATIVE PROBLEM 2: Find the slope and y-intercept of the line whose equation is $y = 3x + 7$.

Solution: Since the equation $y = 3x + 7$ is in the slope-intercept form, $m = 3$ and $b = 7$.

Answer: The slope is $+3$ and the y-intercept is $+7$.

ILLUSTRATIVE PROBLEM 3: Find the slope and the y-intercept of the line whose equation is $3x + 4y = 2$.

Solution: Put the equation into the slope-intercept form by solving for y in terms of x:

$$3x + 4y = 2$$

$$4y = -3x + 2$$

$$y = -\tfrac{3}{4}x + \tfrac{1}{2}$$

Answer: Since $m = -\tfrac{3}{4}$ and $b = \tfrac{1}{2}$, the slope is $-\tfrac{3}{4}$ and the y-intercept is $+\tfrac{1}{2}$.

ILLUSTRATIVE PROBLEM 4: Write an equation of the line with slope $= \frac{1}{2}$ and y-intercept $= -6$.

Solution: Substitute $\frac{1}{2}$ for m and -6 for b in the slope-intercept form of the equation:

$$y = mx + b$$
$$y = \tfrac{1}{2}x + (-6)$$
$$y = \tfrac{1}{2}x - 6 \quad (answer)$$

If we do not wish to have an equation with a fractional coefficient, we can multiply both sides of the equation by 2 to obtain an equivalent equation:

$$2y = x - 12 \quad (answer)$$

Exercises

1. Find the slope and the y-intercept of the graph of each of the following equations:

   a. $y = 4x - 7$ **b.** $y = -6x + 3$ **c.** $y = \frac{2}{5}x + \frac{6}{7}$
   d. $y = -\frac{2}{3}x + 11$ **e.** $y = \frac{2}{3}x$ **f.** $3x = 4y$
   g. $5y = 6x$ **h.** $x + y = 9$ **i.** $x - y = 5$
   j. $2x + 3y = 5$ **k.** $2x - y = 4$ **l.** $x - 2y = 1$
   m. $2y - 3x = -3$ **n.** $x - 5y = 4$

2. In each of the following, write an equation of the line whose slope (m) and y-intercept (b) are given:

   a. $m = 3, b = 2$ **b.** $m = -\frac{1}{2}, b = -2$ **c.** $m = \frac{2}{3}, b = \frac{5}{2}$
   d. $m = -5, b = 0$ **e.** $m = 0, b = 4$ **f.** $m = -\frac{3}{4}, b = 6$
   g. $m = -\frac{5}{6}, b = 0$ **h.** $m = 0, b = -2\frac{1}{2}$ **i.** $m = -1, b = \frac{5}{8}$
   j. $m = 1, b = 0$

3. What do the graphs of $y = 5x - 6$, $y = 5x + 7$, and $y = 5x + 12$ all have in common?

4. What do the graphs of $y = 3x - 9$, $y = \frac{1}{2}x - 9$, and $y = 5x - 9$ all have in common?

5. **a.** Through what point must all lines pass that have zero (0) as their y-intercept?

   b. What must be true of all lines that have zero (0) as their slope?

6. The formula $F = \frac{9}{5}C + 32$ is used to convert Celsius temperature readings (C) to Fahrenheit temperature readings (F).

    a. On separate paper, make a brief table of values for this formula, using $C = -40, 0, 20, 100$.

    b. Make a graph from the table of values, placing C on the horizontal axis and F on the vertical axis. Use a suitable scale for each.

    c. Using the graph, determine the temperature at which the readings are the same.

---

8. SOLVING TWO LINEAR EQUATIONS GRAPHICALLY

Suppose we are asked to solve a problem such as this: Find two numbers whose sum is 7 and whose difference is 3. By letting x and y represent the two numbers, we can obtain the following two equations:

$$x + y = 7$$

$$\text{and } x - y = 3$$

One way to solve such a pair of equations is to draw graphs of them. We know that each of these equations represents a straight line on a graph. We have learned that there is an infinite number of ordered pairs of values that will satisfy each equation. For example, the solution set of $x + y = 7$ includes the ordered pairs (6, 1), (5, 2), (4, 3), (1, 6), (8, −1), $(4\frac{1}{2}, 2\frac{1}{2})$, etc. The solution set of $x - y = 3$ includes the ordered pairs (7, 4), (6, 3), (3, 0), (0, −3), $(4\frac{1}{2}, 1\frac{1}{2})$, etc. We are interested in finding one pair of values that will satisfy both equations at the same time, or *simultaneously*. Hence, two linear equations that have a common solution are called **simultaneous equations**.

To help us in plotting the graph, we can construct a table of values for each equation. Among the points chosen should be the x-intercept and the y-intercept.

$x + y = 7$	
If $x =$	Then $y =$
8	−1
7	0
$4\frac{1}{2}$	$2\frac{1}{2}$
0	7

$x - y = 3$	
If $x =$	Then $y =$
0	−3
3	0
$4\frac{1}{2}$	$1\frac{1}{2}$
6	3
7	4

By using the ordered pairs just listed, we may draw the graphs of both linear equations. These two straight lines intersect at $P(5, 2)$. Since this point is on both straight lines, its coordinates must be solutions of both equations. We may check this by substituting $x = 5$ and $y = 2$ in both equations. We see readily that their sum is 7 and their difference is 3.

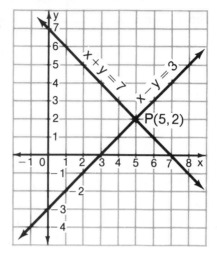

$$x + y = 5 + 2 = 7$$

$$x - y = 5 - 2 = 3$$

Since two non-parallel straight lines can intersect in only one point, it follows that two linear equations in two variables may have only one ordered pair of values as a common solution.

ILLUSTRATIVE PROBLEM: Solve graphically: $3x - y = 1$
$$x + y = 3$$

Solution: Graph $3x - y = 1$, or $y = 3x - 1$.

x	0	2	-1
y	-1	5	-4

Graph $x + y = 3$, or $y = -x + 3$.

x	0	3	2
y	3	0	1

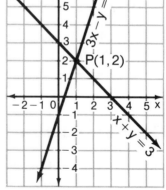

In the figure, we see that point $P(1, 2)$ is the intersection of the graphs.

Check: Substitute 1 for x and 2 for y in the given equations.

$$3x - y = 1 \qquad x + y = 3$$
$$3(1) - 2 = 1 \qquad 1 + 2 = 3$$
$$3 - 2 = 1 \qquad\quad 3 = 3 \checkmark$$
$$1 = 1 \checkmark$$

Answer: $x = 1, y = 2$

Exercises _____

In 1 to 15, solve the pairs of equations graphically and check your answers.

1. $x + y = 9$
$x - y = 1$

2. $x - y = -2$
$2x + y = 5$

3. $x + y = 3$
$7x + y = 9$

4. $y = 2x - 4$
$3x + y = 1$

5. $y = 2x + 1$
$2x + y = 5$

6. $2x + y = 8$
$y = x - 1$

7. $x - y = 1$
$y = 2x + 2$

8. $3x + y = 8$
$2x - 3y = 9$

9. $2x - y = 5$
$x + 2y = 5$

10. $3x + y = 0$
$x - 2y = -7$

11. $3x - y = 1$
$x + 2y = 5$

12. $2x + y = -3$
$x - 3y = -5$

13. $x - 2y = -6$
$4x + y = -6$

14. $2x + 3y = 11$
$4x - 3y = 10$

15. $2x - y = 8$
$y = -1$

16. Draw the graphs of the equations $y = 2x - 1$ and $y = 2x + 3$ on the same set of axes. What are the slopes of these lines? (*Hint:* If the slopes of two lines are equal, then the lines are parallel and there is no common solution to the pair of equations. In other words, there is no point that lies on both lines. Such equations are said to be **inconsistent**.)

17. Draw the graphs of $y = x - 1$ and $2y = 2x - 2$ on the same set of axes. What are the slopes of these lines and their y-intercepts? (*Hint:* If two lines coincide, their equations have an infinite number of common solutions. Such equations are said to be **dependent**.)

_____

Chapter Review Exercises

1. On a graph, the points $(-3, 2)$ and $(-1, 6)$ are joined by a line segment. If this segment is extended so as to intersect the x-axis, at which of the following points does it do so?

  (a) $(0, -4)$ **(b)** $(4, 0)$ **(c)** $(-4, 0)$ **(d)** $(0, 8)$

2. The point P lies on the graph of $x + 2y = 7$. If the ordinate of P is 6, what is its abscissa?

3. The point Q lies on the graph of $5x + 2y = 6$. If the abscissa of Q is 4, what is its ordinate?

In 4 to 7, choose the letter of the choice that best answers the question or completes the sentence.

4. Which ordered pair is a solution of the equation $x - y = -1$?

(a) $(2, -3)$ (b) $(2, 3)$ (c) $(3, -2)$ (d) $(3, 2)$

5. The graph of $x + y = 4$ crosses the y-axis at the point whose coordinates are

(a) $(0, -4)$ (b) $(-4, 0)$ (c) $(0, 4)$ (d) $(4, 0)$

6. The graph of $x = -3$

(a) passes through the origin   (b) is parallel to the x-axis

(c) is parallel to the y-axis   (d) has a y-intercept of -3.

7. Which number pair is *not* in the solution set of $x - 2y = 10$?

(a) $(-4, -7)$ (b) $(0, -5)$ (c) $(5, 0)$ (d) $(6, -2)$

8. The equation of a line is $3x - 2y = 9$. The abscissa of a point on this line is 6. What is the ordinate of this point?

For 9 to 11, use graph paper.

9. Graph the equation $x + 2y = 10$.

10. Graph the equation $2x + 3y = -4$.

11. Graph the equation $3x + 2y = 8$.

In 12 to 14, choose the letter of the choice that best answers the question or completes the sentence.

12. Which ordered pair is the solution of the following pair of equations?

$$6x + 2y = 14$$
$$3x + 2y = 8$$

(a) $(1, 2)$ (b) $(2, 1)$ (c) $(1, 4)$ (d) $(4, -2)$

13. Which of the following is the y-intercept of the line whose equation is $y = 3x - \frac{2}{3}$?

(a) $-\frac{2}{3}$ (b) -2 (c) 3 (d) $\frac{2}{3}$

14. The slope of the line $y - 2x = 1$ is

(a) 1 (b) 2 (c) -2 (d) $\frac{1}{2}$

15. Find the slope of the line whose equation is $3x + 2y = 5$.

16. Solve graphically for x and y and check:

$$x + y = 16$$
$$x - y = 4$$

17. The slope of a line is 3 and its *y*-intercept is 2. Write an equation of this line.

In 18 to 20, solve graphically for *x* and *y* and check.

18. $2x - y = 11$
$x + y = 1$

19. $2x = y + 5$
$x + y = -2$

20. $y = -x$
$2x - y = 3$

CHECK YOUR SKILLS

1. From $7x - 3y$ subtract $5x - 3y$.

2. If one side of a square is represented by $3s + 2$, find the perimeter of the square in terms of *s*.

3. Solve for *m*: $4:6 = m:15$

4. Solve for *c*: $4(c - 3) = 8$

5. A carpenter wishes to cut pieces of board $1\frac{1}{3}$ feet long from a board 9 feet long. What is the maximum number of such pieces she can cut?

6. Solve for *n*: $2n + 3 = 5n - 9$

7. If 50% of a number is 20, then what would 75% of the same number be?

8. Which point is on the graph of $3x - 2y = 7$?
   (a) $(1, -2)$ **(b)** $(2, -1)$ **(c)** $(3, 2)$ **(d)** $(0, -3)$

9. Terry left home on his bicycle, traveling at the rate of 8 miles per hour. One hour later his friend Jan set out to overtake him, traveling at the rate of 10 miles per hour. In how many hours will Jan overtake Terry?

10. What is 46.87 rounded to the nearest tenth?

13

Inequalities

1. MEANING AND SYMBOLS OF INEQUALITY

Thus far, we have dealt only with equations or statements of equality. Frequently, we deal with problems that involve statements indicating that quantities are not equal.

We may say that $5 + 3 \neq 7$, where the symbol $\neq$ is read "is not equal to." Such a statement is called an *inequality.*

To compare two unequal numbers, we use the following symbols:
$>$ is read "is greater than." Thus $5 > 3$ is read "5 is greater than 3."
$<$ is read "is less than." Thus $2\frac{1}{2} < 6$ is read "$2\frac{1}{2}$ is less than 6."

Note that in each case the symbol points to the smaller number.

If we write the inequality $x > 3$, we mean that x may assume any value that is greater than 3; thus, x may be equal to $3\frac{1}{2}$, 4.2, 7, etc. There is an infinite set of numbers that satisfies this inequality, and we call this set the solution set of the inequality.

Note that a temperature of $-20°$ is lower than a temperature of $-10°$. Thus we say that $-20 < -10$.

In the case of the inequality $x < 3$, the solution set would consist of all values of x that are less than 3; thus, members of the solution set would be 2, $1\frac{1}{2}$, 0, -3, -7.4, etc.

Consider the inequality $x + 5 > 11$. What positive integers less than 10 satisfy this inequality? Substitute integers less than 10.

Let $x = 9$. Is $9 + 5 > 11$? Is $14 > 11$? Yes
Let $x = 8$. Is $8 + 5 > 11$? Is $13 > 11$? Yes
Let $x = 7$. Is $7 + 5 > 11$? Is $12 > 11$? Yes
Let $x = 6$. Is $6 + 5 > 11$? Is $11 > 11$? No
Let $x = 5$. Is $5 + 5 > 11$? Is $10 > 11$? No

By continuing this process, we see that 7, 8, and 9 are the only positive integers less than 10 that satisfy the inequality $x + 5 > 11$. Thus, $\{7, 8, 9\}$ is the solution set of $x + 5 > 11$, for x less than 10.

Consider the inequality $2x + 1 < 5$. What integers greater than -5 satisfy this inequality? Substitute integers greater than -5.

Let $x = -4$. Is $2(-4) + 1 < 5$? Is $-8 + 1 < 5$? Is $-7 < 5$? Yes.
Let $x = -3$. Is $2(-3) + 1 < 5$? Is $-6 + 1 < 5$? Is $-5 < 5$? Yes.

By continuing this process, we see that $-4, -3, -2, -1, 0$, and $+1$ are the only integers greater than -5 that can satisfy the inequality $2x + 1 < 5$. We say that the solution set consists of the integers from -4 through $+1$ *inclusive*.

Exercises

For each inequality in 1 to 10, list the integers from 0 to $+5$ inclusive that satisfy the inequality.

1. $y + 4 > 6$ **2.** $x + 2 > 3$ **3.** $p + 2\frac{1}{2} > 5$ **4.** $k - 2 > 1$

5. $4t > 16$ **6.** $a + 6 < 9$ **7.** $3t < 5$ **8.** $3x + 2 > 7$

9. $\dfrac{r}{3} < 1$ **10.** $5m - 3 < 6$

For each inequality in 11 to 20, list the integers from 0 to -5 inclusive that satisfy the inequality.

11. $x + 4 > 1$ **12.** $p + 3 < -1$ **13.** $t + 3\frac{1}{2} < 0$

14. $K + 8\frac{1}{4} > 5$ **15.** $3y - 4 < -13$ **16.** $\dfrac{s}{3} > -1$

17. $5b < -20$ **18.** $2x + 7 < 2$ **19.** $\dfrac{a}{6} < -\frac{1}{2}$

20. $3p + 8 < 8$

In 21 to 30, replace the question mark by one of the symbols $>$, $<$, or $=$ so as to make the statement true.

21. When $a = 4$, $2a + 3$? 7.   **22.** When $x = 0$, $3x - 5$? 8.

23. When $t = 2$, $t - 5$? -2.   **24.** When $y = 2$, $4y - 5$? 3.

25. When $p = 5$, $2p - 7$? 9.   **26.** When $r = 3$, $\dfrac{r}{4}$? 2.

27. When $k = 6$, $3k - 7$? $2\frac{1}{2}$.   **28.** When $b = -2$, $2b - 7$? 5.

29. When $n = 2.4$, $3n - .5$? 6.4.   **30.** When $s = 1.7$, $2s + .4$? 5.2.

2. REPRESENTING INEQUALITIES ON A NUMBER LINE

In previous chapters, we used a number line to illustrate operations on signed numbers. We will find that using a number line to show inequalities gives us a better understanding of them.

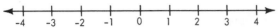

We can use a number line, shown above, to represent all signed numbers, positive and negative. For convenience, we write only the integers, but fractions and decimals are also included. For example, the point on the number line that represents $+2\frac{1}{2}$ lies halfway between $+2$ and $+3$. Similarly, the point on the number line that represents -3.1 lies between -3 and -4 and is closer to -3.

Note that, as we move to the right on the number line, the numbers grow larger. As we move to the left, the numbers grow smaller. Thus, we can verify that

$$-5 < -2 \text{ or } -3\tfrac{1}{2} < 0 \text{ or } -\tfrac{1}{2} > -1, \text{ etc.,}$$

by checking the *order* of these numbers on the number line. For any two numbers on the number line, the smaller number is always to the *left*.

When we represent an inequality with one variable on a number line, we call this the graph of the inequality. The graph of the inequality $x < 2$ is shown below.

The heavy line indicates that the graph of $x < 2$ includes all values of x less than 2. The open circle indicates that 2 is not a member of the solution set. The arrowhead at the left of the heavy line indicates that the solution set continues indefinitely to the left.

Below is the solution set of $x > 1$.

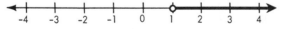

The heavy line indicates all values of x greater than 1. Note the open circle at 1 to indicate it is not included in the solution set. The arrowhead at the right indicates that the graph extends indefinitely to the right.

If we wish to include an endpoint in an inequality, we use a double symbol combining the $=$ symbol and the $<$ or $>$ symbol. For example,

$x \geq 1$ is read "x is greater than or equal to 1." In this case, the open circle on the graph becomes a heavy dot, as indicated below.

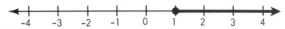

This symbol is frequently convenient. For example, the voting age is now 18 years or over. If V represents the voting age, we may then write $V \geq 18$.

If we wish to indicate that the speed limit (s) is now 55 mph, we may write the inequality $s \leq 55$, which indicates that speeds of 55 mph or less are legal.

Exercises _____

In 1 to 8, use separate paper to graph the inequalities on a number line.

1. $x > 3$ **2.** $x < 4$ **3.** $x < -3$ **4.** $x > -2$

5. $x \leq -2$ **6.** $x \geq -1$ **7.** $x \leq 0$ **8.** $x \geq 1$

9. Express as an inequality in symbols:
    a. n is greater than or equal to $4\frac{1}{2}$.
    b. p is less than or equal to 3.14.
    c. The greatest number of students (n) allowed in a class is 35.
    d. The minimum speed (s) permitted on a highway is 30 mph.
    e. Using a suitable scale on a number line, graph the inequality in **d**.

3. PROPERTIES OF INEQUALITIES

If we start with the inequality $5 > 3$, and add 4 to both sides, we obtain the inequality $5 + 4 > 3 + 4$ or $9 > 7$. This is apparently a true inequality. When the inequality symbol is the same in two inequalities, we say they are of the same **order**, or *the order remains unchanged*.

This illustration suggests the following addition principle of inequality:

Principle: If the same number is added to (or subtracted from) both members of an inequality, the order of the inequality remains unchanged.

In a similar manner, if we start with the inequality $6 > 4$, and multiply both members by 3, we get $6 \times 3 > 4 \times 3$ or $18 > 12$. This is apparently a true inequality.

However, if we start with $6 > 4$, and multiply both members by -3, the left member becomes -18 and the right member becomes -12. Now $-18 < -12$. We say that *the order of the inequality has been reversed.*

These illustrations suggest the following multiplication principle of inequality:

Principle:

1. If both members of an inequality are multiplied (or divided) by the same positive number, the order of the inequality remains unchanged.

2. If both members of an inequality are multiplied (or divided) by the same negative number, the order of the inequality is reversed.

Note that we include division in the multiplication property because dividing by 3 is the same as multiplying by $\frac{1}{3}$ or, in general, division by a number (other than 0) is the same as multiplication by the reciprocal of that number.

4. SOLVING INEQUALITIES

By using the addition and multiplication principles of inequality, we are able to find the solution sets of inequalities.

If you are unfamiliar with sets, review Appendix III on page 441.

ILLUSTRATIVE PROBLEM 1: Solve the inequality $x - 2 > 1$ and graph the solution set.

Solution:

$$
\begin{array}{rcl}
x - \ 2 & > & 1 \\
\underline{+2 \quad +2} & & \text{(add 2 to both members)} \\
x \qquad & > & 3 \quad (answer)
\end{array}
$$

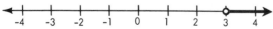

The heavy line indicates all values of x for which $x > 3$, which is also the solution set of the original inequality $x - 2 > 1$. Note that 3 is not included in the graph.

ILLUSTRATIVE PROBLEM 2: Solve the inequality $x + 2 \leq 5$ and graph the solution set. (Read $x + 2$ is less than or equal to 5.)

Solution:

$$
\begin{array}{r}
x + \quad 2 \leq \quad 5 \\
\underline{-2 \quad\quad -2} \quad \text{(add } -2 \text{ to both members)} \\
x \quad\quad\quad \leq \quad 3 \quad \text{(\textit{answer})}
\end{array}
$$

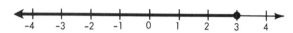

Note that 3 is included in the graph.

ILLUSTRATIVE PROBLEM 3: Solve the inequality $\frac{1}{3}p \geq 1$ and graph the solution set.

Solution:

$$\tfrac{1}{3}p \geq 1$$

$$3 \cdot \tfrac{1}{3}p \geq 1 \cdot 3 \quad \text{(multiply both members by 3)}$$

$$p \geq 3 \quad \text{(\textit{answer})}$$

ILLUSTRATIVE PROBLEM 4: Solve the inequality $-3x + 2 \leq 8$ and graph the solution set.

Solution:

$$
\begin{array}{r}
-3x + \quad 2 \leq \quad 8 \\
\underline{-2 \quad\quad -2} \quad \text{(add } -2 \text{ to both members)} \\
-3x \quad\quad \leq \quad 6 \\
\dfrac{-3x}{-3} \quad\quad \leq \dfrac{6}{-3} \quad \text{(divide both members by } -3) \\
x \quad\quad\quad \geq -2 \quad \text{(\textit{answer})}
\end{array}
$$

In this problem, note that the order of the inequality was *reversed* because we divided by a negative number.

Exercises

In 1 to 18, solve the inequalities. On separate paper, graph their solution sets.

1. $x - 3 > 2$ **2.** $p + 5 > 3$ **3.** $K - 2 < 1$ **4.** $r + \frac{1}{3} > 3$

5. $y - \frac{1}{2} > 0$ **6.** $t + 2.5 < 3.5$ **7.** $s + 2 < 0$ **8.** $4z > 8$

9. $3m \le 9$ **10.** $-2y > 4$ **11.** $-3a \ge -9$ **12.** $.6x \ge -3$

13. $4K - 2 > 2$ **14.** $6 + \dfrac{n}{2} \ge 0$ **15.** $\frac{1}{2}z - 2 < -8$

16. $4x - 3 > 13$ **17.** $3t - 6 \ge 15$ **18.** $-8b - 3b - 2 \le 20$

In 19 to 22, write an inequality and then solve it.

19. Seven times a number is less than 63. What numbers satisfy this condition?

20. A number decreased by 12 is less than 46. Find the set of numbers that satisfies this condition.

21. Twice a number increased by 14 is greater than 6. Find the set of numbers that satisfies this condition.

22. Five times a number decreased by 3 is greater than 7. Find the set of numbers that satisfies this condition.

5. GRAPH OF AN INEQUALITY IN TWO VARIABLES

In Chapter 12, we discussed how to draw the graph of a first-degree equation such as $y = 2x$. Every point on this line has an ordinate (y) that is twice its abscissa (x). Thus, the points $(2, 4)$, $(-1, -2)$, and $(1\frac{1}{2}, 3)$ lie on this line. See Fig. 1.

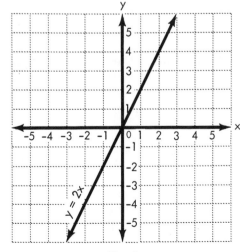

Fig. 1

This infinite line divides the plane of the graph into two **half-planes**. Every point in the half-plane above the line $y = 2x$ has an ordinate which is more than twice its abscissa. For example, the point (1, 4) has an ordinate 4, which is 4 times the abscissa 1. Also, the point $(-2, -3)$ has an ordinate -3, which is more than twice the abscissa -2, since $-3 > 2(-2)$ or $-3 > -4$.

Thus, as shown in Fig. 2, the graph of $y > 2x$ is the half-plane above the line $y = 2x$; this half-plane is shaded in the graph. This half-plane is the set of all points whose ordinates lie above the line $y = 2x$. Note that the line $y = 2x$ appears as a dotted line since it is not actually part of the graph of $y > 2x$. If we were to draw the graph of $y \geq 2x$, we would draw the graph of the line as a solid line since then the line $y = 2x$ would be part of the graph.

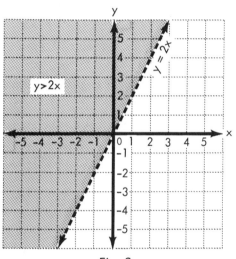

Fig. 2

Every point in the half-plane below the line $y = 2x$ has an ordinate y that is less than twice its abscissa x. For example, the point (2, 3) has an ordinate 3 that is less than twice the abscissa 2 since $3 < 2(2)$ or $3 < 4$. Also, the point $(-1, -3)$ has an ordinate -3 that is less than twice the abscissa -1, since $-3 < 2(-1)$ or $-3 < -2$. Note that both points lie below the line $y = 2x$.

Thus, as shown in Fig. 3, the half-plane below the line $y = 2x$ represents the graph of $y < 2x$. Note that the line $y = 2x$ appears as a dotted line since it is not in the graph of $y < 2x$. If we make the graph of $y \leq 2x$, we would draw the line $y = 2x$ as a solid line to indicate that it is included in the graph.

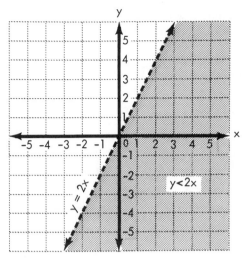

Fig. 3

In order to draw the graph of a first-degree inequality, such as $y > ax + b$ or $y < ax + b$, first draw the line graph of $y = ax + b$. The half-plane above this line is the graph of $y > ax + b$ and the half-plane below this line is the graph of $y < ax + b$.

Let us draw the graph of $y > 2$. First, we draw the line $y = 2$ as a dotted line. The half-plane above this horizontal line represents the set of all points with $y > 2$, that is, all points whose ordinates are greater than 2. We shade the half-plane to complete the graph. See Fig. 4.

To draw the graph of $y < 2$, we shade the region below the dotted line $y = 2$. See Fig. 5.

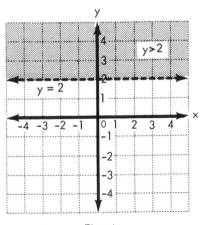

Fig. 4

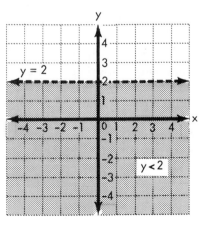

Fig. 5

To draw the graph of $y \geq 2$, we draw the line $y = 2$ as a solid line and shade the half-plane above this line. See Fig. 6.

To draw the graph of $y \leq 2$, we draw the line $y = 2$ as a solid line and shade the half-plane below this line. See Fig. 7.

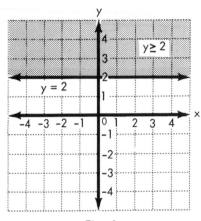

Fig. 6

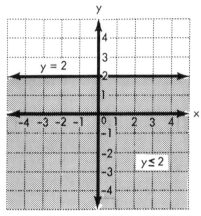

Fig. 7

If we wish to graph the inequality $x > 2$, we draw the dotted vertical line $x = 2$ and shade the half-plane to the right of this line. See Fig. 8. All points in this region have abscissas greater than 2.

For the graph of $x \geq 2$, we would draw the vertical line $x = 2$ as a solid line. The graph would then consist of the shaded area and the solid line.

To draw the graph of $x < 2$, we would shade the region to the left of the vertical dotted line $x = 2$. This half-plane would be the graph of all points in the plane with abscissas less than 2.

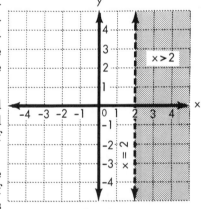

Fig. 8

ILLUSTRATIVE PROBLEM 1: Graph the inequality $y > 3x + 1$.

Solution:

First, graph the line $y = 3x + 1$.

x	0	1	-1
y	1	4	-2

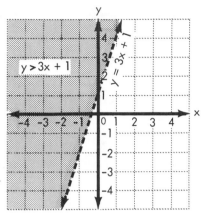

Draw the line dotted since it is not included in the graph. Now shade the half-plane above this line. This shaded half-plane is the graph of $y > 3x + 1$.

ILLUSTRATIVE PROBLEM 2: Graph the inequality $y \leq 2x - 2$.

Solution:

First, graph the line $y = 2x - 2$.

x	0	1	-1
y	-2	0	-4

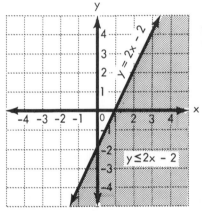

Draw the line solid since it is included in the graph. Now shade the half-plane below this line. This shaded half-plane and the line make up the graph of $y \leq 2x - 2$.

ILLUSTRATIVE PROBLEM 3:

Graph the inequality $y \leq 1$.

Solution: Draw the graph of $y = 1$. We draw the line solid to indicate that it is included in the graph. The half-plane below this line is the graph of $y < 1$. The shaded half-plane and the line make up the graph of $y \leq 1$.

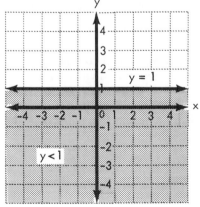

Exercises

In 1 to 18, graph the inequalities. Use graph paper.

1. $y > x$ **2.** $y < 3x$ **3.** $y \geq x + 3$

4. $y \leq 2x + 1$ **5.** $y < x$ **6.** $y \leq x - 2$

7. $y > -2x$ **8.** $y \geq 3x - 4$ **9.** $y < -x + 2$

10. $y \geq \frac{1}{2}x + 3$ **11.** $y \leq 2x + 5$ **12.** $y > -x + 6$

13. $y > \frac{2}{3}x$ **14.** $y \leq \frac{3}{4}x - 2$ **15.** $2y \geq x - 4$

16. $y \geq 3$ **17.** $x \leq -1$ **18.** $y < -2$

In 19 to 26, write an inequality representing the statement. Then graph the inequality.

19. The ordinate of a point is less than its abscissa.

20. The ordinate of a point is greater than or equal to 3 less than its abscissa.

21. The ordinate of a point is greater than 3 times its abscissa.

22. The ordinate of a point is less than 3 more than twice its abscissa.

23. The ordinate of a point is greater than 4 less than half its abscissa.

24. The ordinate of a point is less than one-third its abscissa.

25. The ordinate of a point is greater than -1.

26. The abscissa of a point is negative.

6. SOLVING TWO LINEAR INEQUALITIES GRAPHICALLY

Let us now examine the graph of the solution set of two linear inequalities in two variables.

ILLUSTRATIVE PROBLEM:
Graph the solution set of $x + y \geq 2$

$$y < 1$$

Solution: First, graph $x + y = 2$. By adding $-x$ to both sides of the equation, we can rewrite this equation as $y = -x + 2$.

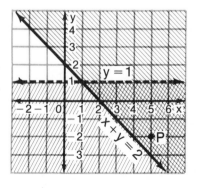

Draw the line solid since it is part of the graph of the inequality $x + y \geq 2$. Shade in the region above the line to represent the graph of $x + y > 2$.

Now draw the line $y = 1$, but make it a dotted line since it is not part of the graph of the inequality $y < 1$. Shade the region below the line to represent the graph of $y < 1$. The doubly shaded (*cross-hatched*) area represents the graph of the solution set of the two inequalities.

Check: As a rough check of this result, select a point within the cross-hatched area, such as $P(5, -2)$. Check to see if $x = 5$, $y = -2$ satisfies both inequalities:

$$x + y \geq 2 \qquad\qquad y < 1$$
$$5 + (-2) \geq 2 \qquad\qquad -2 < 1 \ \rlap{\checkmark}$$
$$3 \geq 2 \ \rlap{\checkmark}$$

Exercises

In 1 to 12, draw the graph of the solution set of the two linear inequalities. Check to see that a point in the cross-hatched region has coordinates that satisfy both inequalities.

1. $y \geq x$
   $x > 3$

2. $y > 2x - 4$
   $y > 2$

3. $x \geq 2$
   $y < -2$

4. $x < 0$
   $y < 0$

5. $y \leq 4x$
   $3x + y > 6$

6. $y > 3x - 2$
   $y \leq x$

7. $y \leq 3x$
   $y > x - 4$

8. $y < 5$
   $y \leq x + 3$

9. $y + 3x > 6$
   $y \leq 2x - 4$

10. $y \geq -x + 1$
   $y < 3x - 3$

11. $y \geq -x + 5$
   $y < 2x + 3$

12. $y \geq -x + 4$
   $y < x + 2$

Chapter Review Exercises

In 1 to 10, choose your answer from the given choices.

1. A value of x that satisfies the inequality $x > -2$ is
 (a) -4 (b) -3 (c) -2 (d) -1

2. A value of y that satisfies the inequality $y - 2 > -3$ is
 (a) 0 (b) -4 (c) -1 (d) -2

3. The solution set of the inequality $2t + 3 > 5$ consists of all values of t for which
 (a) $t > 1$ (b) $t < 1$ (c) $t > 4$ (d) $t < 4$

4. The graph shown is the graph of
 (a) $y > x$
 (b) $y < x$
 (c) $y \le x$
 (d) $y \ge x$

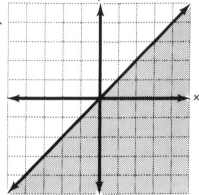

5. The graph below indicates the solution set of
 (a) $x < 0$ (b) $x \le 0$ (c) $x > 0$ (d) $x \ge 0$

6. If $3x - 1 \ge 4$, a member of the solution set is
 (a) 1 (b) 0 (c) 3 (d) -1

7. The solution set of $3x - 3 > 2x + 1$ consists of all values of x for which
 (a) $x < 4$ (b) $x > -2$ (c) $x > 4$ (d) $x > -4$

8. Which number pair does not belong to the solution set of the inequality $y \ge 2x - 3$?
 (a) $(0, -3)$ (b) $(1, -2)$ (c) $(1, 0)$ (d) $(2, 3)$

9. The solution set of $3y - 3 > 12$ is
 (a) $y > 3$ (b) $y < 3$ (c) $y > 5$ (d) $y < 5$

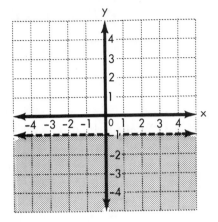

10. Which relation is shown in the graph?
    (a) $y < -1$
    (b) $y > -1$
    (c) $x > -1$
    (d) $x < -1$

11. Solve the inequality: $1 + 2x < 0$

12. Solve the inequality: $2x - 1 < 8$

    For 13 to 18, use separate graph paper.

13. Graph the inequality: $y \geq -x + 1$

14. Graph the inequality: $y < 3x - 6$

15. Graph the set of all points whose ordinates are less than -2.

16. Graph the set of all points whose abscissas are positive or zero.

17. Graph the solution set of the following pair of inequalities:

$$y < 5$$
$$y \leq x + 3$$

18. Using a set of coordinate axes, graph the solution set of the following pair of inequalities:

$$y \leq x - 3$$
$$y > -x + 1$$

19. Choose the letter that best completes the statement:
    The graph of $x = -3$
    (a) has a slope of -3.   (b) passes through the origin.
    (c) is parallel to the y-axis. (d) is parallel to the x-axis.

CUMULATIVE REVIEW

1. Find the numerical value of $|9 - 13|$.

2. Solve for x: $5(x - 4) = 10$

3. What is the product of $5x^3$ and $3x^2$?

4. The length of a rectangle is 6 inches more than the width. If the width is decreased by 2 inches and the length is increased by 4 inches, the perimeter of the new rectangle is 48 inches. Find the original width.

5. Solve the inequality: $3x + 4 < x + 2$

6. Find the solution set of $2n + 3 = 5n - 9$.

7. Graph the inequality $y \le 2x - 5$.

8. Two autos leave a certain place at the same time and travel in opposite directions. The rate of one is three-fourths the rate of the other. If the autos are 364 miles apart after 4 hours, what is the rate of the faster auto?

9. Graph the equation $y = 2x - 1$.

10. What are the coordinates of the point where the graph of the equation $3x + 2y = 12$ intersects the y-axis?

Answers to Cumulative Review

If you get an incorrect answer, refer to the chapter and section shown in brackets for review. For example [5-2] means chapter 5, section 2.

1. 4 [9-3]   2. 6 [11-5]   3. $15x^5$ [10-4]

4. 8 [11-7]   5. $x < -1$ [13-4]   6. 4 [11-4]

7. Graph [13-5]   8. 52 mph [11-9]   9. Graph [12-3]

10. (0, 6) [12-3]

14

Algebraic Solution of Two Linear Equations

1. SOLVING PAIRS OF LINEAR EQUATIONS BY SUBSTITUTION

We are now ready to consider an alternate method of solving simultaneous equations.

ILLUSTRATIVE PROBLEM 1: Solve for x and y:

$$2x + 3y = 24$$

$$y = 2x$$

Solution: Since a quantity may be substituted for an equal quantity, we may substitute the value of y from the second equation wherever y appears in the first equation. The result is

$$2x + 3(2x) = 24$$

We now have only one equation and this equation contains only one variable, x. Now we can solve this equation for x.

$$2x + 6x = 24$$

$$8x = 24$$

$$x = 3$$

To get the corresponding value of y, substitute 3 for x in the equation $y = 2x$, giving $y = 6$. Thus, the solution for the two original equations is $x = 3$, $y = 6$.

Instead of substituting 3 for x in the equation $y = 2x$, we could have substituted 3 for x in the other original equation, $2x + 3y = 24$. This method would have involved a little more work, but it would have also given us $y = 6$.

$$2x + 3y = 24$$
$$2(3) + 3y = 24$$
$$6 + 3y = 24$$
$$3y = 18$$
$$y = 6$$

The solution $x = 3$, $y = 6$ should be checked in *both* of the *original* equations.

Check:

$$2x + 3y = 24 \qquad y = 2x$$
$$2(3) + 3(6) = 24 \qquad 6 = 2(3)$$
$$6 + 18 = 24 \qquad 6 = 6 \ \checkmark$$
$$24 = 24 \ \checkmark$$

Answer: $x = 3$, $y = 6$

A more general situation is illustrated by the following problem.

ILLUSTRATIVE PROBLEM 2: Solve the following pair of equations by substitution and check:

$$2x - 9y = 1$$
$$x - 4y = 1$$

Solution: First change one of the equations to an equivalent equation with x (or y) as one member. By adding $4y$ to both members of the second equation, we obtain

$$x = 4y + 1$$

Substitute this value of x in the first equation.

$$2x - 9y = 1$$
$$2(4y + 1) - 9y = 1 \qquad \text{(substitute } 4y + 1 \text{ for } x\text{)}$$
$$8y + 2 - 9y = 1$$
$$8y - 9y = 1 - 2 \qquad \text{(subtract 2 from both members}$$
$$\qquad\qquad\qquad\qquad \text{and combine like terms)}$$
$$-y = -1$$
$$y = 1 \qquad \text{(divide both members by } -1\text{)}$$

Substitute this value of y in the equation $x = 4y + 1$ and solve for y.

$$x = 4(1) + 1$$
$$x = 4 + 1$$
$$x = 5$$

Check $x = 5$, $y = 1$ in *both original* equations.

Check:

$2x - 9y = 1$	$x - 4y = 1$
$2(5) - 9(1) = 1$	$5 - 4(1) = 1$
$10 - 9 = 1$	$5 - 4 = 1$
$1 = 1$ ✔	$1 = 1$ ✔

Answer: $x = 5$, $y = 1$

Summary: To solve a pair of linear equations in two variables (x and y), proceed as follows:

1. Solve one of the equations for either x or y. Choose the simpler equation, preferably one in which the coefficient of x or y is 1.

2. Substitute this expression for x (or y) in the other equation.

3. The resulting equation is then an equation in one variable (x or y). Solve this equation for the variable.

4. Substitute this value of the variable in the equation found in step (1) or in either of the two original equations and solve for the other variable.

5. Check the values found for x and y in *both original* equations.

Exercises _____

In 1 to 30, solve the following pairs of equations by the substitution method and check.

1. $y = x$	**2.** $x = 3y$	**3.** $3y - 4x = -7$
$\quad 2x + 5y = 21$	$\quad 3y + 2x = 18$	$\quad x = -2$
4. $r = 2s$	**5.** $3p + q = 28$	**6.** $y = -x$
$\quad 5r + 2s = 36$	$\quad q = 4p$	$\quad 3x + 5y = 12$
7. $10a = 4b + 60$	**8.** $t = -4s$	**9.** $m + 3n = 0$
$\quad b = -5a$	$\quad 3s + 7t = 50$	$\quad 2m + 9n = 10$
10. $y = 3 - x$	**11.** $3p + 2q = 8$	**12.** $3c + d = 8$
$\quad 7x + y = 9$	$\quad p + q = 3$	$\quad 2c - 3d = 9$
13. $s = 2r + 1$	**14.** $m = 3n - 7$	**15.** $a = 3b - 1$
$\quad 3r + 2s = 9$	$\quad 2m + 3n = 13$	$\quad 2a + 3b = 8$

16. $3x + 4y = 11$
    $x = 6y$

17. $d = 4 - c$
    $7c + 2d = -2$

18. $p = -3q + 4$
    $2p - q = -6$

19. $5r + 2s = 23$
    $4r + s = 19$

20. $2a + b = 0$
    $2a - b = 1$

21. $2x - 9y = 1$
    $x - 4y = 1$

22. $t - 4u = 1$
    $2t - 9u = 3$

23. $4x + y = 11$
    $x + 2y = 8$

24. $3r - 5s = -9$
    $2r - s = 1$

25. $3s = t + 11$
    $5t = 7s + 1$

26. $3b + c = 26$
    $2b + 5c = 52$

27. $y - 4x = 5$
    $x + y = 15$

28. $2x - 5y = 8$
    $2x = 3y$

29. $2s + 5t = 9$
    $s = -\frac{1}{2}t$

30. $2t + u = 3.9$
    $t = .15u$

---

2. SOLVING PAIRS OF LINEAR EQUATIONS BY ADDITION OR SUBTRACTION

Consider the pair of equations shown below.

$$3x + 2y = 28$$

$$5x - 2y = 20$$

If we try to solve either of these equations for x or y, we get involved in algebraic fractions. In this case, the method of substitution would lead to a rather complicated equation in one variable. Another method that is frequently useful in such cases is the *method of addition*.

Since the coefficients of y are $+2$ and -2, which are additive inverses (opposites), their sum is zero. This leads to the idea of adding the two equations.

$$
\begin{array}{rl}
3x + 2y &= 28 \\
5x - 2y &= 20 \\
\hline
8x &= 48 \\
x &= 6
\end{array}
$$

This value of x may now be substituted in either of the original equations to obtain the value of y. Thus:

$$
\begin{array}{rl}
3x + 2y &= 28 \\
3(6) + 2y &= 28 \\
18 + 2y &= 28 \\
-18 &= -18 \\
\hline
2y &= 10 \\
y &= 5
\end{array}
$$

The solution $x = 6$, $y = 5$ should be checked in *both original* equations.

Check:

$$3x + 2y = 28 \qquad\qquad 5x - 2y = 20$$
$$3(6) + 2(5) = 28 \qquad\qquad 5(6) - 2(5) = 20$$
$$18 + 10 = 28 \qquad\qquad 30 - 10 = 20$$
$$28 = 28 \ \checkmark \qquad\qquad 20 = 20 \ \checkmark$$

In this example, the coefficients of y were *additive inverses*. The same method may be used when the coefficients of x are *additive inverses*. In either case, we eliminate, or get rid of, one of the variables, resulting in a single equation in one variable.

Consider now the pair of equations shown below.

$$5x + 3y = 19$$
$$2x + 3y = 13$$

Here the coefficients of y are equal, and adding both equations will not eliminate one variable. However, if we *subtract* the members of the second equation from the corresponding members of the first equation, the variable y is eliminated and we can find the value of x.

$$3x = 6$$
$$x = 2$$

Substitution of this value of x in either one of the original equations enables us to find y.

$$
\begin{aligned}
5x + 3y &= 19 \\
5(2) + 3y &= 19 \\
10 + 3y &= 19 \\
-10 \qquad &= -10 \\
\hline
3y &= 9 \\
y &= 3
\end{aligned}
$$

Check:

$$5x + 3y = 19 \qquad\qquad 2x + 3y = 13$$
$$5(2) + 3(3) = 19 \qquad\qquad 2(2) + 3(3) = 13$$
$$10 + 9 = 19 \qquad\qquad 4 + 9 = 13$$
$$19 = 19 \ \checkmark \qquad\qquad 13 = 13 \ \checkmark$$

Note that if we had first multiplied both members of the second equation by -1, we could then have proceeded by the method of addition and obtained the same results.

Exercises _____

In 1 to 15, solve each of the following pairs of equations by the addition or subtraction method and check.

1. $x + y = 8$
   $x - y = 2$

2. $r + s = 24$
   $r - s = 12$

3. $2s + t = 13$
   $3s - t = 7$

4. $a + b = 11$
   $a - b = 5$

5. $2p + q = 7$
   $2p - q = 5$

6. $3m + 2n = 9$
   $m + 2n = 7$

7. $3x + 4y = 11$
   $3x + y = 2$

8. $t - 2u = 3$
   $t - u = 5$

9. $2c + 3d = 5$
   $2c - d = 1$

10. $2p + 5q = 20$
   $-2p + 3q = 4$

11. $.3x + .2y = 10$
   $.2x - .2y = 20$

12. $2c + 3d = 17$
   $4c - 3d = 34$

13. $4m - n = 4$
   $4m + 2n = -5$

14. $x + 2y = 0$
   $y - x = 12$

15. $2(2k + m) = 16$
   $4k - m = 4$

3.   SOLVING PAIRS OF LINEAR EQUATIONS BY MULTIPLICATION AND ADDITION

Consider the pair of equations shown below.

$$2x + 3y = 12$$
$$x - 2y = -1$$

If we try to use the addition method directly, the resulting equation will still have two variables. Neither x nor y is eliminated here by using addition or subtraction.

If we multiply both members of the second equation by -2, the result is

$$-2(x - 2y) = -2(-1) \text{ or } -2x + 4y = 2$$

By doing this, the coefficients of the x terms in both equations become additive inverses. We can now add these equations to find y.

$$2x + 3y = 12$$
$$\underline{-2x + 4y = 2}$$
$$7y = 14$$
$$y = 2$$

Now substitute 2 for y in either of the original equations to find x.

$$x - 2(2) = -1$$
$$x - 4 = -1$$
$$x = 3$$

The answer $x = 3$, $y = 2$ should be checked in *both original* equations.

$$2x + 3y = 12 \qquad\qquad x - 2y = -1$$
$$2(3) + 3(2) = 12 \qquad 3 - 2(2) = -1$$
$$6 + 6 = 12 \qquad\qquad 3 - 4 = -1$$
$$12 = 12 \; ✓ \qquad\qquad -1 = -1 \; ✓$$

It may be necessary to find a multiplier for *each* equation to produce a pair of equations so that one variable is eliminated by addition. This is illustrated as follows.

ILLUSTRATIVE PROBLEM: Solve the pair of equations and check:

$$3x + 4y = 1$$
$$2x - 3y = 12$$

Solution: In order to use the addition method, we must make the coefficients of x (or y) additive inverses. This can be accomplished by using -2 as a multiplier for the first equation and 3 as a multiplier for the second equation as shown below.

$$-2(3x + 4y) = 1(-2)$$
$$3(2x - 3y) = 12(3)$$

The following equations are then obtained:

$$-6x - 8y = -2$$
$$6x - 9y = 36$$

Adding these equations, we get

$$-17y = 34$$
$$y = -2$$

Substitute -2 for y in either original equation.

Using the first equation, we obtain

$$3x + 4(-2) = 1$$
$$3x - 8 = 1$$
$$3x = 9$$
$$x = 3$$

Check:

$$3x + 4y = 1 \qquad\qquad 2x - 3y = 12$$
$$3(3) + 4(-2) = 1 \qquad 2(3) - 3(-2) = 12$$
$$9 - 8 = 1 \qquad\qquad 6 + 6 = 12$$
$$1 = 1 \; ✓ \qquad\qquad 12 = 12 \; ✓$$

Answer: $x = 3$, $y = -2$

In using the addition method, transform each equation into one in which the variables are on one side and the constant is on the other side, if the equations are not already in this form.

Consider, for example, the pair of equations shown below.

$$7x = 5 - 2y$$

$$2x + 3y - 16 = 0$$

Rearrange the first equation by adding $2y$ to both members. Rearrange the second equation by adding 16 to both members. The following equations are then obtained:

$$7x + 2y = 5$$

$$2x + 3y = 16$$

We can now proceed to solve this pair of equations, using the addition method as shown in the illustrative problem. By doing this, we find that $x = -1$ and $y = 6$.

Exercises

In 1–21, solve each pair of equations by the multiplication and addition method and check.

1. $3x - 2y = 16$
$5x - y = 22$

2. $3x + y = 6$
$x + 3y = 10$

3. $a + 2b = 4$
$3a + b = 7$

4. $3m + 2n = 23$
$m + 3n = 17$

5. $2r + s = 12$
$r + 2s = 9$

6. $5p + 4q = 27$
$p - 2q = 11$

7. $3c - d = 1$
$c + 2d = 12$

8. $s - 3t = -11$
$3s + t = 17$

9. $x - 2y = -2$
$2x - y = 5$

10. $5a + 8b = 1$
$3a + 4b = -1$

11. $7m - 3n = 23$
$m + 2n = 13$

12. $6r + 5s = 11$
$4r + 3s = 8$

13. $4x - 5y = -2$
$3x - 7y = 5$

14. $4p + 3q = -2$
$6p + 7q = -13$

15. $2t + 7u = 3$
$3t - 5u = 51$

16. $x + y = 1050$
$.03x + .04y = 38$

17. $.25c + .50d = 250$
$c + d = 900$

18. $s + t = 720$
$.02s + .03t = 17.60$

19. $2r - s = 5$
$r + 2(s + 3) = 11$

20. $4x + 5y = 6 + 8y$
$5x + 6y = 27$

21. $8a + b = 8 + 5a$
$2a - 3b = 9$

Chapter Review Exercises

1. Solve by the substitution method and check:

    a. $5x - 3y = 36$ **b.** $2r - s = 1$

        $y = 2 - 3x$ $3r - 5s = -9$

2. Solve by the addition method and check:

    a. $3s + t = 7$ **b.** $m + 2n = -4$

        $2s - t = 8$ $n = -3m + 3$

    c. $5t + 2u = -7$ **d.** $.4r + .3s = 1.5$

        $2t + 3u = 6$ $.5r + .4s = 1.5$

In 3 to 5, solve each of the following pairs of equations algebraically.

3. $x + 3y = 13$ **4.** $4a - 3b = 17$ **5.** $5r - 2s = 8$

    $x + y = 5$ $5a - b = 13$ $3r - 7s = -1$

CHECK YOUR SKILLS

1. Solve for x: $5x + 5 = 6$

2. If 19 is subtracted from three times a number, the result is 110. Find the number.

3. Solve for y: $0.9y + 0.4 = 7.6$.

4. Find the value of $(6a)^2$ when $a = \frac{1}{2}$.

5. Express $\frac{1}{2} + \frac{1}{3}$ as a single fraction.

6. Karl can mow a lawn in 45 minutes. What part of the lawn can he mow in 15 minutes?

7. Round off 87.4562 to the nearest hundredth.

8. Solve for t: $5t - 17 = 4 + 2t$

9. Find the sum of $\frac{2}{3}$ and $\frac{3}{4}$.

10. Solve for y: $5 - y = 2$.

15

Solving Verbal Problems by Using Two Variables

1. INTRODUCTION

We have already seen that many verbal problems can be solved by setting up one equation in one variable. It is frequently more convenient and direct to solve some verbal problems by setting up and solving two equations in two variables.

Recall that algebraic symbols may be expressed in words in several different ways. A few of these ways are listed below:

Algebraic Expression	Word Expression
pq	p times q p multiplied by q the product of p and q
$x + y$	y more than x the sum of x and y x increased by y y added to x
$t - s$	t diminished by s t decreased by s s less than t the difference of t and s $(t > s)$ s subtracted from t
$\dfrac{m}{n}$ or $m \div n$	$\dfrac{1}{n}$ times m m divided by n the quotient of m and n

2. NUMBER PROBLEMS

ILLUSTRATIVE PROBLEM: The sum of two numbers is 42. When twice the smaller number is subtracted from the larger number, the result is 12. What are the two numbers?

Solution:

Let x = the larger number.
Let y = the smaller number.

Then $x + y = 42$.

And $x - 2y = 12$.

Subtract the second equation from the first.

$$\begin{array}{r} x + y = 42 \\ \underline{x - 2y = 12} \\ 3y = 30 \\ y = 10 \end{array}$$

Substitute this value of y for x in the first equation.

$$x + y = 42$$

$$x + 10 = 42$$

$$x = 32$$

Answer: The numbers are 32 and 10.

Note: Check a verbal problem by testing the answers in the original problem. Thus, the sum of 32 and 10 is 42. Also, 2(10) subtracted from 32 is $32 - 20$ or 12.

Exercises

1. If x and y represent numbers, write algebraically:
   a. y more than x **b.** y less than x
   c. y times x **d.** y more than twice x
   e. x decreased by y **f.** x diminished by 5 times y
   g. the product of x and twice y **h.** the quotient of x divided by y
   i. y less than four times x
   j. $\frac{2}{3}$ of x increased by y
   k. y less than the product of 6 and x
   l. x increased by three times y

In 2 to 8, use the given information to set up two equations in two variables. Then solve to find the two numbers.

2. The sum of two numbers is 38 and their difference is 10.

3. The sum of two numbers is 27 and their difference is 12.

4. The sum of two numbers is 28. The larger is 4 more than 5 times the smaller number.

5. The difference between two numbers is 8. The sum of 2 and the larger number is twice the smaller number.

6. The difference between two numbers is 16. Twice the larger is 12 more than three times the smaller.

7. The sum of two numbers is 31. When twice the larger is added to four times the smaller, the result is 89.

8. The sum of two numbers is 18. If 4 times the smaller is subtracted from 3 times the larger, the difference is 12.

---

3. COIN PROBLEMS

Recall the following:

If the value of 3 nickels in cents is 3(5), or 15 cents, then the value of n nickels is $n(5)$, or $5n$ cents.

If the value of 7 dimes in cents is 7(10), or 70 cents, then the value of d dimes is $d(10)$, or $10d$ cents.

If the value of 5 quarters in cents is 5(25), or 125 cents, then the value of q quarters is $q(25)$, or $25q$ cents.

ILLUSTRATIVE PROBLEM: A coin bank contains $6.10, in dimes and nickels. If the number of dimes is 4 less than twice the number of nickels, how many coins of each type does the bank contain?

Solution:

Let d = the number of dimes.
Let n = the number of nickels.

Then $10d$ = the value of the dimes in cents.

And $5n$ = the value of the nickels in cents.

Since the value of the coins is $6.10 (or 610 cents), we may write the equation

$$5n + 10d = 610$$

Since the number of dimes is 4 less than twice the number of nick-els, we can write a second equation

$$d = 2n - 4$$

To solve, substitute $2n - 4$ for d in the first equation.

$$5n + 10d = 610$$

$$5n + 10(2n - 4) = 610 \quad \text{(substitute } 2n - 4 \text{ for } d\text{)}$$

$$5n + 20n - 40 = 610 \quad \text{(multiply to remove parentheses)}$$

$$25n - 40 = 610 \quad \text{(combine similar terms)}$$

$$25n = 650$$

$$n = 26 \quad \text{(divide both members by 25)}$$

To find d, substitute 26 for n in the equation $d = 2n - 4$.

$$d = 2n - 4$$

$$d = 2(26) - 4 \quad \text{(substitute 26 for } n\text{)}$$

$$d = 52 - 4 = 48$$

Check in the original statement of the problem.

$$\text{value of 48 dimes} = \$4.80$$
$$\text{value of 26 nickels} = \underline{\quad 1.30}$$
$$\text{total value} = \$6.10 \ \checkmark$$

The number of dimes (48) should be 4 less than twice the number of nickels (26).

$$48 = 2(26) - 4$$

$$48 = 52 - 4$$

$$48 = 48 \ \checkmark$$

Answer: 26 nickels and 48 dimes

Exercises

1. Represent algebraically the value in cents of n nickels, d dimes, and q quarters.
2. Represent algebraically:
   a. the amount, in cents, that a boy pays for n 15¢ candies and r 9¢ candies.
   b. the amount of change, in cents, that he receives if he pays for the candies with a $5.00 bill.

In 3 to 8, solve each exercise by using two variables. Check your answers.

3. Mort saved nickels and dimes until he had $2.80. The number of nickels was 8 more than the number of dimes. How many coins of each type did he have?

4. Edna has $3.20 in coins, all in nickels and dimes. If she has 40 coins altogether, how many of each kind does she have?

5. A purse contains $3.55, in quarters and dimes. It contains, in all, 25 coins. Find the number of each kind of coin.

6. A mechanic received $340 pay in five-dollar and twenty-dollar bills. The number of $20 bills was 1 less than twice the number of $5 bills. How many bills of each type did he have in his pay?

7. A candy machine contains $5 in nickels and dimes. The number of nickels is twice the number of dimes. How many of each type are there?

8. Bernice bought 50 postage stamps for which she paid $13.00. Some were 20¢ stamps and some were 40¢ stamps. How many of each did she buy?

---

4. BUSINESS AND CONSUMER PROBLEMS

ILLUSTRATIVE PROBLEM: A woman bought 7 ears of corn and 6 oranges for $2.44. A second woman bought 10 ears of corn and 3 oranges for $2.26. Find the price of an ear of corn and the price of an orange.

Solution:

Let x = price of an ear of corn in cents.
Let y = price of an orange in cents.

Then $7x + 6y = 244$.

And $10x + 3y = 226$.

We plan to eliminate the variable y by adding the two equations. This can be done when the coefficients of y are the same number but with opposite signs. To this end, we will multiply the second equation by -2.

$$-2(10x + 3y) = -2(226) \quad \text{or} \quad -20x - 6y = -452$$

We can now add this result to the first equation to find x.

$$7x + 6y = 244$$
$$-20x - 6y = -452$$
$$\overline{-13x = -208}$$
$$x = 16$$

To find y, substitute 16 for x in either original equation.

$$10x + 3y = 226$$
$$10(16) + 3y = 226$$
$$160 + 3y = 226$$
$$3y = 66$$
$$y = 22$$

Check in the original problem.

7 ears of corn @ 16¢ = $1.12
6 oranges @ 22¢   = $1.32
                         $2.44 ✔

10 ears of corn @ 16¢ = $1.60
3 oranges @ 22¢   =   .66
                         $2.26 ✔

Answer: The prices are 16¢ for an ear of corn and 22¢ for an orange.

Exercises

In 1 to 7, solve the problems by using two variables.

1. A dealer sold 200 pairs of ski poles. Some were sold at $6 per pair and the remainder were sold at $11 per pair. Total receipts from this sale were $1600. How many pairs of poles did she sell at $6 each?

2. A school club held a bake sale at which a total of 40 cakes and pies were sold. The cakes sold for $2.50 each and the pies for $2.25 each. The club received $93.75 from the sale. How many cakes were sold?

3. A housewife buys 5 pounds of pork and 3 pounds of ground beef for $8.00. The following week, at the same prices, she pays $6.08 for 4 pounds of pork and 2 pounds of ground beef. What is the price per pound of pork?

4. If 5 pencils and 2 notebooks cost 90¢ and 3 pencils and 1 notebook cost 50¢, what is the cost of a pencil?

5. Mr. Burns took his family to the theater and paid $46 for 1 adult ticket and 4 children's tickets. Mr. Hoffman paid $44 for 2 adult tickets and 2 children's tickets. What is the cost of one child's ticket?

6. Mr. Johnson bought 7 quarts of milk and 2 pints of cream for $11.80. Several days later, at the same prices, he bought 5 quarts of milk and 1 pint of cream for $7.40. What is the cost of one quart of milk?

7. At a Saturday afternoon movie, tickets cost $1.50 for children and $4.00 for adults. If a total of 380 tickets were sold and the receipts were $670, how many children attended?

---

5. INVESTMENT PROBLEMS

If you invest $800 at 10%, the annual interest (income) can be computed as follows:

$$i = p \cdot r \cdot t$$
$$i = \$800(.10)(1)$$
$$i = \$80$$

Note that the $800 investment is called the *principal*. The *rate of interest*, usually given as a percent (in this problem as 10%), is usually rewritten as a decimal. Since *annual* means yearly, the *time* in this problem is $t = 1$.

ILLUSTRATIVE PROBLEM 1: A woman invests $1200 at 8% and $700 at 9%. What is her total yearly income from the two investments?

Solution: $1200(.08) = $96
                     $700(.09) = $63

Total annual income = 96 + 63 = $159   (*answer*)

ILLUSTRATIVE PROBLEM 2: Represent algebraically the total yearly income on $x invested at 10% and $y invested at 12%.

Solution:

yearly income on $x at 10% = $x(.10) = .10x$
yearly income on $y at 12% = $y(.12) = .12y$

Total yearly income = $.10x + .12y$ (*answer*)

ILLUSTRATIVE PROBLEM 3: A man invested a sum of $1500 in two accounts. One account paid simple interest at the rate of 6% a year, and the other paid simple interest at the rate of 8% a year. If the total interest received in one year from both accounts was $103, find the amount invested in *each* account.

Solution: Let x = the number of dollars invested at 6%.
             Let y = the number of dollars invested at 8%.
Then $.06x$ = the annual interest on the 6% investment.
And $.08y$ = the annual interest on the 8% investment.

The total investment is $1500, or $x + y = 1500$.
The total annual interest is $103, or $.06x + .08y = 103$.

To solve these equations, multiply the first equation by -6 and the second equation by 100 to obtain the following equations:

$$-6x - 6y = -9000$$
$$6x + 8y = 10,300$$

Add these two equations to find the value of y.

$$2y = 1300$$
$$y = 650$$

Substitute 650 for y in the first of the original equations to find x.

$$x + y = 1500$$
$$x + 650 = 1500$$
$$x = 850$$

Check these values in the original problem.

The total investment is $1500.
$850 + $650 = $1500
$1500 = $1500 ✔

The total interest is $103.
$.06(850) + .08(650) = 103$
$51 + $52 = $103
$103 = $103 ✔

Answer: The amount invested at 6% is $850, and the amount invested at 8% is $650.

Exercises

1. Represent the annual income when the annual rate is 9% and the principal is:
   a. $400   **b.** $1200   **c.** $p   **d.** $(x + 400)$

2. Represent the total annual income in each of the following:
   a. $800 invested at 7% and $1000 invested at 9%
   b. $1200 invested at 10% and $2000 invested at 11%
   c. x invested at 6% and y invested at 9%
   d. p invested at 8% and $2p$ invested at 10%

In 3 to 11, solve each exercise by using two variables. Check your answers.

3. A sum of $3500 is invested in two parts. One part earns interest at 5% and the other at 8%. The total annual interest is $250. Find the amount invested at *each* rate.

4. Part of $3000 is invested in bonds at 4%, and the remainder in stocks at 6%. The total annual income from both investments is $150. Find the number of dollars invested at each rate.

5. Mrs. Primera invested $6000, from which she received a yearly income of $330. Part of the money is invested at 6% and the remainder at 4%. How much is invested at each rate?

6. Mr. Wong invests $7500, part in bonds that pay 4% a year and the rest in property that yields him 7% a year. If the total income from the two investments is $450, how much did he invest in bonds?

7. A woman invested some money at 5% and twice as much money at 7%. If the total yearly income from the two investments is $760, how much was invested at 7%?

8. Mrs. Crane invested $5000, part at 6% and the remainder at 8%. The annual income from the 8% investment is $260 greater than the annual income from the 6% investment. Find the amount invested at 6%.

9. A man invested an amount of money at 6%. He also invested $400 more than the first amount at 4%. The annual incomes from these investments were equal. How much was invested at 6%?

10. Ms. Rosen invested $3900, part at 6% and the rest at 5%. If the total annual income from the two investments is $215, find the amount invested at 5%.

11. Mr. Martinez invests a certain amount of money at 8%. He invests a sum three times as large as the first at 7%. The total annual income from the two investments is $522. How much did he invest at 8%?

6. MIXTURE PROBLEMS

Many problems in business and industry require the mixing of ingredients that have different unit costs.

If a pound of candy costs 90 cents, what is the cost in cents of 4 pounds of this candy? The cost would be 4(90), or 360 cents ($3.60). The cost of x pounds of this candy would be represented as $90x$ cents.

If a cheaper brand of candy costs 70 cents per pound, the cost of y pounds of this candy would be represented as $70y$ cents.

If the x pounds of candy worth 90 cents a pound were mixed with the y pounds of candy worth 70 cents a pound, we would represent the total value of the mixture as $90x + 70y$ cents.

Assume that the seller must receive the same amount of money after mixing the two brands of candy as he or she would have received without mixing them.

ILLUSTRATIVE PROBLEM 1:
a. What is the total value of a mixture of coffee made up of 20 pounds of coffee costing $3.00 per pound and 25 pounds of coffee costing $4.00 per pound?

b. What is the total value of a mixture of coffee made up of x pounds of coffee costing $3.00 per pound with y pounds of coffee costing $4.00 per pound?

Solution:

a. $20(3.00) + 25(4.00) = 60.00 + 100.00 = \$160$ (*answer*)

b. $x(3.00) + y(4.00) = 3x + 4y$ dollars (*answer*)

ILLUSTRATIVE PROBLEM 2: A grocer wishes to mix grain worth 60 cents per pound with grain worth 90 cents per pound to make a mixture worth 80 cents per pound. How many pounds of each kind should he use to make 75 pounds of the 80-cent mixture?

Solution:

Let x = number of pounds of 60-cent grain.
Let y = number of pounds of 90-cent grain.

Then $x + y$ = total number of pounds of the mixture.

And $60x$ = value of x pounds of 60¢ grain.

And $90y$ = value of y pounds of 90¢ grain.

And $75(80) = 6000$ = total value in cents of the mixture.

Since the total number of pounds is 75, we can write the equation $x + y = 75$.

Since the total value of the mixture is $75(80) = 6000$, we can write the equation $60x + 90y = 6000$.

To solve these two equations, we will multiply the first equation ($x + y = 75$) by -60 to obtain $-60x - 60y = -4500$. We can eliminate x by adding this result to the second equation as shown on the following page.

$$-60x - 60y = -4500$$
$$60x + 90y = 6000$$

$$30y = 1500$$
$$y = 50$$

Since $x + y = 75$
$$x + 50 = 75$$
$$x = 25$$

Check in the original problem.

50 + 25 = 75   (total weight of mixture)
value of 25 lb. at 60¢ per pound = $15.00
value of 50 lb. at 90¢ per pound = $\underline{45.00}$
value of 75 lb. at 80¢ per pound = $60.00 ✔

Answer: The grocer should use 25 pounds of 60¢ grain and 50 pounds of 90¢ grain.

Exercises

1. What is the total value of a mixture of two types of seed, if 30 pounds of seed worth 60 cents per pound are mixed with 20 pounds of seed worth 80 cents per pound?

2. Represent the value, in dollars, of a mixture of two brands of tea, if x pounds of tea worth $2.50 per pound are mixed with y pounds of tea worth $3.00 per pound.

3. A merchant mixes candy worth 65¢ a pound with candy worth 80¢ a pound to make a mixture of 60 pounds worth 70¢ a pound. How many pounds of each kind did she use?

4. A dealer wishes to make a mixture of 100 pounds of seed to sell for 66¢ a pound by mixing seed worth 90¢ a pound with seed worth 50¢ a pound. How many pounds of each should he use?

5. Two kinds of grain, selling separately at $.70 and $.90 per pound, are to be mixed to give 72 pounds selling at $.75 per pound. How many pounds of each are to be used?

6. Teas at $4.50 and $6.00 a kilogram are to be mixed to give 30 kilograms that will sell for $5.40 a kilogram. How many kilograms of each must be used?

7. A merchant wishes to have a mixture of 16 pounds of candy to sell at $.90 a pound. How many pounds of each must he use, if he mixes candy at $.80 a pound with candy at $.96 a pound?

8. A refiner wants to make 50 liters of gasohol to sell at 28¢ a liter by mixing gasoline costing 30¢ a liter with alcohol costing 20¢ a liter. How many liters of each ingredient must be mixed?

9. A grocer wishes to mix peanuts worth $1.00 per pound with cashew nuts worth $1.45 per pound in order to make a 60-pound mixture worth $1.30 per pound. How many pounds of cashews should she use?

10. Walnuts at $4.50 a kilogram and hazel nuts at $2.50 a kilogram are to be mixed to give 32 kilograms worth $3.00 a kilogram. How many kilograms of each are needed?

---

7. DIGIT PROBLEMS

In our number system, which uses ten as its base, we use the symbols 0, 1, 2, 3, 4, 5, 6, 7, 8, 9 to write any number. These number symbols are called *digits*. In a number, the digits have different values depending on their place within the number.

For example, in the number 47, the digit 4 has the value 4(10) or 40. Thus, we may say $47 = 4(10) + 7(1)$.

Likewise, the number 358 may be written as

$$3(100) + 5(10) + 8(1)$$

In the number 358, we refer to the 3 as the hundreds digit, the 5 as the tens digit, and the 8 as the units digit. Each place is given a value that is ten times the value of the place immediately to its right.

In dealing with two-digit numbers, we may represent the tens digit by t and the units digit by u. Thus, the number itself is represented by $t(10) + u(1)$, or $10t + u$. Note that we *cannot* represent the number itself as tu since, in algebraic symbolism, tu means "t times u."

If we *reverse* the digits of the number $10t + u$, then u becomes the tens digit and t becomes the units digit. Thus, the original number with its digits *reversed* becomes $10u + t$.

Do not confuse the representation of the number $10t + u$ with the sum of its digits, which is merely $t + u$. If $t = 5$ and $u = 8$, the number is $10(5) + 8$, or 58, whereas the sum of the digits is $5 + 8 = 13$. The number with digits reversed is $10(8) + 5$, or 85.

ILLUSTRATIVE PROBLEM 1: The tens digit of a two-digit number is 6 and the units digits is 9. Write

a. the number itself.

b. the number with digits reversed.

c. the sum of the digits.

Solution:

a. $10(6) + 9 = 60 + 9 = 69$ (*answer*)

b. $10(9) + 6 = 90 + 6 = 96$ (*answer*)

c. $6 + 9 = 15$ (*answer*)

ILLUSTRATIVE PROBLEM 2: The sum of the digits of a two-digit number is 13. If the digits are interchanged, the new number is 27 more than the original number. Find the original number.

Solution:

Let t = the tens digit.
Let u = the units digit.

Then $10t + u$ = the original number.
And $10u + t$ = the new number with digits interchanged.

The sum of the digits is 13.

$$u + t = 13 \quad \text{(first equation)}$$

The new number is 27 more than the original number.

$$10u + t = 27 + 10t + u$$

Simplify this equation by bringing all the variables to the left member.

$$9u - 9t = 27$$

Divide both sides of this equation by 9.

$$u - t = 3 \quad \text{(second equation)}$$

Using the first equation, $u + t = 13$, and also the second equation, $u - t = 3$, add to solve for u.

$$
\begin{array}{rcl}
u + t &=& 13 \\
u - t &=& 3 \\
\hline
2u \phantom{{}+t} &=& 16 \\
u &=& 8
\end{array}
$$

Substitute 8 for u in the first equation.

$$
\begin{array}{rcl}
u + t &=& 13 \\
8 + t &=& 13 \\
t &=& 5
\end{array}
$$

The original number is $10t + u$ or $10(5) + 8 = 58$.

Check in the original problem.

$$8 + 5 = 13 \qquad\qquad 85 = 27 + 58$$
$$13 = 13 \ \checkmark \qquad 85 = 85 \ \checkmark$$

Answer: 58

Exercises

1. Write the number:
   a. whose tens digit is 7 and units digit is 3.
   b. whose tens digit is 3 and units digit is 7.
   c. whose tens digit is x and units digit is y.

2. a. Represent the three-digit number whose hundreds digit is h, tens digit is t, and units digit is u.
   b. Represent the number obtained by reversing the digits of the number in **a.**

3. The sum of the digits of a two-digit number is 7. If the digits are reversed, the new number is 9 less than the original number. Find the original number.

4. The units digit of a two-digit number is 4 more than the tens digit. If the digits of the number are reversed, a new number is formed that is 10 more than twice the original number. Find the original number.

5. In a two-digit number, the tens digit is 5 more than the units digit. The number is 63 more than the sum of its digits. Find the number.

6. The sum of the digits of a two-digit number is 14. If the tens digit is 4 more than the units digit, what is the number?

7. In a two-digit number, the units digit is 2 more than the tens digit. The number is 30 more than the units digit. Find the number.

8. In a two-digit number, the units digit is twice the tens digit. If the digits are reversed, the resulting number exceeds the original number by 36. Find the original number.

9. The tens digit of a two-digit number exceeds the units digit by 2. If the digits are reversed, the resulting number is 4 times the sum of the digits. Find the original number.

Chapter Review Exercises

1. At a high school game, students paid $.70 admission and adults paid $1.00. If 175 people were in attendance, and the receipts were $145.00, how many students and how many adults attended the game?

2. A person has $10,000 invested, part at 8% and part at 6%. The income from the 8% investment is $240 more than the income from the 6% investment. How much is invested at *each* rate?

3. One month a school store sold 15 banners and 10 shirts for a total of $60. The next month it sold 25 banners and 20 shirts for a total of $110. What was the selling price, in dollars, of one banner?

4. A girl bought 25 candies for $1.66. Some were 6-cent candies and some were 10-cent candies. Find the number of each bought.

5. Six years ago in a certain state park, the deer outnumbered the foxes by 80. Since then the number of deer has doubled and the number of foxes has increased by 20. If there is now a total of 240 deer and foxes in the park, how many foxes were there 6 years ago?

6. The sum of the digits of a two-digit number is 11. The number obtained by interchanging the digits exceeds twice the original number by 34. Find the original number.

7. A bag is filled with quarters and half-dollars. The total value of the coins is $20. If the number of quarters is 8 more than the number of half-dollars, find the number of half-dollars.

8. A dealer mixes two kinds of candy, one worth $.80 a pound and the other worth $1.20 a pound. There are 20 pounds in the mixture and it is to sell at $.96 per pound. Find the number of pounds of each kind in the mixture.

9. The sum of two numbers is 180. One number is 40 more than 3 times the other. Find the numbers.

10. The perimeter of a rectangle is 302 cm. If the length exceeds 5 times the width by 1 cm, what are the dimensions of the rectangle in centimeters?

CUMULATIVE REVIEW

1. Find the slope of the line joining the points $(-2, 3)$ and $(-5, 9)$.

2. Find the slope of the line whose equation is $y = \frac{2}{3}x + 7$.

3. Solve graphically and check.
$$y = 2x - 1$$
$$3x + 2y = 12$$

4. On the same set of coordinate axes, graph the following system of inequalities and label the solution set A. Give the coordinates of a point P in A.
$$x + y > 6$$
$$y \leq 2x - 5$$

5. Solve for m: $12m - 5 = 10m - 3$

6. **a.** What number is the identity element of addition?
   b. What is the additive inverse of -6?
   c. What is the reciprocal of $-\frac{2}{3}$?

7. Two numbers are in the ratio of 5 to 13. If the sum of the two numbers is 36, what is the *larger* number?

8. Solve the following system of equations for x:
$$5x - y = 20$$
$$y = 3x$$

9. Solve the following system of equations for a:
$$2a + 5b = 19$$
$$-3a + 5b = 9$$

10. The difference between two numbers is 9. If 4 times the larger number is 10 times the smaller, what are the numbers?

11. Part of $4000 is deposited in a credit union at 7% interest and the remainder is put in a bank at 6% interest. The total annual interest from both investments is $265. Find the amount deposited at 7%.

12. A merchant wishes to mix peanuts selling for $2.00 a pound with raisins selling for $1.50 a pound. How many pounds of peanuts should she mix to obtain a 120-pound mixture to sell for $1.75 a pound?

Answers to Cumulative Review

If you got an incorrect answer, refer to the chapter and section shown in brackets for review. For example, [2-3] means chapter 2, section 3.

1. -2 [12-6]

2. $\frac{2}{3}$ [12-7]

3. $(2, 3)$ [12-8]

4. For example, $(7, 3)$ [13-6]

5. 1 [11-4]

6. a. 0 b. $+6$ c. $-\frac{3}{2}$ [9-2]

7. 26 [6-2]

8. 10 [14-1]

9. 2 [14-2]

10. 15 and 6 [15-2]

11. $\$2500$ [15-5]

12. 60 [15-6]

16

Multiplying, Dividing, and Factoring Polynomials

1. INTRODUCTION

Recall the following rules:

RULE: To multiply a polynomial by a monomial, multiply each term of the polynomial by the monomial. This is an example of the distributive property of multiplication over addition.

$$x(2x + 3) = x(2x) + x(3)$$
$$= 2x^2 + 3x$$

RULE: To divide a polynomial by a monomial, divide each term of the polynomial by the monomial.

$$\frac{y^3 - 2y^2 + y}{y} = \frac{y^3}{y} - \frac{2y^2}{y} + \frac{y}{y}$$
$$= y^2 - 2y + 1$$

2. MULTIPLYING A POLYNOMIAL BY A BINOMIAL

ILLUSTRATIVE PROBLEM 1: Multiply: $(3x + 2)(x + 5)$

Solution: We may use the distributive property of multiplication over addition if we think of $(3x + 2)$ as a number by which $(x + 5)$ is multiplied.

Thus, we may write

$$(3x + 2)(x + 5) = 3x(x + 5) + 2(x + 5)$$
$$= 3x^2 + 15x + 2x + 10$$
$$= 3x^2 + (15x + 2x) + 10$$
$$= 3x^2 + 17x + 10 \quad (answer)$$

We may sometimes wish to arrange the solution in vertical form in order to multiply. The advantage of the vertical method is that it permits us to arrange like terms in columns.

$$
\begin{array}{r}
3x \;+\; 2 \\
x \;+\; 5 \\
\hline
3x^2 \;+\; 2x \\
+\; 15x \;+\; 10 \\
\hline
3x^2 \;+\; 17x \;+\; 10
\end{array}
$$

We may summarize this process in the following manner:

RULE: The product of any two polynomials is found by multiplying each term of one by each term of the other and adding the results algebraically.

ILLUSTRATIVE PROBLEM 2: Find the product: $(3y^2 - 7y + 1)(2y - 3)$

Solution:

$$
\begin{array}{r}
3y^2 \;-\; 7y \;+\; 1 \\
2y \;-\; 3 \\
\hline
6y^3 \;-\; 14y^2 \;+\; 2y \\
-\; 9y^2 \;+\; 21y \;-\; 3 \\
\hline
6y^3 \;-\; 23y^2 \;+\; 23y \;-\; 3 \quad (answer)
\end{array}
$$

Exercises

In 1 to 40, find the products.

1. $(n + 2)(n + 1)$ 2. $(x + 3)(x + 4)$ 3. $(y - 6)(y - 2)$

4. $(p + 3)(p - 5)$ 5. $(r + 7)(r - 3)$ 6. $(t + 4)(t - 3)$

7. $(m - 8)(m - 2)$ 8. $(k + 8)(k - 9)$ 9. $(s + 7)(s - 5)$

10. $(2b + 1)(b - 3)$ 11. $(3x - 5)(x - 6)$ 12. $(5g - 2)(g - 10)$

13. $(r + 9)(r - 9)$ 14. $(t + 4)(t - 4)$ 15. $(2k + 5)(2k - 5)$

16. $(4a + 3)(a + 2)$ 17. $(3r - 2)(2r - 5)$ 18. $(4p + 1)(p + 4)$

19. $(3x + 5)(x - 3)$ **20.** $(2y + 3)(2y - 3)$ **21.** $(2b - 3)(b - 5)$

22. $(5n + 2)(4n + 3)$ **23.** $(2a - 3)(6a + 9)$ **24.** $(3a + 4b)(a - 2b)$

25. $(3x - y)(x + y)$ **26.** $(r - s)(r + 3s)$ **27.** $(c + 2d)(5c - d)$

28. $(11s - 8t)(2s + t)$ **29.** $(a - 3b)(a + 2b)$ **30.** $(7x - y)(2x + 3y)$

31. $(r + s)(r + s)$ **32.** $(m - n)(m - n)$ **33.** $(2a + b)(2a + b)$

34. $(5k + 2)(5k - 2)$ **35.** $(3g + h)(3g - h)$ **36.** $(5x + y)(5x - y)$

37. $(a^2 + 2a + 1)(a - 1)$ **38.** $(x^2 + 2x + 4)(x - 2)$

39. $(r - 2)(r^2 + 3r + 4)$ **40.** $(2y + 3)(3y^2 + y - 2)$

In 41 to 47, multiply.

41. $x - 5$
   $x + 2$
      42. $3a + 5$
   $2a - 3$
      43. $3y + 4$
   $2y - 5$
      44. $3r - 7$
   $2r + 5$

45. $x^2 + 2x - 5$
         $x - 3$
      46. $y^2 - 3y + 2$
        $y + 4$
      47. $2n^2 + n - 7$
        $3n - 1$

48. Express the area of a rectangle of length $2r + 5$ and width $r - 3$.

49. The side of a square is $2m - 5$. Express the area of the square.

50. The cost of a textbook is $(3b + 2)$ dollars. Represent the cost of $(2b - 1)$ of these books.

51. Express the area of a rectangle whose dimensions are $(2k + 5)$ and $(2k - 5)$.

52. A train travels at a rate of $(r + 30)$ miles per hour for $(3r - 2)$ hours. Represent the distance in miles that it travels.

53. A boy has $(2n - 5)$ coins, each worth $(3n + 4)$ cents. Express the total value of the coins in cents.

54. Express the area of a rectangle whose dimensions are $(2x + 3y)$ and $(x - 3y)$.

---

3. DIVIDING A POLYNOMIAL BY A BINOMIAL

It sometimes becomes necessary to divide a polynomial by a binomial. The process we use here is very similar to the long-division process in arithmetic.

In the following examples, we show the division of 864 by 36 on the left and the division of $x^2 + 8x + 15$ by $x + 3$ on the right. In the center, the steps of both divisions are described.

$36 \overline{)\,864}$ 1. Put the dividend under the division sign and the divisor to the left. $x + 3 \overline{)\,x^2 + 8x + 15}$

$\dfrac{2}{36 \overline{)\,864}}$ 2. Divide the first term of the dividend by the first term of the divisor to get the first term of the quotient. $\dfrac{x}{x + 3 \overline{)\,x^2 + 8x + 15}}$

$\begin{array}{r} 2 \\ 36 \overline{)\,864} \\ 72 \end{array}$ 3. Multiply the complete divisor by the quotient obtained and write the terms of the product under the like terms of the dividend. $\begin{array}{r} x \\ x + 3 \overline{)\,x^2 + 8x + 15} \\ x^2 + 3x \end{array}$

$\begin{array}{r} 2 \\ 36 \overline{)\,864} \\ 72 \\ \hline 144 \end{array}$ 4. Subtract this product from the dividend and bring down the next term of the dividend; consider the remainder as a new dividend and repeat the process. (Remember that algebraic subtraction requires changing the signs of the subtrahend.) $\begin{array}{r} x \\ x + 3 \overline{)\,x^2 + 8x + 15} \\ x^2 + 3x \\ \hline 5x + 15 \end{array}$

$\begin{array}{r} 24 \\ 36 \overline{)\,864} \\ 72 \\ \hline 144 \end{array}$ 5. Divide the first term of the new dividend by the first term of the divisor to get the next term of the quotient. $\begin{array}{r} x + 5 \\ x + 3 \overline{)\,x^2 + 8x + 15} \\ x^2 + 3x \\ \hline 5x + 15 \end{array}$

$\begin{array}{r} 24 \\ 36 \overline{)\,864} \\ 72 \\ \hline 144 \\ 144 \end{array}$ 6. Repeat steps 3 and 4 above with the new term in the dividend. In both cases here, the remainder is zero. $\begin{array}{r} x + 5 \\ x + 3 \overline{)\,x^2 + 8x + 15} \\ x^2 + 3x \\ \hline 5x + 15 \\ 5x + 15 \end{array}$

The quotient is 24.   The quotient is $x + 5$.

We may check in both the arithmetic example and the algebraic example by showing that the product of the quotient and the divisor is equal to the dividend.

$$\begin{array}{r} 24 \\ \times 36 \\ \hline 144 \\ 72 \\ \hline 864 \end{array} \qquad \begin{array}{r} x + 5 \\ x + 3 \\ \hline x^2 + 5x \\ + 3x + 15 \\ \hline x^2 + 8x + 15 \end{array}$$

Before starting the division process in an algebraic problem, we must remember to arrange both the dividend and divisor in either descending or ascending powers of the same variable. The process just described must be continued until the remainder is zero or is of a degree lower than the divisor.

If a particular power of the variable is missing in the dividend, we write the power with a zero coefficient to make the term act as a placeholder. Thus, a dividend of $x^2 + 2$ would be written as $x^2 + 0x + 2$. These ideas are illustrated as follows.

ILLUSTRATIVE PROBLEM 1: Divide $6x^2 - 8 - 11x$ by $3x + 2$.

Solution: Arrange the dividend in descending powers of x.

$$
\begin{array}{r}
2x \;-\; 5 \\
3x + 2\,\overline{)\,6x^2 - 11x - 8} \\
\underline{6x^2 + 4x} \\
- 15x - 8 \\
\underline{- 15x - 10} \\
2
\end{array}
$$

To show the complete result, place the remainder over the divisor and add this fraction to the quotient. Thus,

$$2x - 5 + \frac{2}{3x + 2} \quad (answer)$$

Check:
$$
\begin{array}{l}
2x \;-\; 5 \quad \text{(quotient)} \\
\underline{3x \;+\; 2} \quad \text{(divisor)} \\
6x^2 - 15x \\
\underline{ + 4x - 10} \\
6x^2 - 11x - 10 \\
\underline{ + 2} \quad \text{(remainder)} \\
6x^2 - 11x - 8 \quad \text{(dividend)}
\end{array}
$$

In checking, remember to add the remainder to the product of the quotient and the divisor in order to obtain the dividend.

ILLUSTRATIVE PROBLEM 2: Divide $2y^2 + y^3 + 5$ by $y + 3$.

Solution: Arrange the powers of y in descending order and insert $0y$ for the missing term.

$$
\begin{array}{r}
y^2 \;-\; y + 3 \\
y + 3\,\overline{)\,y^3 + 2y^2 + 0y + 5} \\
\underline{y^3 + 3y^2} \\
- y^2 + 0y \\
\underline{- y^2 - 3y} \\
3y + 5 \\
\underline{3y + 9} \\
- 4
\end{array}
$$

Answer: $y^2 - y + 3 - \dfrac{4}{y + 3}$

Exercises _____

In 1 to 26, divide and check.

1. $x^2 + 11x + 30$ by $x + 6$ **2.** $y^2 - 7y + 12$ by $y - 3$

3. $a^2 - 11a + 30$ by $a - 5$ **4.** $r^2 + 7r + 10$ by $r + 5$

5. $t^2 - 3t + 2$ by $t - 2$ **6.** $p^2 - p - 72$ by $p - 9$

7. $3x^2 - 8x + 4$ by $x - 2$ **8.** $m^2 - 8m - 9$ by $m - 9$

9. $3k^2 + 8k + 4$ by $k + 2$ **10.** $8r^2 + 16rs + 6s^2$ by $4r + 2s$

11. $3n^2 - 2 - n$ by $3n + 2$ **12.** $3t + 2 + t^2$ by $2 + t$

13. $2 + b^2 + 3b$ by $2 + b$ **14.** $2p^2 - 5 + 3p$ by $2p + 5$

15. $y^2 + 10 - 7y$ by $y - 2$ **16.** $s^2 + 6st + 8t^2$ by $s + 4t$

17. $x^2 - 16$ by $x + 4$ **18.** $t^2 - 25$ by $t - 5$

19. $9p^2 - 4q^2$ by $3p - 2q$ **20.** $a^2 - 16b^2$ by $a + 4b$

21. $2r^2 - rs - 6s^2$ by $2r + 3s$ **22.** $x^3 + x^2 + x + 6$ by $x + 2$

23. $y^3 - y + 6$ by $y + 2$ **24.** $t^3 - 4t^2 + 7t - 6$ by $t - 2$

25. $x^3 - 8$ by $x - 2$ **26.** $p^3 - p + 6$ by $p + 2$

In 27 to 32, find the quotient and the remainder.

27. $(x^2 - 5x + 7) \div (x - 2)$ **28.** $(n^2 - 4n + 1) \div (n - 5)$

29. $(3a^2 + 14a + 10) \div (a + 4)$ **30.** $(p^2 + 3p + 5) \div (p + 2)$

31. $(r^3 + r + 10) \div (r - 2)$ **32.** $(t^3 - 1) \div (t - 1)$

33. The area of a rectangle is $s^2 + s - 12$, and its length is $s + 4$. What is its width?

34. One factor of $6y^2 - 11y - 10$ is $2y - 5$. What is the other factor?

35. What is the remainder when $x^2 - 8x - 28$ is divided by $x - 10$?

_____

4. THE MEANING OF FACTORING

Recall that when two or more numbers, or integers, are multiplied to give a product, each number is a *factor* of the product. Thus, the integers 1, 2, 3, 4, 6, and 12 are all factors of 12 since 1×2, 2×6, and 3×4 all equal 12.

What are the factors of the integer 11? The only divisors of 11 are 1 and 11. Such a number is called a ***prime number.*** Other examples of prime numbers are 2, 3, 5, 7, 31, etc.

A prime number is an integer greater than 1 that has only itself and 1 as integral (whole number) factors.

Note that 2 is the only even prime number since all other even numbers have a factor of 2.

It is often necessary to express an integer as the product of prime numbers (prime factors).

For example, the number $20 = 5 \cdot 4 = 5 \cdot 2 \cdot 2$ or $5 \cdot 2^2$.

For any particular integer, there is only one set of prime factors.

ILLUSTRATIVE PROBLEM 1: Find the *greatest common factor* of 60 and 72.

Solution: Factor both numbers into a product of primes.

$$60 = 10 \cdot 6 \qquad\qquad 72 = 2 \cdot 36$$
$$= 5 \cdot 2 \cdot 3 \cdot 2 \qquad\qquad = 2 \cdot 4 \cdot 9$$
$$= 2^2 \cdot 3 \cdot 5 \qquad\qquad = 2 \cdot 2^2 \cdot 3^2$$
$$= 2^3 \cdot 3^2$$

The largest power of 2 *common* to 60 and 72 is 2^2 and the largest power of 3 *common* to both is 3.

Thus, the greatest factor common to both numbers is $2^2 \cdot 3$ or 12. (*answer*)

In a similar manner, we can find the greatest common factor of a pair of monomials.

ILLUSTRATIVE PROBLEM 2: Find the *greatest common factor* of $21c^3d^4$ and $35c^2d^5$.

Solution: The greatest common factor of both 21 and 35 is 7.

The highest power of c that is a factor of both monomials is c^2.

The highest power of d that is a factor of both monomials is d^4.

Thus, the greatest factor common to both monomials is $7c^2d^4$. (*answer*)

When we factor a monomial whose coefficients are integers, we will consider only integral factors.

Exercises _____

1. The numbers 11 and 13 are two prime numbers that differ by 2.
   a. What is the next pair of prime numbers that differ by 2?
   b. What is the next pair of prime numbers after the pair of numbers in **a** that differ by 2?

2. Write each of the following even numbers as the *sum* of two prime numbers.
   a. 8   **b.** 12   **c.** 18   **d.** 28   **e.** 32

3. Express each number below as a product of prime numbers.
   a. 15   **b.** 22   **c.** 26   **d.** 50
   e. 80   **f.** 75   **g.** 110   **h.** 132

4. Find the greatest common factor of each pair of numbers below.
   a. 15 and 20   **b.** 18 and 24   **c.** 32 and 80
   d. 120 and 140   **e.** 60 and 108   **f.** 42 and 70

5. The product of two integers is $54x^3y^5$. Find the second factor if the first factor is:
   a. $9x^2y^2$ **b.** $6x^3y$ **c.** $18x^2y^3$ **d.** $3xy^4$

6. Find the greatest common factor of each pair of monomials below.
   a. $5r, 5s$ **b.** $3x, 6$ **c.** $12y^2, 6y$
   d. $21a^2b, 28a$ **e.** $10c^2d, 15cd^2$ **f.** $6rs^2, -9r^3s$
   g. $28xy, 32x^2y^3$ **h.** $27p^2q, 36p^3q$ **i.** $48b^2c^3d, 42bc^2$
   j. a^3c, a^2c^2 **k.** $16x^2y, 24xy^2$ **l.** $\pi r^2, 2\pi r$

5. FINDING THE GREATEST COMMON MONOMIAL FACTOR OF A POLYNOMIAL

By the distributive property, $3(a + b) = 3a + 3b$.

It thus follows that $3a + 3b = 3(a + b)$.

Therefore, 3 and $(a + b)$ are *factors* of the expression $3a + 3b$. We say that we have *factored* the polynomial $3a + 3b$. We refer to the 3 as the *common monomial factor* and $(a + b)$ as the binomial factor of the polynomial $3a + 3b$.

In factoring a polynomial, we first look for the *greatest common monomial factor* of all the terms of the polynomial. The other polynomial factor is obtained by dividing each term of the original polynomial by the common monomial factor.

ILLUSTRATIVE PROBLEM 1: Factor $x^3 + 4x^2 + x$.

Solution: Note that x is the greatest common factor of all three terms of the polynomial and is, therefore, the common monomial factor. To obtain the trinomial factor, divide each term by x.

$$\frac{x^3}{x} + \frac{4x^2}{x} + \frac{x}{x} = x^2 + 4x + 1$$

Thus, $x^3 + 4x^2 + x = x(x^2 + 4x + 1)$ *(answer)*

ILLUSTRATIVE PROBLEM 2: Evaluate the expression $39(45) + 39(55)$.

Solution: Use factoring to simplify the computation.

$$39(45) + 39(55) = 39(45 + 55)$$
$$= 39(100)$$
$$= 3900 \quad (answer)$$

Exercises———————————————————————————

In 1 to 42, write each polynomial in factored form.

1. $5m + 5n$ **2.** $7r + 7s$ **3.** $9s + 9t$ **4.** $2x - 2y$

5. $6c - 6d$ **6.** $\pi r - \pi s$ **7.** $3a + 6b$ **8.** $4m - 8n$

9. $12f - 18g$ **10.** $ap + aq$ **11.** $kr - ks$ **12.** $cs - ct$

13. $14d - 21e$ **14.** $20x - 15y$ **15.** $50m - 75n$ **16.** $12n - 18$

17. $27p - 36$ **18.** $9y - 12$ **19.** $11t - 11$ **20.** $26 - 13s$

21. $15 - 10z$ **22.** $x^2 + 5x$ **23.** $3m^2 + 4m$ **24.** $8t^2 - 4t$

25. $y^2 - y$ **26.** $r^2 - 2r$ **27.** $a^2 + a$ **28.** $b^2 + 5b$

29. $cx + cy$ **30.** $x^2y + xy^2$ **31.** $c^3d^2 - c^2d^2$

32. $8mn^2 + 24m^2n$ **33.** $4pqr + 20pqr^2$ **34.** $6ab^2 - 30a^2b$

35. $2k^3 + 4k^2 + 6k$ **36.** $2r^3 - 6r^2 - 4r$ **37.** $ar + as + at$

38. $p^3 - p^2 + 5p$ **39.** $4px - 8py - 12pz$ **40.** $at^2 - at + 7a$

41. $6y^5 + 9y^3 - 12y^2$ **42.** $rz + sz - tz$

In 43 to 48, evaluate each expression by using factoring to simplify the computation.

43. $57(78) + 57(22)$ **44.** $29(67) + 29(33)$

45. $83(66) - 83(56)$ **46.** $54(31) - 34(31)$

47. $35(37) + 35(28) + 35(35)$ **48.** $\frac{1}{2}(78) - \frac{1}{2}(58)$

6. MULTIPLYING THE SUM AND DIFFERENCE OF TWO TERMS

Observe the results of the following multiplications.

$$
\begin{array}{ll}
r + s \\
\underline{r - s} \\
r^2 + rs \\
\underline{- rs - s^2} \\
r^2 - s^2
\end{array}
\qquad
\begin{array}{ll}
x + 3 \\
\underline{x - 3} \\
x^2 + 3x \\
\underline{- 3x - 9} \\
x^2 - 9
\end{array}
\qquad
\begin{array}{ll}
3m + 4n \\
\underline{3m - 4n} \\
9m^2 + 12mn \\
\underline{- 12mn - 16n^2} \\
9m^2 - 16n^2
\end{array}
$$

Note that in each case, we are multiplying the sum of two monomials (such as $r + s$) by the difference of these same two monomials (such as $r - s$). The product in each case is the difference between the square of the first monomial and the square of the second monomial in that order.

In general,

$$(x + y)(x - y) = x^2 - y^2$$

ILLUSTRATIVE PROBLEM 1: Find the product: $(a + 6)(a - 6)$

Solution:

$$
\begin{aligned}
(a + 6)(a - 6) &= (a)^2 - (6)^2 \\
&= a^2 - 36 \quad (answer)
\end{aligned}
$$

ILLUSTRATIVE PROBLEM 2: Find the product: $(5x - 7y)(5x + 7y)$

Solution:

$$
\begin{aligned}
(5x - 7y)(5x + 7y) &= (5x)^2 - (7y)^2 \\
&= 25x^2 - 49y^2 \quad (answer)
\end{aligned}
$$

Exercises

In 1 to 22, find the product.

1. $(a + 2)(a - 2)$
2. $(x + 6)(x - 6)$
3. $(p + 1)(p - 1)$
4. $(k + 9)(k - 9)$
5. $(7 - r)(7 + r)$
6. $(12 - t)(12 + t)$
7. $(m + n)(m - n)$
8. $(s + t)(s - t)$
9. $(2b - c)(2b + c)$
10. $(3n + 5)(3n - 5)$
11. $(8x - 7)(8x + 7)$
12. $(6p + 11)(6p - 11)$

13. $(5x + 8y)(5x - 8y)$ **14.** $(2c - 9d)(2c + 9d)$

15. $(rs + 12)(rs - 12)$ **16.** $(ef - 3)(ef + 3)$

17. $(p^2 - 10)(p^2 + 10)$ **18.** $(y^2 + 2)(y^2 - 2)$

19. $(m + .3)(m - .3)$ **20.** $(n + \frac{1}{2})(n - \frac{1}{2})$

21. $(30 - 2)(30 + 2)$ **22.** $(40 - 1)(40 + 1)$

---

7. FACTORING THE DIFFERENCE OF TWO SQUARES

In the previous section, we learned that $(x + y)(x - y) = x^2 - y^2$. The reverse situation tells us that the factors of $x^2 - y^2$ are $(x + y)$ and $(x - y)$.

Thus, the factors of the difference of two *squares* of monomials are the *sum* and the *difference* of the monomials. In general,

$$x^2 - y^2 = (x + y)(x - y)$$

ILLUSTRATIVE PROBLEM 1: Factor $t^2 - 25$.

Solution:

$$t^2 - 25 = (t)^2 - (5)^2$$
$$= (t + 5)(t - 5) \quad (answer)$$

ILLUSTRATIVE PROBLEM 2: Factor $9a^2 - 16b^2$.

Solution:

$$9a^2 - 16b^2 = (3a)^2 - (4b)^2$$
$$= (3a + 4b)(3a - 4b) \quad (answer)$$

Exercises

In 1 to 30, factor the given binomials.

1. $r^2 - s^2$ **2.** $y^2 - 4$ **3.** $b^2 - 9$ **4.** $x^2 - 36$

5. $a^2 - 144$ **6.** $m^2 - 64$ **7.** $k^2 - 81$ **8.** $p^2 - 49$

9. $s^2 - 100$ **10.** $x^2 - 16y^2$ **11.** $4m^2 - n^2$ **12.** $9t^2 - 121$

13. $1 - 4b^2$ **14.** $a^2b^2 - 1$ **15.** $9a^2 - 16$ **16.** $s^2 - 400$

17. $25p^2 - 4q^2$ **18.** $x^2 - \frac{1}{4}$ **19.** $x^4 - y^2$ **20.** $64a^2 - 25b^2$

21. $r^2 - .04$ **22.** $\frac{1}{9}p^2 - q^2$ **23.** $9k^2 - 49n^2$ **24.** $.09c^2 - d^2$

25. $4t^2 - .01$ **26.** $r^4 - 9s^4$ **27.** $x^2y^2 - 9z^2$ **28.** $36a^2 - .25$

29. $x^2 - \frac{1}{25}$ **30.** $s^2 - \frac{4}{25}t^2$

---

8. MULTIPLYING TWO BINOMIALS HORIZONTALLY

Observe the multiplication below.

$$\begin{array}{r}
x + 2 \\
2x + 3 \\
\hline
2x^2 + 4x \\
+ 3x + 6 \\
\hline
2x^2 + 7x + 6
\end{array}$$

Note that the *first* term of the product, $2x^2$, is obtained by multiplying the two *first* terms of the binomials, x and $2x$.

The *last* term of the product, $+6$, is obtained by multiplying the two *last* terms of the binomials, $+2$ and $+3$.

To obtain the *middle* term of the product, let us write the multipliers horizontally.

inner product

$$(x + 2)(2x + 3)$$

outer product

The product of $+2$ and $2x$ is called the *inner product* and equals $+4x$.

The product of x and $+3$ is called the *outer product* and equals $+3x$.

The middle term of the product in the original problem is obtained by adding algebraically the inner and outer products: $4x + 3x = 7x$.

RULE: To find the product of two binomials:

1. Multiply the first terms of the two binomials.

2. Find the inner product and outer product of the two binomials and add them algebraically to get the middle term of the product.

3. Multiply the last terms of the two binomials.

ILLUSTRATIVE PROBLEM 1: Multiply: $(y - 3)(y - 5)$

Solution: Find the first term of the product by multiplying the first terms of the two binomials:

$$y \cdot y = y^2$$

Find the middle term of the product by combining the inner and outer products of the two binomials:

$$\overset{-3y}{\overbrace{(y - 3)(y - 5)}_{-5y}} = -3y - 5y = -8y$$

Find the last term of the product by multiplying the last terms of the two binomials:

$$(-3)(-5) = +15$$

Thus, $(y - 3)(y - 5) = y^2 - 8y + 15$ *(answer)*

ILLUSTRATIVE PROBLEM 2: Multiply: $(2x - 7)(3x + 4)$

Solution:

$$(2x)(3x) = 6x^2 \quad \text{(product of two first terms)}$$

$$\overset{-21x \quad \text{(inner product)}}{\overbrace{(2x - 7)(3x + 4)}_{+8x \quad \text{(outer product)}}} = -21x + 8x = -13x \quad \text{(middle term)}$$

$$(-7)(+4) = -28 \quad \text{(product of two last terms)}$$

Thus, $(2x - 7)(3x + 4) = 6x^2 - 13x - 28$ *(answer)*

Exercises

In 1 to 28, determine the following products.

1. $(x + 5)(x + 2)$
2. $(b + 7)(b + 3)$
3. $(c + 4)(c + 4)$
4. $(y - 2)(y + 1)$
5. $(r - 6)(r - 2)$
6. $(s + 5)(s - 3)$
7. $(n + 7)(n - 6)$
8. $(t - 7)(t - 7)$
9. $(2 + x)(9 + x)$
10. $(3 + d)(11 + d)$
11. $(1 - m)(3 - m)$
12. $(4 - n)(12 - n)$

13. $(c + 15)(c - 6)$ **14.** $(k + 20)(k - 3)$

15. $(p - 1)(p - 1)$ **16.** $(1 - r)(1 - r)$

17. $(9 - y)(5 - y)$ **18.** $(11 - v)(7 + v)$

19. $(2n - 3)(n + 4)$ **20.** $(5x - 3)(x - 6)$

21. $(3p - 5)(4p - 7)$ **22.** $(m - 4)(2m - 7)$

23. $(3z - 2)(3z - 2)$ **24.** $(4y - 5)(4y - 5)$

25. $(2a + 3b)(3a + 4b)$ **26.** $(5x - 6y)(2x - y)$

27. $(4m - 3n)(2m - n)$ **28.** $(7p - 2q)(3p + 2q)$

---

9. FACTORING TRINOMIALS OF THE FORM $x^2 + bx + c$

We have already learned to multiply two binomials by inspection. For example,

$$(x + 3)(x + 4) = x^2 + 7x + 12$$

Thus, the factors of $x^2 + 7x + 12$ are $(x + 3)$ and $(x + 4)$.

In factoring trinomials of this type, we use a system of trial and error and keep checking the factors by inspection to see that they are correct.

ILLUSTRATIVE PROBLEM 1: Factor $x^2 + 7x + 10$.

Solution: We know that the factors are two binomials, the first term of each being x. We write

$$(x + ?)(x + ?)$$

Since the 10 in the trinomial was obtained by multiplying the last terms of the binomial factors, these last terms must be factors of 10. Let us consider all the possible factors of 10. They are 2 and 5, 10 and 1, -2 and -5, and -10 and -1. Since the middle term is positive, it is unlikely that the negative pairs will check out. Let us try as factors 10 and 1:

$$
\begin{array}{c}
\overset{\displaystyle 10x}{\overbrace{\qquad\qquad}} \\[2pt]
(x + 10)(x + 1) \\[2pt]
\underbrace{\qquad\qquad\qquad}_{\displaystyle x}
\end{array}
$$

Since the sum of inner and outer products is $11x$, we reject this choice. Now we try 5 and 2:

$$5x$$

$$\overline{}$$

$$(x + 5)(x + 2)$$

$$\underline{}$$

$$2x$$

The sum of the inner and outer products is now $+7x$, which checks with the original trinomial. Thus,

$$x^2 + 7x + 10 = (x + 5)(x + 2) \quad (answer)$$

ILLUSTRATIVE PROBLEM 2: Factor $y^2 + 10y + 21$.

Solution: $(y + ?)(y + ?)$

The possible factors of 21 are 7 and 3, 21 and 1, and the negatives of these factors. Since the middle term is positive, we may again reject the negative pairs. Let us try 21 and 1.

$$21y$$

$$\overline{}$$

$$(y + 21)(y + 1)$$

$$\underline{}$$

$$y$$

Since the sum of the inner and outer products is $22y$, we reject this choice. Now we try 7 and 3.

$$7y$$

$$\overline{}$$

$$(y + 7)(y + 3)$$

$$\underline{}$$

$$3y$$

The middle term is now $10y$, which checks. Thus,

$$y^2 + 10y + 21 = (y + 7)(y + 3) \quad (answer)$$

ILLUSTRATIVE PROBLEM 3: Factor $m^2 - 2m - 15$.

Solution: The first terms of the factors are each m. Since the middle term is now negative, however, we must allow for a possible negative second term.

$$m^2 - 2m - 15 = (m \quad ?)(m \quad ?)$$

Since the product of the two last terms of the binomial factors must be -15, let us consider all possible pairs of factors of -15: 5 and -3, 3 and -5, 15 and -1, and -15 and 1. If we try these various pairs, we find that only -5 and 3 give us the desired result.

$$-5m$$

$$(m - 5)(m + 3) \; (-5m) + (+3m) = -2m \quad \text{(middle term)}$$

$$+3m$$

Thus,

$$m^2 - 2m - 15 = (m - 5)(m + 3) \quad (answer)$$

In problems such as those above, where the coefficient of the x^2 (or y^2 or m^2) term is 1, we may reduce the number of trials by means of the following observation. In Illustrative Problem 3, the product of the two numerical factors was -15, and their sum was -2, the coefficient of the middle term. Likewise, in Illustrative Problem 2, the product of the two numerical factors was 21, and their sum was 10, the coefficient of the middle term.

RULE: A trinomial of the form $ax^2 + bx + c$, where $a = 1$, may be factored as the product of two binomials if:

1. There are two numbers such that their algebraic sum is the coefficient of the middle term.
2. The product of these same two numbers is equal to the last term of the trinomial.

Exercises

In 1 to 38, factor each trinomial.

1. $x^2 + 4x + 3$
2. $y^2 + 3y + 2$
3. $p^2 + 8p + 7$
4. $t^2 + 10t + 9$
5. $n^2 + 4n + 4$
6. $r^2 + 5r + 6$
7. $c^2 + 6c + 9$
8. $m^2 + 10m + 24$
9. $a^2 + 8a + 15$
10. $b^2 + 9b + 20$
11. $d^2 + 10d + 16$
12. $k^2 + 9k + 14$
13. $t^2 + 13t + 12$
14. $n^2 + 11n + 28$
15. $24 + 11x + x^2$
16. $1 + 2b + b^2$
17. $r^2 - 5r + 4$
18. $p^2 - 5p - 6$
19. $m^2 + 3m - 4$
20. $x^2 + 5x - 6$
21. $t^2 - 6t + 8$
22. $c^2 - c - 2$
23. $y^2 + 5y - 24$
24. $n^2 + 8n - 9$

25. $k^2 - 8k - 9$ **26.** $r^2 - 12r - 45$ **27.** $m^2 - 2m - 24$

28. $s^2 - 15s + 56$ **29.** $d^2 + 8d - 48$ **30.** $x^2 + 2x - 48$

31. $n^2 - 4n - 60$ **32.** $b^2 + 14b + 24$ **33.** $y^2 + 15y + 50$

34. $t^2 + 16t - 80$ **35.** $a^2 - 29a + 100$ **36.** $16 - 10k + k^2$

37. $b^2 - 12b + 20$ **38.** $c^2 - c - 30$

39. The area of a rectangle is $m^2 - 2m - 15$. What are the dimensions of the rectangle if they are factors of the area?

40. The area of a square is $n^2 - 10n + 25$. Represent the side of the square in terms of n.

---

10. FACTORING TRINOMIALS OF THE FORM $ax^2 + bx + c, a \neq 1$

In the previous section, we learned to factor trinomials of the form $ax^2 + bx + c$, where $a = 1$. Let us now consider the case where a is greater than 1.

ILLUSTRATIVE PROBLEM 1: Find the factors of $2x^2 + 7x + 3$.

Solution: The only possible first-degree factors of $2x^2$ are $2x$ and x. Thus, the factors are of the form:

$$(2x + ?)(x + ?)$$

The factors of 3 are $+3$ and $+1$. We need not consider the negatives of these, since all the coefficients are positive. Try 3 and 1:

$$+3x$$
$$\sqcap$$
$$(2x + 3)(x + 1)$$
$$\llcorner_____\lrcorner$$
$$+2x$$

The sum of the inner and outer products is $2x + 3x = 5x$. This is not the desired middle term. Now try 1 and 3:

$$+x$$
$$\sqcap$$
$$(2x + 1)(x + 3)$$
$$\llcorner_____\lrcorner$$
$$+6x$$

Note that $(+6x) + (+x) = +7x$, the desired middle term. Thus,

$$2x^2 + 7x + 3 = (2x + 1)(x + 3) \quad (answer)$$

In some cases, there may be more choices for the first or last terms, so that many more trials will be necessary.

ILLUSTRATIVE PROBLEM 2: Factor $3x^2 - 7x + 4$.

Solution: The first terms of the factors are $3x$ and x.

$$(3x \ ?)(x \ ?)$$

The possible pairs of factors of 4 are:

$$4 \text{ and } 1 \qquad -4 \text{ and } -1$$

$$2 \text{ and } 2 \qquad -2 \text{ and } -2$$

Since the middle term is negative, we must consider the pairs of negative factors. Try -2 and -2:

$$\overset{\displaystyle -2x}{\overbrace{(3x - 2)(x - 2)}}$$
$$\underbrace{}_{-6x}$$

The middle term is $-6x - 2x = -8x$, which is not the desired result. Now try -4 and -1:

$$\overset{\displaystyle -4x}{\overbrace{(3x - 4)(x - 1)}}$$
$$\underbrace{}_{-3x}$$

Now the middle term is $-4x - 3x = -7x$, which is the desired result. Thus,

$$3x^2 - 7x + 4 = (3x - 4)(x - 1) \quad (answer)$$

Exercises

In 1 to 16, factor the given trinomials.

1. $3t^2 + 7t + 2$
2. $2x^2 + 7x + 6$
3. $3y^2 + 5y + 2$
4. $2m^2 + 11m + 15$
5. $10n^2 + 9n + 2$
6. $3p^2 + 10p + 3$
7. $3n^2 - 11n + 6$
8. $5t^2 - 16t + 3$
9. $3k^2 - 11k + 10$

10. $7a^2 - 9a + 2$ **11.** $2b^2 - b - 6$ **12.** $3c^2 + 5c - 12$

13. $4r^2 + 3r - 10$ **14.** $3y^2 - y - 14$ **15.** $3a^2 + ab - 2b^2$

16. $5r^2 + 6rs - 8s^2$

---

11. FACTORING COMPLETELY

Consider the factors of $2x^2 - 8$. Note that 2 is the greatest common monomial factor. Thus:

$$2x^2 - 8 = 2(x^2 - 4)$$

Notice that $x^2 - 4$ is the difference of two squares and may be further factored.

$$2x^2 - 8 = 2(x^2 - 4) = 2(x - 2)(x + 2)$$

Since each of the binomial factors cannot be factored further, we say that the factorization is now complete.

Use the following steps in factoring a polynomial completely:

RULE: To factor a polynomial completely:

1. Look for the greatest common monomial factor first. Express the polynomial in factored form and examine each factor for further factoring.

2. If one of these factors is a binomial, see if it is the difference of two squares. If it is, factor it.

3. If one of these factors is a trinomial, see if it can be factored further. If so, factor it.

4. Check to see that none of the factors can be factored further.

ILLUSTRATIVE PROBLEM 1: Factor completely: $2p^2 + 6p - 8$

Solution: $2p^2 + 6p - 8 = 2(p^2 + 3p - 4)$

$$= 2(p + 4)(p - 1) \quad (answer)$$

ILLUSTRATIVE PROBLEM 2: Factor completely: $a^4 - 81$

Solution: $a^4 - 81 = (a^2 + 9)(a^2 - 9)$

$$= (a^2 + 9)(a + 3)(a - 3) \quad (answer)$$

Exercises———————————————————————————————

In 1 to 26, factor completely.

1. $5a^2 - 5b^2$ 2. $2x^2 - 18$ 3. $2y^2 - 16y + 14$

4. $5c^2 - 35c - 40$ 5. $p^3 - 4p$ 6. $2x^2 - 8x + 8$

7. $p^3 - pq^2$ 8. $\pi r^2 - \pi s^2$ 9. $a^4 - b^4$

10. $2x^4y - 18y^3$ 11. $ay^2 - 4ay - 45a$ 12. $2r^3s^2 - 32rs^4$

13. $2t^2 - 6t - 108$ 14. $6r^2s^3 + 12rs^2$ 15. $4a^5b^3 - 4a^3b^3$

16. $a^3 - 36a$ 17. $5x^2 - 45$ 18. $x^3 - a^2x$

19. $p^3 + 7p^2 + 2p$ 20. $y + 2yr + yr^2$ 21. $n^4 - 1$

22. $3t^2 - 12$ 23. $2a^2y^2 - 3a^2y + a^2$ 24. $8x^2 - 18y^2$

25. $t^4 - t^2$ 26. $m^5 - 9m^3$

Chapter Review Exercises

In 1 to 6, multiply.

1. $(4a + 3)(a + 2)$ 2. $(2r - 5)(3r - 2)$ 3. $(4b + 1)(b + 4)$

4. $(3x + 5)(x - 3)$ 5. $(2y + 3)(2y - 3)$ 6. $(2n - 3)(n - 5)$

7. The length and width of a rectangle are represented by $7x - 8$ and $3x + 5$. What polynomial represents the area?

In 8 to 13, divide.

8. $n^2 - 5n + 6$ by $n - 2$ 9. $2x^2 - 7x - 15$ by $x - 5$

10. $y^2 + 2y - 15$ by $y + 5$ 11. $56x^2 - 11x - 15$ by $7x + 3$

12. $40x^2 + 11xy - 63y^2$ by $8x - 9y$

13. $45x^2 + 69xy - 10y^2$ by $3x + 5y$

14. The area of a square is represented by $9s^2 + 12s + 4$. Represent its side in terms of s.

In 15 to 32, factor completely.

15. $a^2 - 3a - 54$ 16. $6r^2s^2 + 12rs^2$ 17. $9x^2 - 1$

18. $4a^5b^3 + 8a^3b^3$ 19. $a^2 - 36$ 20. $5x^3y^2 + 15x^2y^2$

21. $c^2 - 25$ 22. $4rx + 12r$ 23. $4a^3b + 12a^3$

24. $y^2 + 3y - 10$ **25.** $3x^2 + 2x - 5$ **26.** $4p^2 - 25$

27. $5bx^2 + 20bx$ **28.** $2n^2 + 11n + 15$

29. $3x^2 - 75$ **30.** $3ab^3 - 15ab^2 + 18ab$

31. $3t^2 + 15t - 42$ **32.** $16p^2 - p^2q^4$

CHECK YOUR SKILLS

1. Solve for x: $3x + 2 = -1$

2. Factor: $a^2 - 5a - 24$

3. On a map, if one inch represents 60 miles, how many miles are represented by $2\frac{1}{2}$ inches?

4. If the point $(2, k)$ is located on the graph of the equation $x + 2y = 10$, find the value of k.

5. Solve the following system of equations for x:
$$x + 9y = -6$$
$$2x - 9y = 15$$

6. Solve for m: $0.3m - 2 = 4$

7. What is the numerical value of $3x^2 - 2y$ when $x = -1$ and $y = 0$?

8. Factor: $2x^2 + x$

9. If 75% of x is 24, find the value of x.

10. Solve for y: $3(y - 2) = 12$

17

Algebraic Fractions

1. MEANING OF AN ALGEBRAIC FRACTION

Recall that arithmetic fractions are used in several different ways. The fraction $\frac{4}{5}$ may represent the division of 4 by 5, or the ratio of 4 to 5 (4:5), or 4 parts of a whole that is divided into 5 equal parts.

An **algebraic fraction** is a fraction that contains algebraic expressions in the numerator or denominator or both. Examples are $\frac{y}{5}, \frac{6}{b}, \frac{r}{s}$, $\frac{2x + 3}{x - 2}$, and $\frac{a^2 + 3a + 1}{2a + 5}$. Since division by zero has no meaning, the denominator of a fraction cannot be zero. Thus, $\frac{3}{x - 4}$ is not defined when $x = 4$, since the denominator would then be zero.

Algebraic fractions may be used in the same way as arithmetic fractions. Thus, the fraction $\frac{r}{s}$ may indicate the division of r by s, or the ratio of r to s ($r:s$), or r parts of a whole that is divided into s equal parts.

Exercises

In 1 to 16, express each phrase as a fraction.

1. p divided by 6

2. the ratio of p to q

3. $q \div r$

4. $(x - 5) \div 3$

5. the ratio of the circumference, c, of a circle, to its diameter, d

6. $r \div (r - 7)$

7. $(2t - 8) \div 5t$

8. the cost of 1 shirt if n shirts cost d dollars

9. the length of a rectangle if its area is A and its width is W

10. $(x^2 - y^2) \div (x + y)$

11. $(p^2 + 2p + 1) \div (p + 1)$

12. the rate of a car if it travels m miles in h hours

13. the ratio of the base, b, of a rectangle to its height, h

14. the cost of 1 melon if n melons cost c cents   **15.** $3m \div (n - 2)$

16. the ratio of boys to girls in a class made up of b boys and g girls

In 17 to 28, give the value(s) of the variable for which the fraction has no meaning.

17. $\dfrac{8}{p}$ **18.** $\dfrac{15}{2a}$ **19.** $\dfrac{k + 3}{k - 3}$ **20.** $\dfrac{n}{2 - n}$

21. $\dfrac{p + 5}{p - 7}$ **22.** $\dfrac{r + 5}{r^2}$ **23.** $\dfrac{7c}{c - 9}$ **24.** $\dfrac{14}{p - 8}$

25. $\dfrac{2x + 5}{3x - 1}$ **26.** $\dfrac{m + 5}{m^2 - 9}$ **27.** $\dfrac{2t + 5}{3t - 6}$ **28.** $\dfrac{a - 6}{a^2 - 25}$

2. REDUCING FRACTIONS TO LOWEST TERMS

Recall the following rules:

RULE: A fraction is said to be in *lowest terms* when its numerator and denominator have no common factor other than 1.

Thus, the fractions $\dfrac{3}{4}$ and $\dfrac{a}{b}$ are in lowest terms.

RULE: The numerator and denominator of a fraction may both be *divided* by the same number without changing the value of the fraction.

Thus, $\dfrac{12}{16} = \dfrac{12 \div 2}{16 \div 2} = \dfrac{6}{8}$.

RULE: To reduce a fraction to lowest terms, divide the numerator and denominator by their greatest common factor.

Thus, $\dfrac{12}{16} = \dfrac{12 \div 4}{16 \div 4} = \dfrac{3}{4}$. Here 4 is the greatest common factor of 12 and 16.

If the terms of a fraction are monomials, common factors are relatively easy to obtain.

ILLUSTRATIVE PROBLEM 1: Reduce $\dfrac{6a^2b}{9ab}$ to lowest terms.

Solution: The greatest common factor of $6a^2b$ and $9ab$ is $3ab$. Thus,

$$\frac{6a^2b}{9ab} = \frac{6a^2b \div 3ab}{9ab \div 3ab}$$

We usually show this division by cancelling numbers into numbers and like letter terms into each other as follows:

$$\frac{\overset{2}{\cancel{6}} \cdot \overset{a}{\cancel{a^2}} \cdot \overset{1}{\cancel{b}}}{\underset{3}{\cancel{9}} \cdot \underset{1}{\cancel{a}} \cdot \underset{1}{\cancel{b}}} = \frac{2a}{3} \quad (answer)$$

If the terms of a fraction are polynomials, we must factor them first to see any common factors.

ILLUSTRATIVE PROBLEM 2: Reduce to lowest terms: $\dfrac{3x + 3y}{x^2 - y^2}$

Solution:

$$\frac{3x + 3y}{x^2 - y^2} = \frac{3\overset{1}{\cancel{(x + y)}}}{(x - y)\underset{1}{\cancel{(x + y)}}} = \frac{3}{x - y} \quad (answer)$$

It is important to remember that, in the process of cancellation, we are *dividing* both the entire numerator and the entire denominator of the fraction by a common factor.

The common error shown below should be carefully avoided.

$$(\text{WRONG}) \quad \frac{\cancel{a} + x}{\cancel{a} + y} = \frac{x}{y}$$

ILLUSTRATIVE PROBLEM 3: Reduce to lowest terms: $\dfrac{(r - 1)^2}{r^2 - 9r + 8}$

Solution: Factor the numerator and the denominator and cancel common factors.

$$\frac{(r - 1)^2}{r^2 - 9r + 8} = \frac{(r - 1)\overset{1}{\cancel{(r - 1)}}}{(r - 8)\underset{1}{\cancel{(r - 1)}}} = \frac{r - 1}{r - 8} \quad (answer)$$

Exercises

In 1 to 73, reduce the fraction to lowest terms.

1. $\dfrac{12}{16}$ 2. $\dfrac{20}{30}$ 3. $\dfrac{36}{45}$ 4. $\dfrac{15}{21}$

5. $\dfrac{7p}{7q}$ 6. $\dfrac{ax}{ay}$ 7. $\dfrac{18t}{24t}$ 8. $\dfrac{5n^2}{9n^2}$

9. $\dfrac{by}{cy}$ 10. $\dfrac{p^2q}{p^2r}$ 11. $\dfrac{12cx^2}{15dx}$ 12. $\dfrac{rs^2}{rt^2}$

13. $\dfrac{14mn}{20mn}$ 14. $\dfrac{15prt}{20prt}$ 15. $\dfrac{cd}{c^2}$ 16. $\dfrac{2ab^2}{3a}$

17. $\dfrac{15x}{10x^2}$ 18. $\dfrac{30s^3t}{20st}$ 19. $\dfrac{10rs^4}{14rs}$ 20. $\dfrac{-14r^5}{21r^4}$

21. $\dfrac{-r^2s}{5s}$ 22. $\dfrac{-12a^3b^2c}{18ab^2c}$ 23. $\dfrac{14rs^2}{21r^2s}$ 24. $\dfrac{-40d^2e}{-5de^2}$

25. $\dfrac{-35x^2y^3}{45x^3y^2}$ 26. $\dfrac{-6pq}{9p^2q^3}$ 27. $\dfrac{40s^4t^3}{-32s^2t^4}$ 28. $\dfrac{21ab}{28a^3b^2}$

29. $\dfrac{3(a-2)}{5(a-2)}$ 30. $\dfrac{8(p+q)}{12(p+q)}$ 31. $\dfrac{24(n-1)}{40(n-1)}$ 32. $\dfrac{9a(b-5)}{12a(b-5)}$

33. $\dfrac{r^3(s-t)}{2r(s-t)}$ 34. $\dfrac{18a(2x-y)}{12a^2(2x-y)}$ 35. $\dfrac{6a+6b}{a+b}$ 36. $\dfrac{9x-9y}{6x-6y}$

37. $\dfrac{5c+5d}{3c+3d}$ 38. $\dfrac{r-2}{r^2-4}$ 39. $\dfrac{7b-14}{3b-6}$ 40. $\dfrac{12(x-7)^2}{15(x-7)}$

41. $\dfrac{n^2-25}{2n+10}$ 42. $\dfrac{c^2-49}{3c-21}$ 43. $\dfrac{5k+30}{k^2-36}$ 44. $\dfrac{2p-6}{p^2-9}$

45. $\dfrac{(a-b)^2}{a^2-b^2}$ 46. $\dfrac{9c^2-9d^2}{12(c+d)^2}$ 47. $\dfrac{3p+15}{p^2-25}$ 48. $\dfrac{12}{6x+12}$

49. $\dfrac{3x-21}{x^2-49}$ 50. $\dfrac{5t}{t^2-t}$ 51. $\dfrac{7k+14k^2}{14k}$ 52. $\dfrac{x^3-5x^2}{3x^2}$

53. $\dfrac{a-6}{a^2-7a+6}$ 54. $\dfrac{n^2+3n}{n^2+4n+3}$ 55. $\dfrac{5p-5}{p^2-2p+1}$

56. $\dfrac{y^2-25}{y^2-8y+15}$ 57. $\dfrac{a^2-5a-24}{a^2-64}$ 58. $\dfrac{a^2-b^2}{a^2-2ab+b^2}$

59. $\dfrac{4x^2-9}{6x^2-9x}$ 60. $\dfrac{4x+12}{x^2-4x-21}$ 61. $\dfrac{y^2+4y}{y+4}$

62. $\dfrac{5x - 10}{2x^2 - 3x - 2}$ **63.** $\dfrac{n^2 - 4}{n^2 + 4x + 4}$ **64.** $\dfrac{r^2 - 6rs + 9s^2}{r^2 - 9s^2}$

65. $\dfrac{y^2 - 14y + 45}{y^2 - 8y - 9}$ **66.** $\dfrac{n^2 - 2n + 1}{n^2 + 3n - 4}$ **67.** $\dfrac{p^2 - 7p}{p^2 - 6p - 7}$

68. $\dfrac{c^2 + 2c - 8}{c^2 + c - 6}$ **69.** $\dfrac{t^2 + 7t + 12}{t^2 + 4t}$ **70.** $\dfrac{x^2 - 2x - 8}{(x + 2)^2}$

71. $\dfrac{n^2 - 5n - 14}{n^2 - 9n + 14}$ **72.** $\dfrac{(2y + 5)^2}{2y^2 - 9y - 35}$ **73.** $\dfrac{14n^3 - 21n^2 + 7n}{7n^3 - 7n^2}$

---

3. MULTIPLYING FRACTIONS

Recall the following:

To multiply one fraction by another, first multiply the numerators together to form the numerator of the product. Then multiply the denominators together to form the denominator of the product. Finally, reduce the product to lowest terms. Thus,

$$\frac{3}{7} \times \frac{5}{6} = \frac{3 \times 5}{7 \times 6} = \frac{15}{42} = \frac{15 \div 3}{42 \div 3} = \frac{5}{14}$$

We usually simplify the computation by dividing any number in a numerator and any number in a denominator by their greatest common factor. Thus, using 3 as a divisor,

$$\frac{\overset{1}{\cancel{3}}}{7} \times \frac{5}{\underset{2}{\cancel{6}}} = \frac{5}{14}$$

Multiplication of algebraic fractions is similar. Remember to factor polynomials in numerators and denominators before performing cancellations.

ILLUSTRATIVE PROBLEM 1: Multiply: $\dfrac{5x^2}{6} \cdot \dfrac{3}{2x}$

Solution: Dividing by 3 and by x,

$$\frac{5\overset{x}{\cancel{x^2}}}{\underset{2}{\cancel{6}}} \cdot \frac{\overset{1}{\cancel{3}}}{2\cancel{x}} = \frac{5x}{4} \quad (answer)$$

ILLUSTRATIVE PROBLEM 2: Multiply: $\dfrac{3y^2}{y^2 - 9} \cdot \dfrac{y + 3}{y}$

Solution: Dividing by $(y + 3)$ and by y,

$$\frac{3y^2}{y^2 - 9} \cdot \frac{y + 3}{y} = \frac{3\overset{y}{\cancel{x}^2}}{\cancel{(y + 3)}(y - 3)} \cdot \frac{\overset{1}{\cancel{y + 3}}}{\underset{1}{\cancel{x}}} = \frac{3y}{y - 3} \quad (answer)$$

ILLUSTRATIVE PROBLEM 3: Multiply: $\dfrac{a^2 - 25}{a^2 + a - 20} \cdot \dfrac{a - 4}{a^2 - 5a}$

Solution: Factor all polynomials in the fractions and divide by common factors.

$$\frac{\cancel{(a - 5)}\cancel{(a + 5)}}{\cancel{(a - 4)}\cancel{(a + 5)}} \cdot \frac{\cancel{(a - 4)}}{a\cancel{(a - 5)}} = \frac{1}{a} \quad (answer)$$

Exercises

In 1 to 43, multiply and find each product in simplest form.

1. $\dfrac{4}{5} \cdot \dfrac{3}{7}$ **2.** $\dfrac{14}{9} \cdot \dfrac{18}{35}$ **3.** $\dfrac{3}{4} \cdot 28$ **4.** $\dfrac{7}{p} \cdot \dfrac{p^2}{2}$

5. $\dfrac{8}{x} \cdot \dfrac{xy}{6}$ **6.** $\dfrac{32}{m} \cdot \dfrac{m^2}{24}$ **7.** $\dfrac{6t}{a} \cdot \dfrac{b}{3t}$ **8.** $\dfrac{10r^3}{x} \cdot \dfrac{3x^2}{5r}$

9. $\dfrac{p}{cd} \cdot \dfrac{cd^2}{p^2}$ **10.** $\dfrac{x^3}{3y} \cdot \dfrac{6y}{x^2}$ **11.** $\dfrac{pq}{r} \cdot \dfrac{r^2}{q}$ **12.** $r^2 s \cdot \dfrac{7}{rs^2}$

13. $\dfrac{n^3}{15} \cdot \dfrac{10}{3n}$ **14.** $\dfrac{12a^2}{b^2} \cdot \dfrac{ab}{15a^3}$ **15.** $8k^2 \cdot \dfrac{3k}{4}$

16. $\dfrac{2x - 10}{21} \cdot \dfrac{14}{x^2 - 25}$ **17.** $\dfrac{5p + 5q}{20pq} \cdot \dfrac{30p^2q^2}{p^2 - q^2}$ **18.** $\dfrac{3m^2}{m^2 - 9} \cdot \dfrac{m + 3}{m}$

19. $\dfrac{x^2 - 1}{x} \cdot \dfrac{4x^2}{x + 1}$ **20.** $\dfrac{n^2 - 9}{2n - 6} \cdot \dfrac{1}{n + 3}$ **21.** $\dfrac{3t^2}{t^2 - 4} \cdot \dfrac{t + 2}{6t}$

22. $\dfrac{3b}{a + b} \cdot 5a^2$ **23.** $\dfrac{a + b}{a - b} \cdot \dfrac{a}{a + b}$ **24.** $\dfrac{(x + 5)^2}{25} \cdot \dfrac{5}{x + 5}$

25. $\dfrac{12p^2}{p^2 - 4} \cdot \dfrac{p - 2}{8p}$ **26.** $\dfrac{d + 4}{c - 1} \cdot \dfrac{c^3 - c}{d^2 - 16}$ **27.** $\dfrac{x}{3} \cdot \dfrac{6}{x^2 + x}$

28. $\dfrac{r^2 - 16}{8} \cdot \dfrac{10}{5r + 20}$

29. $\dfrac{x^2 - 7x + 12}{x^2 - 4} \cdot \dfrac{2x + 4}{x - 3}$

30. $\dfrac{y^2 - 9}{y^2 - 3y - 18} \cdot \dfrac{y - 6}{y^2 - 3y}$

31. $\dfrac{a^2 + 2ab + b^2}{a^2 - b^2} \cdot \dfrac{6a}{3a + 3b}$

32. $\dfrac{a - 1}{2a^2 + 4a + 2} \cdot \dfrac{(a + 1)^2}{a - 1}$

33. $\dfrac{3x + 6}{x^2 - 7x + 10} \cdot \dfrac{x - 5}{2 + x}$

34. $\dfrac{y^2 - 25}{3y + 9} \cdot \dfrac{y + 3}{y^2 + 10y + 25}$

35. $\dfrac{n - 4}{n^2 - 5n} \cdot \dfrac{n^2 - 25}{n^2 + n - 20}$

36. $\dfrac{p + 2}{3p + 3} \cdot \dfrac{p^2 + 5p + 4}{2p + 4}$

37. $\dfrac{r^2 + r - 6}{r^2 - 1} \cdot \dfrac{r - 1}{r - 2}$

38. $\dfrac{t^2 - 6t + 8}{t^2 - 3t} \cdot \dfrac{t^2 - 9}{t - 4}$

39. $\dfrac{c^2 - 6c + 9}{c^2 - 4c + 4} \cdot \dfrac{c^2 - 4}{c - 3}$

40. $\dfrac{6m^2}{m^2 - 4m + 4} \cdot \dfrac{5m - 10}{12m}$

41. $\dfrac{x^2 - 4}{x + 2} \cdot \dfrac{6}{2x - 4}$

42. $\dfrac{5}{2a - 2b} \cdot \dfrac{a^2 - b^2}{10}$

43. $\dfrac{a^2 + 8a + 16}{b^2} \cdot \dfrac{6b^3}{2a + 8}$

4. DIVIDING FRACTIONS

Recall the following rule:

RULE: To divide one fraction by another, *invert* the divisor (the fraction after the division sign) and *multiply*.

$$\text{Thus, } \frac{2}{5} \div \frac{3}{7} = \frac{2}{5} \cdot \frac{7}{3} = \frac{14}{15}.$$

ILLUSTRATIVE PROBLEM 1: Divide: $\frac{3}{5} \div 9$

Solution: Since $9 = \frac{9}{1}$, the reciprocal of 9 is $\frac{1}{9}$. Thus,

$$\frac{3}{5} \div \frac{9}{1} = \frac{3}{5} \cdot \frac{1}{9} = \frac{\overset{1}{\cancel{3}}}{5} \cdot \frac{1}{\underset{3}{\cancel{9}}} = \frac{1}{15} \quad (answer)$$

Note that for any integer x, $x = \dfrac{x}{1}$.

ILLUSTRATIVE PROBLEM 2: Divide: $\frac{5}{9} \div \frac{15}{18}$

Solution: Invert the divisor, and then factor and cancel.

$$\frac{5}{9} \div \frac{15}{18} = \frac{5}{9} \cdot \frac{18}{15}$$

$$\frac{\overset{1}{\cancel{5}}}{9} \cdot \frac{\overset{1}{\cancel{9}} \cdot 2}{\underset{1}{\cancel{5}} \cdot 3} = \frac{2}{3} \quad (answer)$$

The same principle applies to division of algebraic fractions. Division by a fraction is equivalent to multiplication by the reciprocal of the fraction. Thus,

$$\frac{p}{q} \div \frac{r}{s} = \frac{p}{q} \cdot \frac{s}{r} = \frac{ps}{qr}$$

ILLUSTRATIVE PROBLEM 3: Divide: $\dfrac{5p^2}{q} \div \dfrac{15p^3}{q^2}$

Solution: Invert the divisor and cancel.

$$\frac{5p^2}{q} \div \frac{15p^3}{q^2} = \frac{\overset{1}{\cancel{5p^2}}}{\cancel{q}} \cdot \frac{\overset{q}{\cancel{q^2}}}{\underset{3p}{\cancel{15p^3}}} = \frac{q}{3p} \quad (answer)$$

ILLUSTRATIVE PROBLEM 4: Divide: $\dfrac{s^2 - t^2}{4t^3} \div \dfrac{s - t}{2t^4}$

Solution: Invert, factor, and cancel.

$$\frac{s^2 - t^2}{4t^3} \div \frac{s - t}{2t^4} = \frac{s^2 - t^2}{4t^3} \cdot \frac{2t^4}{s - t}$$

$$= \frac{(s + t)\cancel{(s - t)}}{\underset{2}{\cancel{4t^3}}} \cdot \frac{\overset{t}{\cancel{2t^4}}}{\cancel{s - t}}$$

$$= \frac{t(s + t)}{2} \quad (answer)$$

Exercises

In 1 to 28, divide and write the quotient in lowest terms.

1. $\dfrac{9}{16} \div \dfrac{3}{10}$

2. $18 \div \dfrac{3}{5}$

3. $\dfrac{1}{3} \div \dfrac{4}{6}$

4. $\dfrac{3}{20} \div 6$

5. $\dfrac{3p}{8} \div \dfrac{6p^3}{2}$

6. $\dfrac{14}{k^2} \div \dfrac{7}{10k}$

7. $\dfrac{8x}{3y} \div \dfrac{4x^3}{9y^2}$

8. $\dfrac{4a^2b}{9pq^3} \div \dfrac{8b^4}{6p^2q^5}$

9. $\dfrac{3m^2n}{8x} \div \dfrac{9mn^3}{16x^4}$

10. $9cd \div \dfrac{6d^3}{c}$

11. $\dfrac{12x^2y^3}{5p} \div 8xy$

12. $\dfrac{x^3}{3y} \div \dfrac{x^2}{9y}$

13. $\dfrac{t}{n^2} \div \dfrac{t^2}{n^3}$

14. $\dfrac{(-2b)^2}{4} \div \dfrac{1}{b}$

15. $\dfrac{3m^2}{m^2 - 4} \div \dfrac{6m}{m + 2}$

16. $\dfrac{r^2 - t^2}{4rt} \div \dfrac{r - t}{2r^2}$

17. $\dfrac{x + 1}{2y - 2} \div \dfrac{x + 1}{y - 1}$

18. $\dfrac{s^2 - t^2}{10st} \div \dfrac{s - t}{-5s}$

19. $\dfrac{2n - 3}{3n - 2} \div \dfrac{4n - 6}{9n - 6}$

20. $\dfrac{x - 2}{x^2 + 4x + 4} \div \dfrac{1}{x^2 - 4}$

21. $\dfrac{2a - 1}{a^3 + 3a} \div \dfrac{a + 1}{a^2 + 3}$

22. $\dfrac{c^3 - 4c^2 + 3c}{c + 2} \div (c - 3)$

23. $\dfrac{p^2 - 2p + 1}{5p + 5} \div \dfrac{p - 1}{20}$

24. $\dfrac{(k + 3)^2}{k^2 + k - 6} \div \dfrac{k + 3}{k - 2}$

25. $\dfrac{m^2 - n^2}{m^2 + 2mn + n^2} \div \dfrac{m - n}{2m + 2n}$

26. $\dfrac{(b + 2)^2}{4b^2 - 16} \div \dfrac{8}{3b - 6}$

27. $\dfrac{3x - 12}{2x} \div \dfrac{x^2 - x - 12}{2x^3}$

28. $\dfrac{y^2 - 9}{2y - 6} \div (y + 3)$

5. COMBINING FRACTIONS THAT HAVE THE SAME DENOMINATOR

Recall the following rule:

RULE: Fractions that are to be combined by addition (or subtraction) must have the same denominator. The numerators are then added (or subtracted) and the result is placed over the common denominator.

$$\text{Thus, } \frac{2}{7} + \frac{3}{7} = \frac{2+3}{7} = \frac{5}{7}$$

and

$$\frac{4}{5} - \frac{1}{5} = \frac{4-1}{5} = \frac{3}{5}$$

The same rule applies to algebraic fractions. Thus,

$$\frac{8}{y} + \frac{3}{y} - \frac{4}{y} = \frac{8+3-4}{y} = \frac{7}{y}$$

and

$$\frac{7x}{5} - \frac{2x}{5} = \frac{7x-2x}{5} = \frac{5x}{5} = x$$

ILLUSTRATIVE PROBLEM 1: Combine: $\dfrac{9}{2n} + \dfrac{3}{2n} - \dfrac{5}{2n}$

Solution: $\dfrac{9}{2n} + \dfrac{3}{2n} - \dfrac{5}{2n} = \dfrac{9+3-5}{2n} = \dfrac{7}{2n}$ *(answer)*

ILLUSTRATIVE PROBLEM 2: Express as a single fraction in simplest form:

$$\frac{5x+2}{3} - \frac{x+1}{3}$$

Solution: To avoid errors in sign, we enclose the polynomials in the numerators within parentheses.

$$\frac{5x+2}{3} - \frac{x+1}{3} = \frac{(5x+2)-(x+1)}{3}$$

$$= \frac{5x+2-x-1}{3} = \frac{4x+1}{3} \quad (answer)$$

Exercises

In 1 to 38, combine fractions and reduce results to lowest terms.

1. $\dfrac{3}{8} + \dfrac{2}{8}$ 2. $\dfrac{15}{19} - \dfrac{2}{19}$ 3. $\dfrac{13}{24} - \dfrac{7}{24}$ 4. $\dfrac{8}{11} + \dfrac{3}{11}$

5. $\dfrac{7}{a} + \dfrac{3}{a}$ 6. $\dfrac{9}{x} - \dfrac{3}{x}$ 7. $\dfrac{12}{c} + \dfrac{3}{c} - \dfrac{8}{c}$ 8. $\dfrac{4p}{9} + \dfrac{5p}{9}$

9. $\dfrac{7}{2r} + \dfrac{3}{2r} - \dfrac{5}{2r}$

10. $\dfrac{9}{2n} + \dfrac{5}{2n}$

11. $\dfrac{8c}{3d} + \dfrac{4c}{3d} - \dfrac{5c}{3d}$

12. $\dfrac{2m}{a} + \dfrac{8m}{a}$

13. $\dfrac{13p}{p-3} - \dfrac{9p}{p-3}$

14. $\dfrac{5r}{x} - \dfrac{3r}{x} + \dfrac{2r}{x}$

15. $\dfrac{7n}{2k} + \dfrac{5n}{2k}$

16. $\dfrac{4p}{y} - \dfrac{3q}{y}$

17. $\dfrac{7a}{5m} + \dfrac{8a}{5m} - \dfrac{4a}{5m}$

18. $\dfrac{10b}{7n} + \dfrac{3c}{7n}$

19. $\dfrac{8}{x-5} + \dfrac{3}{x-5}$

20. $\dfrac{9}{p-q} - \dfrac{4}{p-q}$

21. $\dfrac{p}{p+2} + \dfrac{7}{p+2}$

22. $\dfrac{3n}{2n+5} + \dfrac{4n}{2n+5}$

23. $\dfrac{k}{k^2-9} - \dfrac{3}{k^2-9}$

24. $\dfrac{a}{a^2-b^2} - \dfrac{b}{a^2-b^2}$

25. $\dfrac{2c+3}{7} + \dfrac{3c+5}{7}$

26. $\dfrac{8n-3}{5} + \dfrac{2n+7}{5}$

27. $\dfrac{2m}{15} - \dfrac{m+7}{15}$

28. $\dfrac{7}{2x} + \dfrac{x-3}{2x}$

29. $\dfrac{y-3}{y} - \dfrac{y+6}{y}$

30. $\dfrac{x+2}{5x} - \dfrac{x-3}{5x}$

31. $\dfrac{2x+1}{9} - \dfrac{2(x-6)}{9}$

32. $\dfrac{2a+1}{3} - \dfrac{5a+1}{3}$

33. $\dfrac{3m+5}{m-5} + \dfrac{2m-3}{m-5}$

34. $\dfrac{3n+4}{2n-1} - \dfrac{n+5}{2n-1}$

35. $\dfrac{p^2-3pq}{p-q} + \dfrac{q^2+pq}{p-q}$

36. $\dfrac{7t-8}{t^2-4} - \dfrac{6t-10}{t^2-4}$

37. $\dfrac{r^2-4r}{r-2} + \dfrac{r+2}{r-2}$

38. $\dfrac{7b^2+5b}{b+1} - \dfrac{5b+7}{b+1}$

---

6. COMBINING FRACTIONS THAT HAVE DIFFERENT DENOMINATORS

Recall the following rules used to combine arithmetic fractions that have different denominators:

RULE 1: To add (or subtract) unlike fractions, change the fractions to equivalent fractions that have the same denominator. This *least common denominator* (L.C.D.) is the smallest number that is exactly divisible by each of the denominators.

RULE 2: The numerator and denominator of a fraction may both be multiplied by the same number without changing the value of the fraction.

These rules are demonstrated below.

ILLUSTRATIVE PROBLEM 1: Add $\frac{1}{2}$ and $\frac{2}{3}$.

Solution: The L.C.D. is obviously 6.

$$\frac{1}{2} = \frac{1 \cdot 3}{2 \cdot 3} = \frac{3}{6}$$

$$\frac{2}{3} = \frac{2 \cdot 2}{3 \cdot 2} = \frac{4}{6}$$

Thus, $\dfrac{1}{2} + \dfrac{2}{3} = \dfrac{3}{6} + \dfrac{4}{6} = \dfrac{7}{6} = 1\frac{1}{6}$ (*answer*)

Here we can see readily that 6 is the L.C.D. for denominators 2 and 3. In many problems, the L.C.D. is not quite so obvious.

Suppose we wish to add $\frac{7}{60} + \frac{5}{54}$. The L.C.D. of 60 and 54 will contain all of the various prime factors, each one as many times as it occurs in any of the denominators. Thus

$$60 = 12 \cdot 5 = 4 \cdot 3 \cdot 5 = 2^2 \cdot 3 \cdot 5$$

$$54 = 6 \cdot 9 = 2 \cdot 3 \cdot 3 \cdot 3 = 2 \cdot 3^3$$

The L.C.D. must contain 2, 3, and 5 as factors, but each factor must be used the greatest number of times it appears in any denominator. Thus, the L.C.D. of 60 and 54 $= 2^2 \cdot 3^3 \cdot 5 = 540$.

To change $\frac{7}{60}$ to an equivalent fraction that has a denominator of 540, we must multiply both the numerator and the denominator by 9 ($540 \div 60 = 9$).

$$\frac{7}{60} = \frac{7 \cdot 9}{60 \cdot 9} = \frac{63}{540}$$

To change $\frac{5}{54}$ to an equivalent fraction that has a denominator of 540, we must multiply both the numerator and the denominator by 10 ($540 \div 54 = 10$).

$$\frac{5}{54} = \frac{5 \cdot 10}{54 \cdot 10} = \frac{50}{540}$$

Thus, $\dfrac{7}{60} + \dfrac{5}{54} = \dfrac{63}{540} + \dfrac{50}{540} = \dfrac{113}{540}.$

To determine the L.C.D. of several fractions:

1. Write each denominator as a product of prime factors.
2. The L.C.D. is the product of the highest powers of each of the various prime factors of the denominators.
3. To find the multiplier needed to convert each fraction to an equivalent one, divide the L.C.D. by the denominator of the fraction.

These same principles apply to the addition of algebraic fractions.

ILLUSTRATIVE PROBLEM 2: Combine: $\dfrac{7a}{40} - \dfrac{5a}{18}$

Solution:

$$40 = 10 \cdot 4 = 5 \cdot 2 \cdot 2 \cdot 2 = 5 \cdot 2^3$$

$$18 = 6 \cdot 3 = 2 \cdot 3 \cdot 3 = 2 \cdot 3^2$$

The L.C.D. $= 5 \cdot 2^3 \cdot 3^2 = 5 \cdot 8 \cdot 9 = 360$.

The terms of $\dfrac{7a}{40}$ must be multiplied by 9 since $360 \div 40 = 9$.

$$\frac{7a}{40} = \frac{7a \cdot 9}{40 \cdot 9} = \frac{63a}{360}$$

The terms of $\dfrac{5a}{18}$ must be multiplied by 20 since $360 \div 18 = 20$.

$$\frac{5a}{18} = \frac{5a \cdot 20}{18 \cdot 20} = \frac{100a}{360}$$

Thus, $\dfrac{7a}{40} - \dfrac{5a}{18} = \dfrac{63a}{360} - \dfrac{100a}{360} = \dfrac{-37a}{360}$ *(answer)*

ILLUSTRATIVE PROBLEM 3: Combine: $\dfrac{3}{x^2y} + \dfrac{4}{xy^2}$

Solution: Taking the product of the highest powers of the factors of the denominators, we see that the L.C.D. $= x^2y^2$. The terms of $\dfrac{3}{x^2y}$ must be multiplied by y since $x^2y^2 \div x^2y = y$.

$$\frac{3 \cdot y}{x^2y \cdot y} = \frac{3y}{x^2y^2}$$

The terms of $\dfrac{4}{xy^2}$ must be multiplied by x since $x^2y^2 \div xy^2 = x$.

$$\frac{4 \cdot x}{xy^2 \cdot x} = \frac{4x}{x^2y^2}$$

Thus, $\dfrac{3}{x^2y} + \dfrac{4}{xy^2} = \dfrac{3y}{x^2y^2} + \dfrac{4x}{x^2y^2} = \dfrac{3y + 4x}{x^2y^2}$ (*answer*)

ILLUSTRATIVE PROBLEM 4: Combine: $\dfrac{a + 1}{2a} + \dfrac{a - 1}{3a}$

Solution: The factors of $2a$ are 2 and a.
The factors of $3a$ are 3 and a.
The L.C.D. is $2 \cdot 3 \cdot a = 6a$.

Note that the terms of the first fraction must be multiplied by 3 since $6a \div 2a = 3$. The terms of the second fraction must be multiplied by 2 since $6a \div 3a = 2$.

Thus,

$$\frac{a + 1}{2a} + \frac{a - 1}{3a} = \frac{3(a + 1)}{6a} + \frac{2(a - 1)}{6a}$$

$$= \frac{3(a + 1) + 2(a - 1)}{6a}$$

$$= \frac{3a + 3 + 2a - 2}{6a} = \frac{5a + 1}{6a}$$ (*answer*)

ILLUSTRATIVE PROBLEM 5: Combine: $\dfrac{2}{n + 1} + \dfrac{3}{n^2 - 1}$

Solution: The number $(n + 1)$ is prime and has no factors other than itself and 1.

The factors of $n^2 - 1$ are $(n + 1)(n - 1)$.

The L.C.D. is $(n + 1)(n - 1)$. Thus,

$$\frac{2}{n + 1} = \frac{2(n - 1)}{(n + 1)(n - 1)} = \frac{2n - 2}{n^2 - 1}$$

The second fraction need not be changed.

$$\frac{2}{n + 1} + \frac{3}{n^2 - 1} = \frac{2n - 2}{n^2 - 1} + \frac{3}{n^2 - 1}$$

$$= \frac{2n - 2 + 3}{n^2 - 1} = \frac{2n + 1}{n^2 - 1}$$ (*answer*)

Exercises _____

In 1 to 13, find the L.C.D. for two fractions, the denominators of which are as follows:

1. 15 and 20   **2.** 24 and 36   **3.** $12x$ and $18x$

4. $21a^2b$ and $14ab^2$ **5.** mp and pq

6. $15r^2s$ and $25rs^3$ **7.** $n + 2$ and $n - 2$

8. $6(p + 3)$ and $8(p + 3)$ **9.** $(2a - 4)$ and $(3a - 6)$

10. k and $k - 1$ **11.** $c^2 - 4$ and $5c - 10$

12. $(p^2 - 25)$ and $(p + 5)$ **13.** $(a^2 - b^2)$ and $4a - 4b$

In 14 to 60, combine fractions and reduce to lowest terms.

14. $\dfrac{5}{12} + \dfrac{2}{3}$ **15.** $\dfrac{7}{8} - \dfrac{2}{5}$ **16.** $\dfrac{7}{10} + \dfrac{2}{5} - \dfrac{1}{3}$

17. $\dfrac{5p}{8} - \dfrac{p}{3}$ **18.** $\dfrac{8x}{9} + \dfrac{x}{6}$ **19.** $5n - \dfrac{8n}{3}$

20. $\dfrac{5r^2}{3} + \dfrac{3r^2}{4}$ **21.** $\dfrac{x^2}{5} - \dfrac{2x^2}{6}$ **22.** $\dfrac{7s}{6} - \dfrac{5s}{9}$

23. $\dfrac{c}{5} + \dfrac{d}{10}$ **24.** $\dfrac{a}{8} - \dfrac{b}{12}$ **25.** $\dfrac{r}{2} + \dfrac{s}{3} - \dfrac{t}{4}$

26. $\dfrac{4}{a} + \dfrac{8}{b}$ **27.** $\dfrac{3}{2x} - \dfrac{5}{3x}$ **28.** $\dfrac{1}{a} + \dfrac{2}{b} + \dfrac{3}{c}$

29. $\dfrac{5}{y^2} - \dfrac{3}{y}$ **30.** $2 + \dfrac{3}{x^2}$ **31.** $\dfrac{9}{rs} - 1$

32. $\dfrac{7}{pq} + \dfrac{3}{qr}$ **33.** $\dfrac{8}{r^2s} + \dfrac{5}{rs^2}$ **34.** $\dfrac{3}{a^2} + \dfrac{2}{a} - \dfrac{1}{a^3}$

35. $\dfrac{a}{4} + \dfrac{b}{8}$ **36.** $\dfrac{2x}{5} - \dfrac{x + 7}{15}$ **37.** $\dfrac{y}{5} - \dfrac{y}{6} + \dfrac{y}{4}$

38. $\dfrac{x + 2}{3} + \dfrac{x - 3}{7}$ **39.** $\dfrac{2}{3x} + \dfrac{a}{2x}$ **40.** $\dfrac{n + 7}{3} + \dfrac{n - 3}{4}$

41. $\dfrac{y + 4}{y} + \dfrac{y - 4}{4}$ **42.** $\dfrac{2a + 1}{3} + \dfrac{3a - 1}{2}$ **43.** $\dfrac{r - 4}{3} + \dfrac{r}{4}$

44. $\dfrac{3a}{a^2 - 9} + \dfrac{3}{a - 3}$ **45.** $\dfrac{x - 3}{3x} + \dfrac{x + 2}{2x}$

46. $\dfrac{3}{(a + 4)^2} - \dfrac{2}{a(a + 4)}$ **47.** $\dfrac{n + 2}{3} - \dfrac{n - 2}{5}$

48. $\dfrac{2y + 3}{3y} + \dfrac{y - 2}{2y}$ **49.** $\dfrac{x + y}{3} + \dfrac{x - y}{4}$ **50.** $\dfrac{3a - 2}{5} - \dfrac{2a + 3}{15}$

51. $\dfrac{6r + s}{15} - \dfrac{2r - s}{5}$ **52.** $\dfrac{m + 3}{4} + \dfrac{2m - 3}{5}$ **53.** $\dfrac{5s + 2t}{st^2} - \dfrac{2s + 3t}{s^2 t}$

54. $\dfrac{5}{2p - 8} - \dfrac{p}{p - 4}$ **55.** $\dfrac{1}{x + y} + \dfrac{1}{x - y}$ **56.** $\dfrac{n}{n^2 - 49} - \dfrac{n}{n + 7}$

57. $\dfrac{5}{a + b} - \dfrac{3}{a^2 + ab}$ **58.** $\dfrac{7}{m + n} - \dfrac{3}{m - n}$

59. $\dfrac{4}{x^2 y + xy^2} - \dfrac{6}{x^2 - y^2}$ **60.** $\dfrac{6}{a^2 - 2a - 8} - \dfrac{a}{a + 2}$

---

7. CHANGES IN THE VALUE OF A FRACTION

Consider the set of fractions $\frac{2}{7}, \frac{3}{7}, \frac{4}{7}, \frac{5}{7}$, etc. Note that these fractions are increasing in value. This set of fractions illustrates that the value of a fraction such as $\dfrac{x}{7}$ increases as x increases. Two principles concerning fractions with positive numerators and denominators are summarized below.

Principle 1: If the denominator of a fraction remains fixed and the numerator increases, then the value of the fraction increases. In like manner, we may conclude that if the numerator of a fraction decreases and the denominator remains fixed, the value of the fraction decreases.

Now consider the set of fractions $\frac{2}{6}, \frac{2}{5}, \frac{2}{4}, \frac{2}{3}, \frac{2}{2}$, etc. Note that these fractions are increasing in value: $\frac{2}{6} = \frac{1}{3}, \frac{2}{4} = \frac{1}{2}, \frac{2}{2} = 1$. This leads to another principle concerning fractions with positive numerators and denominators.

Principle 2: If the numerator of a fraction remains fixed and the denominator decreases, then the value of the fraction increases. In like manner, we may conclude that, if the numerator of a fraction remains fixed and the denominator increases, then the value of the fraction decreases.

Thus, the value of a fraction such as $\dfrac{3}{x}$ increases as x decreases and the value of $\dfrac{3}{x}$ decreases as x increases.

ILLUSTRATIVE PROBLEM: If $y = \dfrac{9}{x + 3}$, does the value of y increase or decrease as x decreases from 6 to 0?

Solution: In the equation $y = \dfrac{9}{x + 3}$, substitute the values 6, 3, and 0 for x and solve for y.

$$\text{If } x = 6, y = \frac{9}{6 + 3} = \frac{9}{9} = 1$$

$$\text{If } x = 3, y = \frac{9}{3 + 3} = \frac{9}{6} = 1\frac{1}{2}$$

$$\text{If } x = 0, y = \frac{9}{0 + 3} = \frac{9}{3} = 3$$

Answer: We see that the value of y *increases* as x decreases, as stated in Principle 2.

Exercises

1. If x is positive and $y = \dfrac{7}{x}$, does y increase or decrease as x increases?

2. If $y = \dfrac{2x}{3}$ and x is positive, does y increase or decrease as x decreases?

3. If x is positive, does the value of the fraction $\dfrac{8}{x + 2}$ increase or decrease as x increases?

4. If x is positive, does $y = \dfrac{5}{2x + 3}$ increase or decrease as x decreases?

5. If $y = \dfrac{5x + 1}{10}$ and x is positive, does y increase or decrease as x increases?

6. If $y = \dfrac{x^2 - 2}{3}$ and x is positive, does y increase or decrease as x increases?

7. If $y = \dfrac{2}{x + 3}$ and x is positive, does y increase or decrease as x increases?

8. $\dfrac{p + 1}{q + 2}$ is a fraction in which p and q are positive whole numbers. If p remains fixed and q increases, does the value of the original fraction increase, decrease, or remain the same?

9. $\dfrac{p}{q}$ is a fraction in which p and q are positive whole numbers. The value of the fraction will remain the same if
   (a) p and q are both increased by the same number.
   (b) p and q are both decreased by the same number.
   (c) p and q are both multiplied by the same number.

10. In the equation $y = 2 - \dfrac{1}{x}$, does y increase or decrease as positive x increases?

11. In the expression $y = \dfrac{5}{2 + x^2}$, does y increase or decrease as x increases?

Chapter Review Exercises

In 1 to 4, reduce to lowest terms.

1. $\dfrac{8x^2y^3}{12xy^2}$　　2. $\dfrac{a^2 - b^2}{a + b}$　　3. $\dfrac{p^2 + 3p}{p^2 - p - 12}$　　4. $\dfrac{y^2 - 4}{y + 2}$

In 5 to 10, multiply.

5. $3\frac{3}{4} \times 1\frac{1}{5}$

6. $\dfrac{4r^3s}{a^2} \cdot \dfrac{5a^3}{2rs}$

7. $\dfrac{x^2 - 9}{4} \cdot \dfrac{8}{x - 3}$

8. $\dfrac{1}{d + 4} \cdot \dfrac{d^2 - 16}{d}$

9. $\dfrac{t^2 - 25}{3} \cdot \dfrac{6}{2t + 10}$

10. $\dfrac{5}{2x - 2y} \cdot \dfrac{x^2 - y^2}{10}$

In 11 to 16, divide.

11. $\dfrac{a^2 - 9}{2} \div \dfrac{6a - 18}{4}$

12. $\dfrac{x^2 - 9}{5} \div \dfrac{2(x + 3)}{10}$

13. $\dfrac{y^2 - y - 42}{9} \div \dfrac{y - 7}{3}$ **14.** $\dfrac{(n - 3)^2}{n^2 - n - 6} \div \dfrac{n - 3}{n + 2}$

15. $\dfrac{m^2 - 3m - 18}{m - 6} \div \dfrac{m^2 - 9}{2m - 6}$ **16.** $\dfrac{6k^5}{(x - y)^2} \div \dfrac{9k}{x^2 - y^2}$

In 17 to 28, combine.

17. $\dfrac{a}{b} - \dfrac{b}{a}$ **18.** $\dfrac{x}{7} - \dfrac{x}{14}$ **19.** $\dfrac{2x}{3y} + \dfrac{5x}{2y}$

20. $\dfrac{3a}{5b} + \dfrac{7a}{2b}$ **21.** $\dfrac{3t}{5} - \dfrac{t}{6}$ **22.** $\dfrac{r + s}{3} + \dfrac{r - s}{4}$

23. $\dfrac{2x + 3}{3x} + \dfrac{x - 2}{2x}$ **24.** $\dfrac{4}{x} + \dfrac{1}{y}$ **25.** $\dfrac{1}{a} + \dfrac{1}{2a}$

26. $\dfrac{c + 3}{4} + \dfrac{2c - 3}{5}$ **27.** $\dfrac{3m + n}{4} + \dfrac{2m - n}{3}$ **28.** $\dfrac{c}{d} - \dfrac{d}{c}$

CHECK YOUR SKILLS

1. Express as a trinomial the product of $x - 5$ and $x + 3$.
2. From $7x + 8y$ subtract $2x - 3y$.
3. Find the product of $5x^4y^3$ and $2x^2y^2$.
4. What is the additive inverse of 7?
5. What is the sum of $\dfrac{x + 7}{3}$ and $\dfrac{x - 2}{4}$?
6. What is the y-intercept of the graph of the equation $3x + 2y = 12$?
7. A grocer sold p pounds of butter at c cents per pound. What was the number of cents the grocer received for the sale, in terms of p and c?
8. Divide and express the answer as a fraction in lowest terms: $\dfrac{x^2 + 3x - 4}{2x + 2} \div \dfrac{x + 4}{x + 1}$
9. Solve for y: $\dfrac{2y}{3} = 6$
10. Solve for x in terms of b and c: $2x + b = c$

18

Fractional and Literal Equations

1. SOLVING EQUATIONS THAT HAVE FRACTIONAL COEFFICIENTS

Consider the following problem:

Two-thirds of a certain number is 28 more than three-eighths of the number. Find the number.

Let n = the number.

Then, $\frac{2}{3}n = \frac{3}{8}n + 28$.

A good way to solve an equation involving fractions is to transform the equation into an equivalent equation that has no fractions. This may be done by multiplying both members of the equation by the L.C.D. of all fractions in the equation.

In the preceding equation, the L.C.D. = 24. Thus:

$$\frac{2}{3}n = \frac{3}{8}n + 28$$

$$24(\tfrac{2}{3}n) = 24(\tfrac{3}{8}n + 28) \qquad \text{(multiply by L.C.D., 24)}$$

$$16n = 24(\tfrac{3}{8}n) + 24(28) \qquad \text{(distributive property)}$$

$$16n = 9n + 672$$

$$7n = 672 \qquad \text{(subtract } 9n)$$

$$n = 96$$

This method of getting rid of fractions is referred to as "clearing the equation of fractions." It consists simply of multiplying both members of the equation by the L.C.D. of all fractions in the equation and then solving the resulting equation.

Note: We can get rid of fractions only when we are dealing with equations.

ILLUSTRATIVE PROBLEM 1: Solve and check: $\dfrac{2x}{3} + \dfrac{x}{5} = 65$

Solution: The L.C.D. of $\dfrac{2x}{3}$ and $\dfrac{x}{5}$ is $5 \cdot 3 = 15$. Multiply both members of the equation by 15 (L.C.D.).

$$15\left(\dfrac{2x}{3} + \dfrac{x}{5}\right) = 65(15)$$

$$15\left(\dfrac{2x}{3}\right) + 15\left(\dfrac{x}{5}\right) = 975$$

$$10x + 3x = 975$$

$$13x = 975$$

$$x = 75 \quad (answer)$$

Check in the original equation.

$$\dfrac{2x}{3} + \dfrac{x}{5} = 65$$

$$\dfrac{2(75)}{3} + \dfrac{75}{5} = 65$$

$$50 + 15 = 65$$

$$65 = 65 \; ✔$$

ILLUSTRATIVE PROBLEM 2: Solve and check: $\dfrac{y + 5}{10} + \dfrac{2y + 5}{5} = \dfrac{23}{2}$

Solution: The L.C.D. is 10. Multiply both members of the equation by 10 (L.C.D.).

$$10\left(\dfrac{y + 5}{10} + \dfrac{2y + 5}{5}\right) = 10\left(\dfrac{23}{2}\right)$$

$$10\left(\dfrac{y + 5}{10}\right) + 10\left(\dfrac{2y + 5}{5}\right) = 115$$

$$1(y + 5) + 2(2y + 5) = 115$$

$$y + 5 + 4y + 10 = 115$$

$$5y + 15 = 115$$

$$5y = 100$$

$$y = 20$$

Answer: $y = 20$

Check:

$$\dfrac{y + 5}{10} + \dfrac{2y + 5}{5} = \dfrac{23}{2}$$

$$\dfrac{20 + 5}{10} + \dfrac{40 + 5}{5} = \dfrac{23}{2}$$

$$2\tfrac{1}{2} + 9 = 11\tfrac{1}{2}$$

$$11\tfrac{1}{2} = 11\tfrac{1}{2} \; ✔$$

ILLUSTRATIVE PROBLEM 3: Solve the system of equations and check:

$$\frac{x}{5} - \frac{y}{10} = 1 \quad \text{(equation 1)}$$

$$\frac{x}{3} - \frac{y}{5} = 1 \quad \text{(equation 2)}$$

Solution: Clear the equations of fractions and then solve the resulting pair of equations. The L.C.D. in the first equation is 10; the L.C.D. in the second equation is 15.

Multiply equation 1 by 10. $\quad 10\left(\dfrac{x}{5} - \dfrac{y}{10}\right) = 1 \cdot 10$

$$10\left(\frac{x}{5}\right) - 10\left(\frac{y}{10}\right) = 10$$

Equation 1 becomes: $\qquad\qquad 2x - y = 10 \quad$ (new equation 1)

Multiply equation 2 by 15. $\quad 15\left(\dfrac{x}{3} - \dfrac{y}{5}\right) = 1 \cdot 15$

$$15\left(\frac{x}{3}\right) - 15\left(\frac{y}{5}\right) = 15$$

Equation 2 becomes: $\qquad\qquad 5x - 3y = 15 \quad$ (new equation 2)

To eliminate y, multiply both sides of the new equation 1 by -3 to get $-6x + 3y = -30$. Then add this result to the new equation 2.

$$\begin{aligned} -6x + 3y &= -30 \\ 5x - 3y &= 15 \\ \hline -x &= -15 \\ x &= 15 \end{aligned}$$

Substitute 15 for x in new equation 1 and solve for y.

$$\begin{aligned} 2x - y &= 10 \\ 2(15) - y &= 10 \\ 30 - y &= 10 \\ -y &= -20 \\ y &= 20 \end{aligned}$$

Check in both original equations.

$$\frac{x}{5} - \frac{y}{10} = 1 \qquad\qquad \frac{x}{3} - \frac{y}{5} = 1$$

$$\frac{15}{5} - \frac{20}{10} = 1 \qquad\qquad \frac{15}{3} - \frac{20}{5} = 1$$

$$3 - 2 = 1 \qquad\qquad 5 - 4 = 1$$

$$1 = 1 \ \checkmark \qquad\qquad 1 = 1 \ \checkmark$$

Answer: $x = 15, y = 20$

Exercises _____

In 1 to 28, solve and check each equation.

1. $\dfrac{r}{2} + \dfrac{r}{3} = 1$

2. $\dfrac{2y}{3} = 6$

3. $\dfrac{n}{3} + \dfrac{n}{2} = 5$

4. $\dfrac{3x}{4} = 7 - x$

5. $\dfrac{p}{4} + \dfrac{p}{3} = 7$

6. $\dfrac{k}{3} + \dfrac{k}{5} = 8$

7. $\dfrac{t}{4} - 2 = \dfrac{t}{6}$

8. $\dfrac{a}{2} - \dfrac{a}{3} = 5$

9. $\dfrac{2y}{3} + \dfrac{y}{2} = 7$

10. $\dfrac{3c}{7} - \dfrac{c}{3} = 4$

11. $\dfrac{3k}{2} - \dfrac{k}{3} = 4$

12. $\dfrac{2x}{5} - 4 = \dfrac{2x}{3}$

13. $\dfrac{n+5}{8} = \dfrac{n-1}{4}$

14. $\dfrac{m-5}{4} = \dfrac{m}{5}$

15. $\dfrac{s+1}{9} = \dfrac{5}{3}$

16. $\dfrac{y}{8} - \dfrac{y}{10} = 3$

17. $\dfrac{x+2}{3} - \dfrac{x-2}{5} = 2$

18. $\dfrac{m}{2} + \dfrac{3m}{4} = 5$

19. $\dfrac{x}{3} + \dfrac{x}{4} = \dfrac{7}{12}$

20. $\dfrac{2d}{3} - \dfrac{d}{2} = 1$

21. $\dfrac{5z}{2} - \dfrac{2z}{3} = 11$

22. $\dfrac{5x-1}{2} - \dfrac{2x+5}{3} = \dfrac{7x+3}{6}$

23. $\dfrac{x+3}{3} - \dfrac{x-2}{4} = 6$

24. $\dfrac{n}{3} - \dfrac{3n-4}{16} = 2$

25. $\dfrac{6w-3}{2} - \dfrac{w+2}{5} = \dfrac{37}{10}$

26. $\dfrac{3y+4}{4} - \dfrac{10y+16}{8} = \dfrac{2y-17}{3}$

27. $\dfrac{2p+1}{9} - \dfrac{2(p-6)}{3} = 1$

28. $\dfrac{3n-4}{3} = \dfrac{2n+4}{6}$

In 29 to 33, solve and check each pair of equations.

29. $\dfrac{x-y}{2} = 1$

$\dfrac{x+y}{2} = 4$

30. $\dfrac{x}{4} + \dfrac{y}{2} = 7$

$\dfrac{3x}{4} - \dfrac{y}{2} = 1$

31. $\dfrac{3x+2}{2} + \dfrac{5y-2}{3} = 8$

$3x - 5y = 7$

32. $\dfrac{3r+2}{7} + \dfrac{5s-7}{2} = 6$

$2r - 5s = -7$

33. $2c + 3d = 18$

$\dfrac{3c-d}{2} = 8$

34. If $p - q = 3$ and $\dfrac{p}{5} + \dfrac{q}{5} = 1$, what is the value of pq?

35. If $\frac{3}{5}$ of a certain number is diminished by $\frac{1}{4}$ of the number, the result is 28. Find the number.

---

2. SOLVING FRACTIONAL EQUATIONS

An equation such as $\dfrac{9}{x} = \dfrac{3}{5}$, in which a variable appears in the *denominator* of one or more terms, is called a **fractional equation.**

To solve such an equation, we may use the same method of "clearing of fractions" that we used with equations having fractional coefficients.

ILLUSTRATIVE PROBLEM 1: Solve and check: $\dfrac{9}{x} = \dfrac{3}{5}$

Solution: The L.C.D. is $5x$.
Multiply both members by $5x$.

$$\frac{9}{x}(5x) = \frac{3}{5}(5x)$$

$$45 = 3x$$

$$x = 15 \quad (answer)$$

Check in the original equation.

$$\frac{9}{x} = \frac{3}{5}$$

$$\frac{9}{15} = \frac{3}{5}$$

$$\frac{3}{5} = \frac{3}{5} \; ✔$$

ILLUSTRATIVE PROBLEM 2: Solve and check: $\dfrac{1}{2} - \dfrac{1}{p} = \dfrac{1}{3}$

Solution: The L.C.D. is $6p$.

$$6p\left(\frac{1}{2} - \frac{1}{p}\right) = \frac{1}{3}(6p)$$

$$6p\left(\frac{1}{2}\right) - 6p\left(\frac{1}{p}\right) = 2p$$

$$3p - 6 = 2p$$

$$3p - 2p = 6$$

$$p = 6 \quad (answer)$$

Check in the original equation.

$$\frac{1}{2} - \frac{1}{p} = \frac{1}{3}$$

$$\frac{1}{2} - \frac{1}{6} = \frac{1}{3}$$

$$\frac{3}{6} - \frac{1}{6} = \frac{1}{3}$$

$$\frac{2}{6} = \frac{1}{3}$$

$$\frac{1}{3} = \frac{1}{3} \; ✔$$

ILLUSTRATIVE PROBLEM 3: Solve and check: $\dfrac{9y - 12}{y - 2} = 3 + \dfrac{6}{y - 2}$

Solution: The L.C.D. is $(y - 2)$. Multiply both sides of the equation by $(y - 2)$.

$$(y - 2)\,\frac{9y - 12}{y - 2} = (y - 2)\left(3 + \frac{6}{y - 2}\right)$$

$$9y - 12 = (y - 2) \cdot 3 + (y - 2) \cdot \frac{6}{y - 2}$$

$$9y - 12 = 3y - 6 + 6$$

$$9y - 3y = 12 - 6 + 6$$

$$6y = 12$$

$$y = 2$$

Check: When we substitute 2 for y in the original equation, we see that the denominators both become zero. Since division by zero is not defined, the fractions in the equation become meaningless. Thus, $y = 2$ is not a solution. In this case, the solution set is the **empty set** ($\varnothing$).

Answer: There is no value of y that satisfies the equation.

Note: When both members of an equation are multiplied by a variable expression that may represent zero, the resulting equation may not be equivalent to the given equation. Each solution, therefore, must be checked in the given equation.

Exercises _____

In 1 to 23, solve and check each equation.

1. $\dfrac{24}{h} = \dfrac{16}{4}$

2. $\dfrac{3}{4x} = 6$

3. $\dfrac{5}{y} = \dfrac{30}{18}$

4. $\dfrac{x}{x + 2} = \dfrac{3}{5}$

5. $\dfrac{w}{6} = \dfrac{7}{2}$

6. $\dfrac{6}{x} + \dfrac{x - 3}{2x} = 2$

7. $\dfrac{20}{12} = \dfrac{5}{y}$

8. $\dfrac{6}{4} = \dfrac{15}{m}$

9. $\dfrac{p}{p - 3} = \dfrac{7}{2}$

10. $\dfrac{11}{3p} = \dfrac{1}{6}$

11. $\dfrac{n - 4}{n + 3} = \dfrac{9}{2}$

12. $\dfrac{1}{2r} + \dfrac{2}{3r} = \dfrac{7}{24}$

13. $\dfrac{x+1}{2x-3} = \dfrac{8}{11}$

14. $\dfrac{5}{a} + \dfrac{3}{2a} = \dfrac{13}{18}$

15. $\dfrac{2}{5y} + \dfrac{11}{40} = \dfrac{3}{2y}$

16. $\dfrac{5}{x} = \dfrac{9}{x+4}$

17. $\dfrac{12}{2t-1} = 3$

18. $\dfrac{3-k}{3+k} = \dfrac{-5}{11}$

19. $\dfrac{7}{4-3p} = \dfrac{14}{5}$

20. $\dfrac{9}{m} = \dfrac{5}{m-2}$

21. $\dfrac{s+3}{2s} + \dfrac{5}{11} = \dfrac{12}{s}$

22. $\dfrac{8}{x} + 2 = \dfrac{x+34}{3x}$

23. $\dfrac{23+y}{y} - 3 = \dfrac{9}{2y}$

In 24 to 27, solve and check each pair of equations.

24. $\dfrac{r}{s+1} = \dfrac{2}{3}$

$r + s = 9$

25. $3p - 2q = 6$

$\dfrac{q-1}{p} = \dfrac{1}{2}$

26. $3x - 2y = 19$

$\dfrac{y-1}{x} = -\dfrac{3}{5}$

27. $d - n = 3$

$\dfrac{n+4}{d+4} = \dfrac{2}{3}$

28. Solve the equation and check.

$$\dfrac{7x-12}{x-3} = 3 + \dfrac{9}{x-3}$$

3. SOLVING EQUATIONS THAT HAVE DECIMAL COEFFICIENTS

Recall that decimals are simply fractions with powers of 10 in their denominators (such as 10, 100, 1000, etc.). Thus,

$$.3 = \dfrac{3}{10} \qquad .47 = \dfrac{47}{100} \qquad .835 = \dfrac{835}{1000}$$

To solve equations that have decimal coefficients, we may use the same method (clearing of fractions) used with equations that had fractional coefficients. To completely clear an equation of decimals, multiply both members of the equation by the highest power of 10 that appears among the decimal fractions in the equation. The L.C.D. is this power of 10.

When the equation has been cleared of decimals, it may then be solved by following the usual procedures for solving equations containing integers only.

ILLUSTRATIVE PROBLEM 1: Solve and check: $.3b - .6 = 2.1$

Solution: Multiply both sides of the equation by the L.C.D., 10.

$$10(.3b - .6) = 10(2.1)$$
$$10(.3b) - 10(.6) = 21$$
$$3b - 6 = 21$$
$$3b = 27$$
$$b = 9 \quad (answer)$$

Check in the original equation.

$$.3b - .6 = 2.1$$
$$.3(9) - .6 = 2.1$$
$$2.7 - .6 = 2.1$$
$$2.1 = 2.1 \; \checkmark$$

Note: In checking, we must take care to work the arithmetic of each side separately and then see that the results match. We should not repeat the method of the solution, which was to multiply both sides by a particular number.

ILLUSTRATIVE PROBLEM 2: Solve and check:
$$.05x + .08(600 - x) = 40.5$$

Solution: Multiply both sides of the equation by 100.

$$100[.05x + .08(600 - x)] = 100(40.5)$$
$$5x + 8(600 - x) = 4050$$
$$5x + 4800 - 8x = 4050$$
$$-3x = -750$$
$$x = 250$$

Check:

$$.05x + .08(600 - x) = 40.5$$
$$.05(250) + .08(600 - 250) = 40.5$$
$$12.50 + 28.00 = 40.50$$
$$40.50 = 40.50 \; \checkmark$$

Again we have worked each side of the equation separately.

Answer: $x = 250$

Exercises

In 1 to 15, solve and check.

1. $.01t = 3$ **2.** $.3n = 1.2$ **3.** $.2a + 3 = 9$

4. $.7x + .14x = .42$ **5.** $.07z = .014$ **6.** $0.4x = 2.4$

7. $.05y - 4 = 6$ **8.** $2.3p + 4.5 = 16$ **9.** $.08n = 1.68$

10. $.25a + .5 = 2$ **11.** $.06x + .08(800 - x) = 57$

12. $.2p - .06p = 7.84$ **13.** $1.8 + .05n = 5 - .35n$

14. $.06(x - 10) = .08x$ **15.** $1.8y + 16.5 = .3y$

In 16 to 19, solve and check each pair of equations.

16. $.75r + .5s = 8$ **17.** $1.5c + .75d = 2.25$
$\quad\ \ r - .4s = 0$ $\quad\ \ .25c - .5d = 1$

18. $.6p - q = .2$ **19.** $m + n = 4000$
$\quad\ \ p - q = 9.6$ $\quad\ \ .06m + .08n = 290$

4. SOLVING WORK PROBLEMS

Problems involving **rates of work** by machines or persons usually require the solution of equations containing fractions.

ILLUSTRATIVE PROBLEM 1: One machine can do a piece of work in 12 hours and a newer machine can do the same work in 6 hours. How many hours will it take both machines to do the work if they operate together?

Solution: In such problems, consider the rate of work of each machine. If the older machine can do the job in 12 hours, then it does $\frac{1}{12}$ of the job in one hour. In 2 hours it does $\frac{2}{12}$ of the job, in 3 hours it does $\frac{3}{12}$, etc. In n hours it does $\frac{n}{12}$ of the job.

Likewise, the newer machine does $\frac{1}{6}$ of the job in one hour. In 2 hours it does $\frac{2}{6}$ of the job, in 3 hours it does $\frac{3}{6}$, etc. In n hours it does $\frac{n}{6}$ of the job.

Let n = the number of hours for both machines to complete the job.

In that time, the older machine does $\frac{n}{12}$ of the job and the newer one does $\frac{n}{6}$ of the job. **The complete job is represented by 1.**

Thus, $\dfrac{n}{12} + \dfrac{n}{6} = 1$. The L.C.D. is 12.

$$12\left(\dfrac{n}{12} + \dfrac{n}{6}\right) = 12(1)$$

$$12\left(\dfrac{n}{12}\right) + 12\left(\dfrac{n}{6}\right) = 12(1)$$

$$n + 2n = 12$$

$$3n = 12$$

$$n = 4$$

Check in the original problem.
In 4 hours, the older machine does $\frac{4}{12}$ or $\frac{1}{3}$ of the job. The newer machine does $\frac{4}{6}$ or $\frac{2}{3}$ of the job. The total amount of work done is 1.

$$\tfrac{1}{3} + \tfrac{2}{3} = 1$$

$$1 = 1 \ \checkmark$$

Answer: Both machines working together can complete the job in 4 hours.

ILLUSTRATIVE PROBLEM 2: Ted can shovel the snow from a long driveway in 3 hours, while Jerry can do it in 2 hours. If they work together, how long should it take to clear the snow from the driveway?

Solution:

Let x = the number of hours it takes them together.

Ted's rate = $\frac{1}{3}$ of the job per hour.

Jerry's rate = $\frac{1}{2}$ of the job per hour.

$\dfrac{x}{3}$ = the part of the job Ted does in x hours.

$\dfrac{x}{2}$ = the part of the job Jerry does in x hours.

1 = the whole job done by both in x hours.

Then:

$$\dfrac{x}{3} + \dfrac{x}{2} = 1 \quad (\text{L.C.D.} = 6)$$

$$6\left(\dfrac{x}{3} + \dfrac{x}{2}\right) = 6(1)$$

$$2x + 3x = 6$$

$$5x = 6$$

$$x = 1\tfrac{1}{5}$$

Check in the original problem.

In $1\frac{1}{5}$ hours, Ted does $\dfrac{1\frac{1}{5}}{3}$ part of the job, and Jerry does $\dfrac{1\frac{1}{5}}{2}$ part of the job.

$$\frac{1\frac{1}{5}}{3} = \frac{6}{5} \div \frac{3}{1}$$

$$= \frac{\overset{2}{\cancel{6}}}{5} \cdot \frac{1}{\underset{1}{\cancel{3}}} = \frac{2}{5} \quad \text{(Ted's part)}$$

$$\frac{1\frac{1}{5}}{2} = \frac{6}{5} \div \frac{2}{1}$$

$$= \frac{\overset{3}{\cancel{6}}}{5} \cdot \frac{1}{\underset{1}{\cancel{2}}} = \frac{3}{5} \quad \text{(Jerry's part)}$$

Ted's part of the job plus Jerry's part equals the whole job.

$$\frac{2}{5} + \frac{3}{5} = \frac{5}{5} = 1 \ \checkmark$$

Answer: $1\frac{1}{5}$ hours

Exercises

1. Anita can type a report in 5 hours. At this rate, what part of the report does she type in **a.** 1 hour? **b.** 3 hours? **c.** x hours?

2. A pipe can fill a pool in 40 minutes. At this rate, what part of the pool does it fill in **a.** 1 minute? **b.** 10 minutes? **c.** 15 minutes? **d.** n minutes?

3. Rudy takes x days to complete a job. At this rate, what part of the job does he do **a.** each day? **b.** in 3 days? **c.** in d days?

4. Teresa can paint a room in 6 hours. What portion of the room can she paint **a.** each hour? **b.** in 2 hours? **c.** in n hours?

5. Marvin can wax a car in 6 hours alone, while his friend Tony can do the same job in 4 hours alone. How many hours will it take the two boys working together?

6. A farmer can plow a field in 30 minutes, while his helper takes 50 minutes to plow the same field. How many minutes will it take if they both work together, using their own plows?

7. Mr. Werner can paint his house in 8 hours, while his son Carl takes 10 hours to do the same job. How many hours will it take them, working together?

8. A machine can make 1000 copies of a report in 1 hour. A newer machine can produce the same number of copies in 40 minutes. If both machines are run at the same time, how many minutes will it take them to produce the 1000 copies?

9. Maria and her brother can paint a fence in 6 hours. If Maria works alone, it takes her 10 hours. How many hours will it take her brother to paint the fence alone?

10. A carpenter can build a cottage in 20 days. His helper takes 24 days to build the same cottage. In how many days could they build the cottage, working together?

11. It takes 50 minutes to fill a pool and 90 minutes to empty it. If the inlet pipe and outlet pipe are both open, how many minutes will it take to fill the pool?

12. A mason works twice as fast as his helper. They require 7 hours to build a concrete walk, working together. How many hours will it take the mason, working alone?

---

5. VERBAL PROBLEMS INVOLVING EQUATIONS WITH FRACTIONS

Many of the types of problems that we have already learned to solve lead to the solution of equations involving fractions. Some examples of these problems are shown below.

ILLUSTRATIVE PROBLEM 1: The numerator of a fraction is 10 less than the denominator. If 3 is added to both the numerator and denominator, the value of the resulting fraction is $\frac{1}{2}$. Find the original fraction.

Solution:

Let d = the denominator of the original fraction.

Then $d - 10$ = the numerator of the original fraction.

And $\dfrac{d - 10}{d}$ = the original fraction.

When we add 3 to both numerator and denominator, the resulting fraction equals $\frac{1}{2}$.

$$\frac{d - 10 + 3}{d + 3} = \frac{1}{2}$$

$$\frac{d - 7}{d + 3} = \frac{1}{2} \quad \text{(L.C.D. = } 2(d + 3))$$

$$2(d + 3) \left(\frac{d - 7}{d + 3} \right) = 2(d + 3) \frac{1}{2}$$

$$2(d - 7) = (d + 3) 1$$

$$2d - 14 = d + 3$$

$$2d - d = 14 + 3$$

$$d = 17$$

$$d - 10 = 17 - 10 = 7$$

Check in the original problem. 7 is 10 less than 17 and

$$\frac{7 + 3}{17 + 3} = \frac{10}{20} = \frac{1}{2} \quad \checkmark$$

Answer: The original fraction is the numerator over the denominator, or $\frac{7}{17}$.

ILLUSTRATIVE PROBLEM 2: Mr. Garcia drove 20 miles at a certain speed and then drove the next 10 miles at twice the speed. If the whole trip took 50 minutes, at what rate did he travel the first 20 miles?

Solution: Recall the relationship: distance = rate × time. This relationship can be expressed as the formula $d = rt$ or $t = \frac{d}{r}$. When d is in miles and t is in hours, r is in miles per hour.

Let r = rate in miles per hour for first 20 miles.

Then $2r$ = rate in miles per hour for next 10 miles.

And $\frac{20}{r}$ = time in hours for first 20 miles.

And $\frac{10}{2r} = \frac{5}{r}$ = time in hours for next 10 miles.

And $\frac{50}{60} = \frac{5}{6}$ = time in hours for total trip.

$$\frac{20}{r} + \frac{5}{r} = \frac{5}{6}$$

$$6r\left(\frac{20}{r} + \frac{5}{r}\right) = 6r\left(\frac{5}{6}\right)$$

$$120 + 30 = 5r$$
$$150 = 5r$$
$$r = 30$$
$$2r = 60$$

Check: Time for first 20 miles $= \frac{20}{30} = \frac{2}{3}$ hour.

Time for next 10 miles $= \frac{10}{60} = \frac{1}{6}$ hour.

$\frac{2}{3} + \frac{1}{6} = \frac{4}{6} + \frac{1}{6} = \frac{5}{6}$ hour

$= \frac{5}{6} \times 60 = 50$ minutes ✔

Answer: Mr. Garcia traveled the first 20 miles at 30 mph and the next 10 miles at 60 mph.

Exercises

1. The numerator of a fraction is 5 less than the denominator. If 6 is added to the numerator and 9 is added to the denominator, the resulting fraction has a value of $\frac{1}{2}$. Find the original fraction.

2. The denominator of a fraction is 5 more than the numerator. If 1 is added to the numerator and 2 is subtracted from the denominator, the resulting fraction has a value of $\frac{2}{3}$. Find the original fraction.

3. Mrs. Caso has a sum of money invested in bonds at 9%, and $2500 more than this in stocks at 6%. If the annual income from the two investments is $600, how much is invested in bonds?

4. The numerator of a fraction is 3 less than the denominator. If 4 is added to both numerator and denominator, the value of the resulting fraction is $\frac{2}{3}$. Find the original fraction.

5. Phil drove from home to school, a distance of 25 miles. He drove the first 10 miles at 50 miles per hour. If the whole trip took 32 minutes, at what rate did he travel the rest of the trip?

6. A roast beef loses 25% of its weight in cooking. If a roast weighs $4\frac{1}{2}$ pounds after cooking, how many pounds did it weigh before cooking?

7. Maria buys a television set for $224 after a 30% discount. What was the original price of the set?

8. A coat was marked for sale at $360. This was a discount of 20% off the original sale price. Find the original sale price.

9. Sue ran at a rate of 8 miles per hour from her home to school. She walked back home at 4 miles per hour. If the round trip took 3 hours, how many miles is it from her home to the school?

10. A man invested $3200 for one year, part at 6%, and the remainder at 7%. If the total income from the investments was $212, find the amount invested at 6%.

11. At a sale of fur coats, Mrs. Ortiz paid $570 for a coat that had been reduced 40%. What was the original price?

12. What number added to 12% of itself is equal to 78.4?

13. A suit costing $112 is sold for a profit that is 30% of the selling price. How much is the selling price?

14. The width of a rectangle is $\frac{3}{4}$ of its length. If the perimeter of the rectangle is $87\frac{1}{2}$, find its length.

15. The numerator of a fraction is 1 less than the denominator. If 5 is added to the denominator, the result is $\frac{1}{2}$. Find the original fraction.

---

6. TRANSFORMING FORMULAS

In the last section, we saw that it was desirable to work with the distance formula, $d = rt$, in another form, $t = \dfrac{d}{r}$. We often find it necessary to rewrite a formula in a more useful way.

We will now learn how to rewrite formulas (equations with many variables) so that a particular letter becomes the subject, or stands alone on one side of the equation.

ILLUSTRATIVE PROBLEM 1: Solve the formula $p = 3s$ for s. (This is the formula for the perimeter of an equilateral triangle of side s.)

Solution: Solving for s means that we want s (not p) to stand alone as the subject of the equation.

$$p = 3s$$

$$\frac{p}{3} = \frac{3s}{3} \quad \text{(dividing both members by 3 produces the desired result)}$$

$$\frac{p}{3} = s \quad \text{(answer)}$$

ILLUSTRATIVE PROBLEM 2: Solve the formula $A = bh$ for h. (This is the formula for the area of a rectangle with sides b and h.)

Solution: Solving for h means that we want h to stand alone as the subject.

$$A = bh$$

$$\frac{A}{b} = \frac{bh}{b} \quad \text{(dividing both members by } b \text{ produces the desired result)}$$

$$\frac{A}{b} = h \quad (answer)$$

ILLUSTRATIVE PROBLEM 3: Solve the formula $V = \frac{1}{3}Bh$ for B. (This is the formula for the volume of a cone.)

Solution: Solving for B means that we want B to stand alone as the subject.

$$V = \frac{1}{3}Bh$$

$$3V = Bh \quad \text{(multiply both members by 3)}$$

$$\frac{3V}{h} = \frac{Bh}{h} \quad \text{(divide both members by } h\text{)}$$

$$\frac{3V}{h} = B \quad (answer)$$

ILLUSTRATIVE PROBLEM 4: Solve the formula $A = \frac{1}{2}h(b + c)$ for c. (This is the formula for the area of a trapezoid with height h and with bases b and c.)

Solution: To solve for c:

$$A = \frac{1}{2}h(b + c)$$

$$2A = h(b + c) \quad \text{(multiply both members by 2)}$$

$$2A = hb + hc \quad \text{(use the distributive property)}$$

$$2A - hb = hc \quad \text{(subtract } hb \text{ from both members)}$$

$$\frac{2A - hb}{h} = c \quad \text{(dividing both members by } h \text{ produces the desired result)}$$

Exercises

In 1 to 34, solve each equation for the indicated variable.

1. $p = 4s$ for s　　　**2.** $h = 16t$ for t　　　**3.** $d = 40h$ for h

4. $d = rt$ for r　　　**5.** $c = 2\pi r$ for r　　　**6.** $E = IR$ for I

7. $P = EI$ for E　　　**8.** $A = \frac{1}{2}bh$ for b　　　**9.** $v = gt$ for t

10. $p = 2l + 2w$ for w　　**11.** $I = PRT$ for R　　**12.** $s = \frac{1}{2}gt^2$ for g

13. $ax - b = c$ for x　　**14.** $dy + c = h$ for y　　**15.** $\dfrac{px}{q} = r$ for x

16. $S = 2\pi rh$ for r　　　　　　**17.** $rz - s = o$ for z

18. $2n - b = a + b$ for n　　　　**19.** $3a + 2 = b$ for a

20. $F = \frac{9}{5}C + 32$ for C　　　　**21.** $S = \dfrac{n}{2}(a + b)$ for a

22. $A = \frac{1}{2}h(b + c)$ for h　　　　**23.** $A = \pi ab$ for a

24. $S = 2\pi rh + 2\pi r^2$ for h　　　**25.** $h = vt - 16t^2$ for v

26. $P = I^2R$ for R　　　　　　**27.** $3bx = a$ for x

28. $CW + a = b$ for W　　　　**29.** $ay = by + 3$ for y

30. $an + b = 4$ for n　　　　　**31.** $3z + 6m = 7m$ for z

32. $ax - 2(x + b) = 3a$ for x　　**33.** $\dfrac{PV}{T} = k$ for V

34. $A = p + prt$ for p

Chapter Review Exercises

In 1 to 13, solve and check.

1. $\dfrac{18}{12} = \dfrac{x}{2}$　　　　　**2.** $\dfrac{a}{2} + \dfrac{a}{6} = 2$　　　　**3.** $.02x = 15$

4. $1.2x - .35 = .5x + 5.25$　　　**5.** $\frac{1}{3}y - 12 = 4$

6. $\dfrac{c + 2}{3} - \dfrac{c - 1}{4} = 4$　　　　**7.** $\frac{5}{9}(F - 32) = 37$

8. $\dfrac{x}{3} - \dfrac{x}{4} = 1$　　　　　**9.** $\dfrac{n - 1}{2} + \dfrac{n - 2}{3} - \dfrac{n - 3}{4} = 6$

10. $3 + \dfrac{15}{d} = \dfrac{27}{4}$

11. $.1y + .01y = 2.2$

12. $\dfrac{2y}{4} - 4 = \dfrac{2y}{3}$

13. $\dfrac{x + 5}{8} = \dfrac{x - 1}{4}$

14. Solve the given pair of equations and check.

$$\frac{x}{y + 1} = \frac{2}{3}$$

$$x + y = 9$$

In 15 to 20, solve each equation for the variable indicated.

15. $a + bx = c$ for x **16.** $x + ay = m$ for y **17.** $pt - q = 0$ for t

18. $pn + q = r$ for n **19.** $A = \dfrac{h}{2}(b + c)$ for b **20.** $I = \dfrac{E}{r + s}$ for s

21. A test was failed by 15% of a class. If 6 students failed, how many students were in the class?

22. Mrs. Perez invested some money at 8% and twice as much at 10%. If the total yearly income from the two investments is $1120, how much was invested at 8%?

23. If 25% of a number is 80, find the number.

24. The denominator of a fraction is 4 more than the numerator. If 3 is added to both the numerator and the denominator, the resulting fraction has a value of $\frac{5}{7}$. Find the original fraction.

25. A mason's helper takes twice as long as the mason to build a wall. Together they can build the wall in 6 hours. How many hours will it take the mason alone to build the wall?

26. What number must be subtracted from the numerator and the denominator of $\frac{12}{19}$ so that the resulting fraction has a value of $\frac{1}{2}$?

27. If x is 80% of y, what percent is y of x?

28. What number increased by 20% of itself is equal to 180?

29. Ricardo buys a camera for $126 after a 30% discount. What was the original price of the camera?

CUMULATIVE REVIEW

1. Express the product $(x + 14)(x - 3)$ as a trinomial.

2. Factor: $x^2 - a^2$ 3. Solve for x: $\dfrac{10}{3x} = \dfrac{2}{3}$

4. Solve for x in terms of a and b: $2x + b = a$

5. Multiply and express the answer in simplest form:
$$\frac{x^2 - 3x - 40}{x^2 - 25} \cdot \frac{2x - 10}{x - 8}$$

6. Subtract and express as a single fraction in simplest form:
$$\frac{x + y}{3} - \frac{x}{4}$$

7. Write an equation that can be used to solve the following problem. Working alone, Maria can mow a lawn in 2 hours. Richard can do the same job in 3 hours. How long will it take them to do the job if they work together?

8. Divide and express as a single fraction in simplest form:
$$\frac{5}{x - 3} \div \frac{5}{x}$$

9. Reduce $\dfrac{x^2 + 2x - 3}{x^2 + 3x}$ to a fraction in simplest form.

10. Solve for n and check: $n - 4 = \dfrac{n - 1}{4}$

Answers to Cumulative Review

If you get an incorrect answer, refer to the chapter and section shown in brackets for review. For example, [5-2] means chapter 5, section 2.

1. $x^2 + 11x - 42$ [16-8]   2. $(x + a)(x - a)$ [16-7]   3. 5 [18-2]

4. $\dfrac{a - b}{2}$ [18-6]   5. 2 [17-3]   6. $\dfrac{x + 4y}{12}$ [17-6]   7. $\dfrac{x}{2} + \dfrac{x}{3} = 1$ [18-4]

8. $\dfrac{x}{x - 3}$ [17-4]   9. $\dfrac{x - 1}{x}$ [17-2]   10. 5 [18-1]

19

Roots and Radicals

1. INTRODUCTION

The numbers with which you are familiar consist of positive and negative integers, fractions, decimals, percents, and zero. Examples of these numbers are 8, $-\frac{2}{3}$, .6, and -25%. Each of these numbers can be expressed as a fraction with an integer for its numerator and denominator. Remember that 8 can be expressed as $\frac{8}{1}$, .6 as $\frac{6}{10}$ or $\frac{3}{5}$, and -25% as $-\frac{1}{4}$. Numbers that can be expressed in this form are called **rational numbers.**

Definition: Numbers that can be expressed in the form of $\frac{p}{q}$, where p and q are integers and $q \neq 0$, are called **rational numbers.**

The set of rational numbers includes the set of positive and negative integers, as well as zero and all fractions whose numerators and denominators can be expressed as integers. Thus far we have dealt only with rational numbers. In this chapter, we shall see a need for another type of number.

2. FINDING A ROOT OF A NUMBER

To begin, let us consider a situation that leads to an operation other than $+$, $-$, $\times$, $\div$.

If we let s stand for the side of a square, the formula for the area of a square is $A = s^2$. We use this formula to find the area, A, when a side, s, is given. If a square has a side measuring 4 inches, then the area is 4×4, or 16 square inches.

Let us now consider the reverse situation. If the area of a square is given as 25 square units, what is the value of s? By substituting 25 for A in the formula $A = s^2$, we obtain the equation $25 = s^2$. Thus, s is a number that, when multiplied by itself, equals 25. We see that $s = 5$, since $5 \cdot 5 = 5^2 = 25$.

Using this example, we can say that 5 is a **square root** of 25 or $\sqrt{25} = 5$. To indicate the square root of a number, we use a radical sign, $\sqrt{}$. The number under the radical sign is called the **radicand.**

Finding the square root of a number is the inverse of the operation of squaring a number.

Definition: The square root of a number is one of the two equal factors of the number.

Based on this definition, note that -5 is also a square root of 25, since $(-5)(-5) = 25$. Since $(+5)(+5) = 25$ and $(-5)(-5) = 25$, both $+5$ and -5 are square roots of 25. This example illustrates the truth of the following principle.

Principle: Every positive number has two square roots that are additive inverses of each other.

The positive square root of a number is called the **principal square root.** Although $+9$ and -9 are both square roots of 81, $+9$ is called the principal square root.

To indicate that the principal square root of a number is to be found, a radical sign, $\sqrt{}$, is placed over the number. For example:

$$\sqrt{36} = 6 \qquad \sqrt{\tfrac{4}{25}} = \tfrac{2}{5} \qquad \sqrt{.25} = .5$$

To indicate that the negative square root of a number is to be found, we place a minus sign in front of the radical. For example:

$$-\sqrt{36} = -6 \qquad -\sqrt{\tfrac{4}{25}} = -\tfrac{2}{5} \qquad -\sqrt{.25} = -.5$$

We have already learned that to cube a number means to use it as a factor *three* times. For example, the cube of 2 is $2 \times 2 \times 2 = 8$, or $2^3 = 8$.

Finding a cube root of a number means finding one of the *three* equal factors of that number. For example, the cube root of 8 is 2, since $2^3 = 2 \times 2 \times 2 = 8$. We can write this statement as $\sqrt[3]{8} = 2$. The small symbol ^{3} in $\sqrt[3]{8}$ is called the **index** and it indicates that we are seeking one of the *three* equal factors whose product is 8.

What is $\sqrt[3]{27}$? What is $\sqrt[3]{64}$?

Note that $\sqrt[3]{-8} = -2$, since $(-2)(-2)(-2) = (+4)(-2) = -8$. We see that a negative number *does* have a cube root.

In like manner, the index of $\sqrt[4]{16}$ is 4 and it indicates that we are seeking one of *four* equal factors of 16. Apparently, $\sqrt[4]{16} = 2$ since $2^4 = 2 \cdot 2 \cdot 2 \cdot 2 = 16$.

Note that, when we write the square root in radical form and the index does not appear, it is understood to be 2. Thus $\sqrt{36} = 6$ means $\sqrt[2]{36} = 6$. Note also that since squaring a number can never yield a negative result, a negative number does *not* have a square root.

ILLUSTRATIVE PROBLEM 1: Find the principal square root of 121.

Solution: Since $11^2 = 11 \cdot 11 = 121$, then $\sqrt{121} = 11$ (*answer*)

ILLUSTRATIVE PROBLEM 2: Find the cube root of 64.

Solution: Since $4^3 = 4 \cdot 4 \cdot 4 = 64$, then $\sqrt[3]{64} = 4$ (*answer*)

ILLUSTRATIVE PROBLEM 3: Find the value of $\sqrt{\frac{9}{25}}$.

Solution: Since $\frac{3}{5} \cdot \frac{3}{5} = \frac{9}{25}$, then $\sqrt{\frac{9}{25}} = \frac{3}{5}$ (*answer*)

ILLUSTRATIVE PROBLEM 4: Solve the equation $x^2 = 64$.

Solution:
$$x^2 = 64$$
$$x = \pm\sqrt{64}$$
$$x = \pm 8 \quad (answer)$$

Notice that both a positive *and* a negative result are obtained.

Exercises

In 1 to 12, find the principal square root of the number.

1. 9	**2.** 1	**3.** 0	**4.** 144
5. 400	**6.** 900	**7.** 225	**8.** $\frac{1}{4}$
9. $\frac{4}{9}$	**10.** .09	**11.** .64	**12.** 1.21

In 13 to 17, state the index and the radicand.

13. $\sqrt[3]{-1}$ **14.** $\sqrt[4]{81}$ **15.** $\sqrt{4}$ **16.** $\sqrt[3]{27}$ **17.** $\sqrt{\frac{9}{16}}$

In 18 to 49, find the value of each expression.

18. $\sqrt{9}$ **19.** $\sqrt{49}$ **20.** $-\sqrt{1}$ **21.** $-\sqrt{144}$

22. $\pm\sqrt{36}$ **23.** $-\sqrt{\frac{4}{25}}$ **24.** $\sqrt{.04}$ **25.** $\sqrt{0}$

26. $\pm\sqrt{\frac{9}{64}}$ **27.** $\sqrt[3]{-27}$ **28.** $-\sqrt{.36}$ **29.** $\sqrt[3]{125}$

30. $\pm\sqrt{\frac{64}{81}}$ **31.** $\sqrt[4]{16}$ **32.** $-\sqrt[3]{64}$ **33.** $\sqrt{(5)^2}$

34. $\sqrt{(\frac{1}{3})^2}$ **35.** $\sqrt{(-3)^2}$ **36.** $\sqrt{(\frac{1}{2})^2}$ **37.** $\sqrt[3]{1}$

38. $(\sqrt{9})^2$ **39.** $(\sqrt{49})^2$ **40.** $(\sqrt{25})^2$ **41.** $\sqrt[3]{-125}$

42. $\sqrt{81} + \sqrt{4}$ **43.** $\sqrt{100} - \sqrt{36}$ **44.** $\sqrt{121} + \sqrt{1}$

45. $\sqrt{9 + 16}$ **46.** $\sqrt{100 - 64}$ **47.** $\sqrt{(11)^2} + \sqrt{16}$

48. $3\sqrt{16} - \sqrt{9}$ **49.** $2\sqrt{100} + 4\sqrt{25}$

In 50 to 58, solve for the variable.

50. $p^2 = 81$ **51.** $x^2 = 1$ **52.** $y^2 = 400$

53. $t^2 = \frac{4}{9}$ **54.** $m^2 = \frac{36}{25}$ **55.** $2k^2 = 18$

56. $r^2 - 16 = 0$ **57.** $n^2 - 100 = 0$ **58.** $3z^2 - 48 = 0$

---

3. ESTIMATING THE SQUARE ROOT OF A NUMBER

In the previous section, we dealt only with numbers whose square roots were rational numbers. Numbers that have rational square roots are called **perfect squares.**

ILLUSTRATIVE PROBLEM 1: The area of a square is 7 square centimeters. What is the length in centimeters of each side of the square?

Solution:
$$A = s^2$$
$$s^2 = 7$$
$$s = \sqrt{7} \quad (answer)$$

We know that $\sqrt{4} = 2$ and $\sqrt{9} = 3$. Therefore, we expect that the value of $\sqrt{7}$ will be between 2 and 3, since 7 is between 4 and 9. It has been proven mathematically that there is no rational number whose square is equal to 7. We say that $\sqrt{7}$ is an **irrational number.**

Definition: An **irrational number** is any number that cannot be expressed as the ratio of two integers.

It has been proven that if the square root of an integer is *between* two consecutive integers, the root is irrational. Thus, $\sqrt{3}$, $\sqrt{10}$, and $\sqrt{17}$ are irrational numbers. The number π, which we use in the formulas for the circumference and area of a circle, is also an irrational number. The rational numbers $\frac{22}{7}$ and 3.14 are approximations of π.

In general, if a number, x, is not a perfect square, then $\sqrt{x}$ is an irrational number. Thus, $\sqrt{\frac{1}{2}}$, $-\sqrt{.3}$, and $\sqrt{.07}$ are irrational numbers. However, we may find rational numbers to approximate these square roots to as many decimal places as we wish.

ILLUSTRATIVE PROBLEM 2: Is $\sqrt{15}$ rational or irrational?
Solution: Since $3^2 = 9$ and $4^2 = 16$, then $\sqrt{15}$ is between 3 and 4. That is, $3 < \sqrt{15} < 4$.
Answer: Since 15 is not a perfect square, $\sqrt{15}$ is *irrational*.

ILLUSTRATIVE PROBLEM 3: Between which two consecutive integers is $\sqrt{29}$?
Solution: Since $5^2 = 25$ and $6^2 = 36$, then $\sqrt{29}$ is between 5 and 6.
Answer: $5 < \sqrt{29} < 6$

Exercises

In 1 to 15, determine whether the number is rational or irrational.

1. $\sqrt{121}$ 2. $\sqrt{75}$ 3. $\sqrt{84}$ 4. $-\sqrt{40}$

5. $\sqrt{32}$ 6. $\sqrt{225}$ 7. $\sqrt{59}$ 8. $-\sqrt{400}$

9. $\sqrt{78}$ 10. $\sqrt{128}$ 11. $\sqrt{\frac{3}{4}}$ 12. $-\sqrt{\frac{49}{81}}$

13. $\sqrt{.38}$ 14. $-\sqrt{.36}$ 15. $\sqrt{.9}$

In 16 to 25, between which two consecutive integers is each number?

16. $\sqrt{3}$ 17. $\sqrt{37}$ 18. $\sqrt{69}$ 19. $\sqrt{87}$

20. $\sqrt{145}$ 21. $\sqrt{210}$ 22. $-\sqrt{90}$ 23. $\sqrt{17}$

24. $-\sqrt{70}$ 25. $\sqrt{135}$

26. The area of a square is 19 square meters. Between which two consecutive integral values (in meters) does the value of a side of the square lie?

27. For any two-digit number, how many digits are there in the integral part of its square root?

28. **a.** What is the largest positive integer whose square root is a one-digit integer?
    b. What is the smallest positive integer whose square root is a two-digit integer?

4. APPROXIMATING THE SQUARE ROOT OF A NUMBER

There are several ways of approximating the square root of a number. One simple method is known as **successive division.**

ILLUSTRATIVE PROBLEM: Approximate $\sqrt{21}$ to the nearest tenth.

Solution:

STEP 1. Estimate the square root to the nearest integer. Since $4^2 = 16$ and $5^2 = 25$, $\sqrt{21}$ is between 4 and 5. Thus, $4 < \sqrt{21} < 5$. Note that $\sqrt{21}$ is closer to $\sqrt{25}$ than it is to $\sqrt{16}$. Thus, $\sqrt{21} \approx 5$ (to the nearest integer).

STEP 2. Divide by the estimated value. Since $\sqrt{21}$ is closer to 5, divide 21 by 5, carrying the quotient to one more decimal place than there is in the divisor.

$$\begin{array}{r} 4.2 \\ 5\overline{)21.0} \end{array}$$ Do not round off the quotient.

STEP 3. Average the divisor and quotient above. Since the quotient 4.2 is less than 5, our estimate of 5 was too large. Therefore, $4.2 < \sqrt{21} < 5$. Find the average of 4.2 and 5.

$$\frac{4.2 + 5}{2} = \frac{9.2}{2} = 4.6$$

STEP 4. Now *divide* 21 by 4.6, carrying the quotient to one more decimal place than there is in the divisor.

$$\begin{array}{r} 4.56 \\ 4.6\overline{)21.000} \\ \underline{18\ 4} \\ 2\ 60 \\ \underline{2\ 30} \\ 300 \\ \underline{276} \\ 24 \end{array}$$

Thus, $4.56 < \sqrt{21} < 4.6$

STEP 5. Average. $\dfrac{4.56 + 4.6}{2} = \dfrac{9.16}{2} = 4.58$

Our estimate is 4.58.

Check: $4.58 \times 4.58 = 20.98$ (very close to 21)

These steps may be continued to obtain as close an approximation as we please. Thus, $\sqrt{21}$ is approximately equal to 4.6 to the nearest tenth, or $\sqrt{21} \approx 4.6$ *(answer).*

Exercises _____

In 1 to 10, approximate the square root to the *nearest integer.*

1. $\sqrt{6}$ 2. $\sqrt{11}$ 3. $\sqrt{17}$ 4. $\sqrt{50}$

5. $-\sqrt{70}$ 6. $\sqrt{120}$ 7. $\sqrt{8.5}$ 8. $\sqrt{450}$

9. $-\sqrt{1000}$ 10. $\sqrt{84.75}$

In 11 to 20, approximate the square root to the *nearest tenth.*

11. $\sqrt{5}$ 12. $\sqrt{13}$ 13. $\sqrt{19}$ 14. $\sqrt{23.4}$

15. $-\sqrt{28}$ 16. $\sqrt{37}$ 17. $\sqrt{74}$ 18. $\sqrt{85}$

19. $\sqrt{115}$ 20. $-\sqrt{18.6}$

In 21 to 30, approximate the square root to the *nearest hundredth.*

21. $\sqrt{6}$ 22. $\sqrt{15}$ 23. $\sqrt{31}$ 24. $\sqrt{18.2}$

25. $\sqrt{71}$ 26. $-\sqrt{18}$ 27. $\sqrt{90}$ 28. $-\sqrt{110}$

29. $\sqrt{19.8}$ 30. $\sqrt{41.4}$

---

5. USING A TABLE OF SQUARES AND SQUARE ROOTS

People working in business and industry often find it convenient to use a table to find the squares and square roots of numbers. Such a table appears on page 485.

To find the square of a number, locate the number in the column headed "No." Then read across that row to the column headed "Square" to find the square of the number. Thus, you will find that $(31)^2 = 961$.

To find the square root of a number, locate the number in the column headed "No." Then read across that row to the column headed "Square Root." Thus, you will find that $\sqrt{31} \approx 5.568$. The square roots are given correct to thousandths, but may be rounded off to give a square root to the nearest hundredth or the nearest tenth. Thus, $\sqrt{31} \approx 5.57$ to the nearest hundredth or $\sqrt{31} \approx 5.6$ to the nearest tenth.

The table may also be used to find the square root of any perfect square up to 10,000. For example, determine the square root of 3249. Look for the number 3249 in the columns headed "Square." Then read across the row to the left under "No." and you see that $\sqrt{3249} = 57$.

Note that the table permits you to find the squares and square roots of all numbers from 1 to 150.

Exercises

In 1 to 5, use the table on page 485 to find each of the following:

1. $(23)^2$ **2.** $(37)^2$ **3.** $(51)^2$ **4.** $(67)^2$ **5.** $(93)^2$

In 6 to 15, use the table on page 485 to find the square root to the *nearest hundredth.*

6. $\sqrt{11}$ **7.** $\sqrt{42}$ **8.** $\sqrt{62}$ **9.** $-\sqrt{71}$

10. $\sqrt{45}$ **11.** $\sqrt{18}$ **12.** $-\sqrt{28}$ **13.** $\sqrt{65}$

14. $\sqrt{84}$ **15.** $\sqrt{97}$

In 16 to 25, use the table on page 485 to find the square root to the *nearest tenth.*

16. $\sqrt{7}$ **17.** $\sqrt{15}$ **18.** $\sqrt{27}$ **19.** $-\sqrt{35}$

20. $\sqrt{41}$ **21.** $\sqrt{53}$ **22.** $-\sqrt{60}$ **23.** $\sqrt{68}$

24. $\sqrt{82}$ **25.** $\sqrt{91}$

In 26 to 35, use the table on page 485 to find the square root of each perfect square.

26. $\sqrt{121}$ **27.** $\sqrt{225}$ **28.** $\sqrt{625}$ **29.** $\sqrt{841}$

30. $\sqrt{1089}$ **31.** $\sqrt{1369}$ **32.** $\sqrt{2809}$ **33.** $\sqrt{4096}$

34. $\sqrt{5476}$ **35.** $\sqrt{7921}$

---

6. MULTIPLICATION OF RADICALS

Note that $\sqrt{16} \cdot \sqrt{9} = 4 \cdot 3 = 12$. If we first multiply the radicands, we obtain $\sqrt{16} \cdot \sqrt{9} = \sqrt{16 \cdot 9} = \sqrt{144} = 12$. It thus appears that $\sqrt{16} \cdot \sqrt{9} = \sqrt{16 \cdot 9}$.

Let us try another example.

$$\sqrt{4} \cdot \sqrt{25} = 2 \cdot 5 = 10$$
$$\sqrt{4} \cdot \sqrt{25} = \sqrt{4 \cdot 25} = \sqrt{100} = 10$$

Thus, $\sqrt{4} \cdot \sqrt{25} = \sqrt{4 \cdot 25}$.

To generalize, if c and d are positive numbers, then

$$\sqrt{c} \cdot \sqrt{d} = \sqrt{cd}$$

and we can say that the square root is *distributive over multiplication* of positive numbers.

To multiply $2\sqrt{5}$ by $4\sqrt{3}$, we may use the associative and commutative properties of multiplication as follows:

$$(2\sqrt{5})(4\sqrt{3}) = (2 \cdot 4)(\sqrt{5} \cdot \sqrt{3}) = 8\sqrt{15}$$

To generalize, if c and d are positive numbers and p and q are rational numbers, then

$$p\sqrt{c} \cdot q\sqrt{d} = pq\sqrt{cd}$$

Note that $\sqrt{9 + 16} = \sqrt{25} = 5$. However, $\sqrt{9 + 16} \neq \sqrt{9} + \sqrt{16} = 3 + 4 = 7$. Therefore, the square root is *not* distributive over addition. In general,

$$\sqrt{a + b} \neq \sqrt{a} + \sqrt{b}$$

ILLUSTRATIVE PROBLEM 1: Multiply: $2\sqrt{5} \cdot 3\sqrt{7}$

Solution:
$$2\sqrt{5} \cdot 3\sqrt{7} = 2 \cdot 3\sqrt{5 \cdot 7}$$
$$= 6\sqrt{35} \quad (answer)$$

ILLUSTRATIVE PROBLEM 2: Multiply: $3\sqrt{p} \cdot 4\sqrt{q}$

Solution:
$$3\sqrt{p} \cdot 4\sqrt{q} = 3 \cdot 4\sqrt{p \cdot q}$$
$$= 12\sqrt{pq} \quad (answer)$$

Exercises _____

In 1 to 24, find the products.

1. $\sqrt{2} \cdot \sqrt{3}$ **2.** $\sqrt{5} \cdot \sqrt{3}$ **3.** $\sqrt{7} \cdot \sqrt{6}$ **4.** $\sqrt{2} \cdot \sqrt{11}$

5. $\sqrt{8} \cdot \sqrt{2}$ **6.** $\sqrt{13} \cdot \sqrt{3}$ **7.** $\sqrt{x} \cdot \sqrt{y}$ **8.** $\sqrt{5} \cdot \sqrt{t}$

9. $\sqrt{50} \cdot \sqrt{2}$ **10.** $\sqrt{7} \cdot \sqrt{7}$ **11.** $\sqrt{10} \cdot \sqrt{7}$ **12.** $\sqrt{2a} \cdot \sqrt{b}$

13. $2\sqrt{5} \cdot 4\sqrt{3}$ **14.** $3\sqrt{7} \cdot \sqrt{5}$ **15.** $2\sqrt{3} \cdot 8\sqrt{3}$

16. $3\sqrt{18} \cdot \sqrt{8}$ **17.** $-3\sqrt{2} \cdot 5\sqrt{2}$ **18.** $4\sqrt{m} \cdot 3\sqrt{n}$

19. $2\sqrt{5} \cdot 3\sqrt{45}$ **20.** $3\sqrt{5} \cdot 3\sqrt{5}$ **21.** $5\sqrt{27} \cdot 2\sqrt{\frac{1}{3}}$

22. $5\sqrt{2} \cdot 2\sqrt{32}$ **23.** $-4\sqrt{27} \cdot \sqrt{3}$ **24.** $a\sqrt{r} \cdot b\sqrt{s}$

7. SIMPLIFYING RADICALS

In the preceding section, we concluded that, if the radicands are positive numbers, then

$$\sqrt{ab} = \sqrt{a} \cdot \sqrt{b}$$

We can use this relationship to find the square root of a number that is not included in the table.

ILLUSTRATIVE PROBLEM 1: Find $\sqrt{300}$ to the nearest tenth.

Solution:

$$\sqrt{300} = \sqrt{100 \cdot 3} = \sqrt{100} \cdot \sqrt{3}$$
$$\sqrt{300} = 10\sqrt{3}$$

From the square root table, we find $\sqrt{3} \approx 1.732$. Thus,

$$\sqrt{300} \approx 10(1.732) \approx 17.3 \quad (answer)$$

Note that we could not find $\sqrt{300}$ in the table, since the table lists only the square roots of numbers from 1 to 150. We say that $10\sqrt{3}$ is the **simplest form** of $\sqrt{300}$, since the radicand 3 has no factor, other than 1, that is a perfect square.

To express the square root of an integer in simplest form, write it as an integer or as the product of an integer and the square root of an integer having no square factor other than 1. To simplify the square root of an integer, look for perfect square factors other than 1.

ILLUSTRATIVE PROBLEM 2: Find $\sqrt{160}$ in simplest form.

Solution:

$$\sqrt{160} = \sqrt{16 \cdot 10} = 4\sqrt{10} \quad (answer)$$

Since 10 has no perfect square factors other than 1, then $4\sqrt{10}$ is in simplest form.

In simplifying radicals, try to find the largest possible perfect square factor of the integral radicand.

ILLUSTRATIVE PROBLEM 3: Simplify: $\sqrt{72}$

Solution:

$$\sqrt{72} = \sqrt{36 \cdot 2} = \sqrt{36} \cdot \sqrt{2}$$
$$\sqrt{72} = 6\sqrt{2} \quad (answer)$$

We may apply the same procedure to simplifying the square roots of monomial products where the radicands are non-negative numbers.

Recall that, since $a \cdot a = a^2$, then $\sqrt{a^2} = a$ ($a \geq 0$). Likewise, $\sqrt{a^4} = a^2$, $\sqrt{a^6} = a^3$, etc.

ILLUSTRATIVE PROBLEM 4: Simplify: $\sqrt{9x^2y^4}$

Solution:

$$\sqrt{9x^2y^4} = \sqrt{9} \cdot \sqrt{x^2} \cdot \sqrt{y^4}$$
$$= 3xy^2 \quad (answer)$$

ILLUSTRATIVE PROBLEM 5: Simplify: $\sqrt{48r^3}$

Solution:

$$\sqrt{48 \cdot r^3} = \sqrt{16 \cdot 3 \cdot r^2 \cdot r} = \sqrt{16} \cdot \sqrt{r^2} \cdot \sqrt{3r}$$
$$= 4r\sqrt{3r} \quad (answer)$$

Note that by choosing the largest perfect square factor of the radicand, we can reduce the number of steps needed to simplify the radical.

Exercises

In 1 to 40, simplify each square root.

1. $\sqrt{18}$	**2.** $\sqrt{12}$	**3.** $\sqrt{20}$	**4.** $\sqrt{50}$
5. $\sqrt{80}$	**6.** $\sqrt{75}$	**7.** $-\sqrt{27}$	**8.** $-\sqrt{98}$
9. $\sqrt{54}$	**10.** $\sqrt{108}$	**11.** $\sqrt{150}$	**12.** $-3\sqrt{8}$
13. $5\sqrt{32}$	**14.** $-\sqrt{72}$	**15.** $\sqrt{128}$	**16.** $\frac{1}{3}\sqrt{90}$
17. $\frac{2}{5}\sqrt{125}$	**18.** $\sqrt{84}$	**19.** $\sqrt{147}$	**20.** $\sqrt{180}$
21. $\sqrt{2m^2}$	**22.** $\sqrt{8r^2}$	**23.** $\sqrt{x^3}$	**24.** $\sqrt{18y^4}$
25. $\sqrt{p^5}$	**26.** $\sqrt{48c^3}$	**27.** $\sqrt{9n^6}$	**28.** $\sqrt{2x^4y^8}$
29. $\sqrt{25r^2s}$	**30.** $\sqrt{4a^6}$	**31.** $\sqrt{12t^4}$	**32.** $\sqrt{16y^5}$
33. $\sqrt{a^5n^4}$	**34.** $\sqrt{72b^3}$	**35.** $\sqrt{3u^2}$	**36.** $\sqrt{49xy^3}$
37. $3\sqrt{50p^2}$	**38.** $-5\sqrt{18r^4s^3}$	**39.** $7\sqrt{a^2b}$	**40.** $\sqrt{72x^4y^6}$

In 41 to 48, use the square root table on page 485 to approximate the square root to the nearest tenth.

41. $\sqrt{124}$　　　**42.** $\sqrt{200}$　　　**43.** $\sqrt{126}$　　　**44.** $\sqrt{116}$

45. $\sqrt{245}$　　　**46.** $\sqrt{360}$　　　**47.** $\frac{1}{2}\sqrt{196}$　　　**48.** $\frac{1}{3}\sqrt{405}$

49. Express in terms of x and y a side of a square the area of which is $9x^2y$.

50. Find, to the nearest tenth, a side of a square the area of which is 192.

---

8. DIVISION OF RADICALS

Note that $\dfrac{\sqrt{36}}{\sqrt{9}} = \dfrac{6}{3} = 2$ and that $\sqrt{\dfrac{36}{9}} = \sqrt{4} = 2$.

It thus appears that $\dfrac{\sqrt{36}}{\sqrt{9}} = \sqrt{\dfrac{36}{9}}$.

We may generalize this relationship for positive radicands:

$$\frac{\sqrt{p}}{\sqrt{q}} = \sqrt{\frac{p}{q}}$$

ILLUSTRATIVE PROBLEM 1: Divide and simplify: $\dfrac{\sqrt{32}}{\sqrt{8}}$

Solution:

$$\frac{\sqrt{32}}{\sqrt{8}} = \sqrt{\frac{32}{8}} = \sqrt{4} = 2 \quad (answer)$$

ILLUSTRATIVE PROBLEM 2: Divide and simplify: $\sqrt{54} \div \sqrt{3}$

Solution:

$$\frac{\sqrt{54}}{\sqrt{3}} = \sqrt{\frac{54}{3}} = \sqrt{18} = \sqrt{9 \cdot 2}$$
$$= \sqrt{9} \cdot \sqrt{2}$$
$$= 3\sqrt{2} \quad (answer)$$

ILLUSTRATIVE PROBLEM 3: Divide and simplify: $6\sqrt{12} \div 2\sqrt{3}$

Solution:

$$\frac{6\sqrt{12}}{2\sqrt{3}} = \frac{6}{2}\sqrt{\frac{12}{3}} = 3\sqrt{4}$$

$$= 3 \cdot 2 = 6 \quad (answer)$$

ILLUSTRATIVE PROBLEM 4: Divide and simplify: $9\sqrt{x^3y^2} \div 6\sqrt{xy}$

Solution:

$$\frac{9\sqrt{x^3y^2}}{6\sqrt{xy}} = \frac{9}{6}\sqrt{\frac{x^3y^2}{xy}} = \frac{3}{2}\sqrt{x^2y}$$

$$= \frac{3}{2}\sqrt{x^2}\sqrt{y}$$

$$= \frac{3}{2}x\sqrt{y} \quad (answer)$$

Exercises

In 1 to 24, divide and simplify.

1. $\sqrt{75} \div \sqrt{3}$

2. $\sqrt{144} \div \sqrt{9}$

3. $\sqrt{50} \div \sqrt{2}$

4. $\sqrt{72} \div \sqrt{6}$

5. $\sqrt{81} \div \sqrt{36}$

6. $\sqrt{48} \div \sqrt{3}$

7. $\sqrt{56} \div \sqrt{7}$

8. $\sqrt{125} \div \sqrt{5}$

9. $\sqrt{60} \div \sqrt{3}$

10. $12\sqrt{30} \div 4\sqrt{5}$

11. $15\sqrt{2} \div 5\sqrt{2}$

12. $35\sqrt{6} \div 7\sqrt{6}$

13. $6\sqrt{8} \div 3\sqrt{2}$

14. $9\sqrt{21} \div 6\sqrt{7}$

15. $27\sqrt{80} \div 9\sqrt{10}$

16. $\sqrt{25x} \div \sqrt{x}$

17. $3\sqrt{18a} \div 6\sqrt{2a}$

18. $\sqrt{n^5} \div \sqrt{n^3}$

19. $\sqrt{27t^3} \div \sqrt{9t}$

20. $10\sqrt{x^3y} \div 5\sqrt{xy}$

21. $9\sqrt{5p} \div 3\sqrt{p}$

22. $\dfrac{\sqrt{15m^5}}{\sqrt{5m^3}}$

23. $\dfrac{8\sqrt{18r^4s^2}}{12\sqrt{2r^2s^2}}$

24. $\dfrac{2\sqrt{s^5t^2}}{6\sqrt{st}}$

9. RATIONALIZING THE DENOMINATOR

An expression such as $\dfrac{1}{\sqrt{2}}$ has an irrational number in the denominator. If the denominator is changed to its approximate decimal value from the square root table, the expression becomes $\dfrac{1}{\sqrt{2}} \approx \dfrac{1}{1.414}$.

The process of evaluating this fraction then becomes an exercise in long division. The original fraction can be evaluated more easily by first changing the denominator to a rational number.

If the numerator and denominator of $\dfrac{1}{\sqrt{2}}$ are each multiplied by $\sqrt{2}$, another fraction having the same value is obtained. We are in effect multiplying the fraction by 1, the identity element for multiplication. Thus,

$$\frac{1}{\sqrt{2}} = \frac{1}{\sqrt{2}} \cdot \frac{\sqrt{2}}{\sqrt{2}} = \frac{\sqrt{2}}{2} \approx \frac{1.414}{2} \approx .707$$

Changing a fraction with an irrational number in its denominator to an equivalent fraction with a rational denominator is called ***rationalizing the denominator.***

To do this in the simplest way, multiply the numerator and denominator by the *smallest* radical that will make the denominator of the resulting fraction a rational number.

ILLUSTRATIVE PROBLEM 1: Rationalize the denominator of $\dfrac{6}{\sqrt{3}}$.

Solution:

$$\frac{6}{\sqrt{3}} = \frac{6}{\sqrt{3}} \cdot \frac{\sqrt{3}}{\sqrt{3}} = \frac{6\sqrt{3}}{3} = 2\sqrt{3} \quad (answer)$$

ILLUSTRATIVE PROBLEM 2: Rationalize the denominator of $\dfrac{2}{\sqrt{8}}$.

Solution:

$$\frac{2}{\sqrt{8}} = \frac{2}{\sqrt{8}} \cdot \frac{\sqrt{2}}{\sqrt{2}} = \frac{2\sqrt{2}}{\sqrt{16}} = \frac{2\sqrt{2}}{4} = \frac{\sqrt{2}}{2} \quad (answer)$$

Suppose we had not multiplied by $\dfrac{\sqrt{2}}{\sqrt{2}}$, the smallest appropriate radical,

but had chosen, instead, to multiply by $\dfrac{\sqrt{8}}{\sqrt{8}}$, the most obvious choice

for the given problem. Then,

$$\frac{2}{\sqrt{8}} \cdot \frac{\sqrt{8}}{\sqrt{8}} = \frac{2\sqrt{8}}{8} = \frac{\sqrt{8}}{4}.$$

This result, $\dfrac{\sqrt{8}}{4}$, looks very different from the previously obtained an-

swer of $\dfrac{\sqrt{2}}{2}$. Upon further inspection, we should realize that $\dfrac{\sqrt{8}}{4}$ can be

simplified to obtain the same result.

$$\frac{\sqrt{8}}{4} = \frac{\sqrt{4}\,\sqrt{2}}{4} = \frac{2\sqrt{2}}{4} = \frac{\sqrt{2}}{2} \quad (answer)$$

Note that some time spent at the outset in looking for the smallest radical to be used avoids extra work later.

Exercises

In 1 to 25, rationalize the denominator and simplify the resulting fraction.

1. $\dfrac{3}{\sqrt{3}}$ 2. $\dfrac{1}{\sqrt{7}}$ 3. $\dfrac{10}{\sqrt{5}}$ 4. $\dfrac{6}{\sqrt{2}}$ 5. $\dfrac{2}{\sqrt{6}}$

6. $\dfrac{5}{\sqrt{5}}$ 7. $\dfrac{3}{\sqrt{11}}$ 8. $\dfrac{12}{\sqrt{3}}$ 9. $\dfrac{20}{\sqrt{10}}$ 10. $\dfrac{7}{\sqrt{7}}$

11. $\dfrac{10}{\sqrt{8}}$ 12. $\dfrac{6}{\sqrt{12}}$ 13. $\dfrac{15}{\sqrt{50}}$ 14. $\dfrac{3}{\sqrt{18}}$ 15. $\dfrac{6}{\sqrt{27}}$

16. $\dfrac{8}{\sqrt{20}}$ 17. $\dfrac{4}{3\sqrt{8}}$ 18. $\dfrac{18}{7\sqrt{12}}$ 19. $\dfrac{20}{3\sqrt{5}}$ 20. $\dfrac{12}{\sqrt{32}}$

21. $\dfrac{\sqrt{3}}{\sqrt{5}}$ 22. $\dfrac{\sqrt{2}}{\sqrt{3}}$ 23. $\dfrac{\sqrt{3}}{\sqrt{10}}$ 24. $\dfrac{x}{\sqrt{x}}$ 25. $\dfrac{rs}{\sqrt{s}}$

In 26 to 30, find the value of the fraction to the *nearest tenth*.

26. $\dfrac{3}{\sqrt{2}}$ 27. $\dfrac{6}{\sqrt{6}}$ 28. $\dfrac{9}{\sqrt{3}}$ 29. $\dfrac{3}{\sqrt{12}}$ 30. $\dfrac{15}{\sqrt{5}}$

10. ADDITION AND SUBTRACTION OF RADICALS

Radicals having the same index and the same radicand are called **like radicals.** Thus, $4\sqrt{3}$ and $5\sqrt{3}$ are like radicals. Do you see why? In both $4\sqrt{3}$ and $5\sqrt{3}$, the index is understood to be 2 and the radicand is 3. However, $\sqrt{2}$ and $\sqrt{5}$ are unlike radicals because they have different radicands (2 and 5). Also, $\sqrt[3]{5}$ and $\sqrt{5}$ are unlike radicals because they have different indexes (3 and 2).

Like radicals may be added or subtracted in the same way that we add or subtract like algebraic terms. Thus,

$$4\sqrt{3} + 5\sqrt{3} = (4 + 5)\sqrt{3} = 9\sqrt{3}$$
$$7\sqrt{2} - 3\sqrt{2} = (7 - 3)\sqrt{2} = 4\sqrt{2}$$

To add or subtract like radicals, add or subtract their coefficients and place the result in front of the common radical.

Unlike radicals such as $2\sqrt{5}$ and $3\sqrt{2}$ cannot be combined into a single term. However, some radicals that appear to be unlike may turn out to be like radicals when simplified. They may then be combined by the rule above. For example, the radicals $3\sqrt{8} + 7\sqrt{2}$ appear to be unlike.

But $3\sqrt{8} = 3\sqrt{4 \cdot 2} = 3\sqrt{4} \cdot \sqrt{2} = 3 \cdot 2 \cdot \sqrt{2} = 6\sqrt{2}$

Then $6\sqrt{2} + 7\sqrt{2} = (6 + 7)\sqrt{2} = 13\sqrt{2}.$

Thus, before we can determine whether radicals are like or unlike, we must first simplify them. If they turn out to be like radicals, then we can use the rule above for addition or subtraction.

ILLUSTRATIVE PROBLEM 1: Combine $4\sqrt{3} + 5\sqrt{3} - \sqrt{3}$

Solution:

$$4\sqrt{3} + 5\sqrt{3} - \sqrt{3} = (4 + 5 - 1)\sqrt{3} = 8\sqrt{3} \quad (answer)$$

ILLUSTRATIVE PROBLEM 2: Find the sum of $2\sqrt{18}$ and $\sqrt{50}$.

Solution:
$$2\sqrt{18} + \sqrt{50}$$
$$= 2\sqrt{9 \cdot 2} + \sqrt{25 \cdot 2}$$
$$= 2 \cdot 3\sqrt{2} + 5\sqrt{2}$$
$$= 6\sqrt{2} + 5\sqrt{2}$$
$$= (6 + 5)\sqrt{2} = 11\sqrt{2} \quad (answer)$$

ILLUSTRATIVE PROBLEM 3: Find the difference of $5\sqrt{12}$ and $\dfrac{6}{\sqrt{3}}$.

Solution:

$$5\sqrt{12} - \frac{6}{\sqrt{3}}$$

$$= 5\sqrt{4 \cdot 3} - \frac{6}{\sqrt{3}} \cdot \frac{\sqrt{3}}{\sqrt{3}}$$

$$= 5 \cdot 2\sqrt{3} - \frac{6\sqrt{3}}{3}$$

$$= 10\sqrt{3} - 2\sqrt{3}$$

$$= (10 - 2)\sqrt{3} = 8\sqrt{3} \quad (answer)$$

Exercises

In 1 to 42, combine radicals.

1. $4\sqrt{5} + 7\sqrt{5}$ **2.** $6\sqrt{2} + 3\sqrt{2}$ **3.** $\sqrt{7} + 8\sqrt{7}$

4. $9\sqrt{6} - 4\sqrt{6}$ **5.** $8\sqrt{10} - \sqrt{10}$ **6.** $3\sqrt{11} - 2\sqrt{11}$

7. $3\sqrt{2} + 4\sqrt{2} + \sqrt{2}$ **8.** $6\sqrt{3} - 2\sqrt{3} + \sqrt{3}$

9. $\sqrt{6} + 5\sqrt{6} - 2\sqrt{6}$ **10.** $8\sqrt{3} - 2\sqrt{3} + 5\sqrt{2}$

11. $3\sqrt{7} + 5\sqrt{2} - 2\sqrt{2}$ **12.** $\sqrt{5} + 4\sqrt{5} - \sqrt{10}$

13. $4\sqrt{x} + 3\sqrt{x}$ **14.** $8\sqrt{n} - 3\sqrt{n}$

15. $\sqrt{a} + 5\sqrt{a} - 2\sqrt{a}$ **16.** $\sqrt{3} + \sqrt{48}$

17. $\sqrt{8} - \sqrt{2}$ **18.** $\sqrt{75} - \sqrt{12}$

19. $\sqrt{27} + \sqrt{12}$ **20.** $5\sqrt{3} - \sqrt{27}$

21. $3\sqrt{2} + \sqrt{98}$ **22.** $4\sqrt{2} - \sqrt{18}$

23. $5\sqrt{2} - \sqrt{32}$ **24.** $\sqrt{48} - \sqrt{12}$

25. $2\sqrt{3} + \sqrt{27}$ **26.** $5\sqrt{2} + \sqrt{8}$

27. $\sqrt{125} - \sqrt{45}$ **28.** $3\sqrt{50} - 5\sqrt{18}$

29. $\sqrt{80} - \sqrt{45}$ **30.** $3\sqrt{20} - \sqrt{45}$

31. $\sqrt{72} - \sqrt{50}$ **32.** $5\sqrt{27} - \sqrt{108}$

33. $3\sqrt{2} + \sqrt{50} - \sqrt{72}$ **34.** $\sqrt{12} + 2\sqrt{27} - \sqrt{75}$

35. $3\sqrt{45} + 5\sqrt{20} - 6\sqrt{5}$ **36.** $5\sqrt{6} + \dfrac{12}{\sqrt{6}}$

37. $7\sqrt{12} - \dfrac{9}{\sqrt{3}}$

38. $\sqrt{8} - \dfrac{4}{\sqrt{2}}$

39. $\sqrt{25y} - \sqrt{9y}$

40. $\sqrt{2n^2} + \sqrt{8n^2}$

41. $\sqrt{27r^3} - \sqrt{12r^3}$

42. $\sqrt{a^3} - \sqrt{a}$

Chapter Review Exercises

1. Using the division method, find the following square roots to the *nearest tenth*:

    a. $\sqrt{34}$ **b.** $\sqrt{60}$ **c.** $\sqrt{85}$ **d.** $\sqrt{110}$ **e.** $\sqrt{150}$

2. Using the table of square roots on page 485, check the square roots found in problem 1.

3. Multiply or divide as indicated:

    a. $\sqrt{5} \cdot \sqrt{7}$ **b.** $\sqrt{18} \div \sqrt{2}$ **c.** $\sqrt{3} \cdot \sqrt{11}$

    d. $\sqrt{12} \div \sqrt{2}$ **e.** $\sqrt{2x} \cdot \sqrt{8x}$ **f.** $\sqrt{3ab} \div \sqrt{3b}$

    g. $\sqrt{rs} \div \sqrt{r}$ **h.** $\sqrt{50n} \cdot \sqrt{2n}$

4. Simplify the following radicals:

    a. $5\sqrt{12}$ **b.** $\sqrt{108}$ **c.** $\sqrt{125}$ **d.** $\sqrt{81n^4}$

    e. $\sqrt{a^3}$ **f.** $\sqrt{48x^2}$ **g.** $\dfrac{15}{\sqrt{10}}$ **h.** $\dfrac{9}{\sqrt{3}}$

    i. $\dfrac{6}{\sqrt{8}}$ **j.** $\dfrac{t}{\sqrt{t}}$

5. Approximate the following square roots to the *nearest tenth*:

    a. $\sqrt{84}$ **b.** $\sqrt{500}$ **c.** $\dfrac{6}{\sqrt{3}}$ **d.** $\dfrac{8}{\sqrt{12}}$ **e.** $\sqrt{240}$

6. Combine radicals in each of the following:

    a. $\sqrt{27} - \sqrt{12}$ **b.** $\sqrt{6} + \sqrt{24}$ **c.** $5\sqrt{7} + 2\sqrt{7} - 3\sqrt{7}$

    d. $\sqrt{200} + \sqrt{50}$ **e.** $\sqrt{162} - \sqrt{18}$ **f.** $\sqrt{54} + \sqrt{24} - 2\sqrt{72}$

    g. $\sqrt{45} - \dfrac{10}{\sqrt{5}}$ **h.** $\sqrt{a^3} + \sqrt{9a}$ **i.** $\sqrt{2a^2} + \sqrt{8a^2}$

CHECK YOUR SKILLS

1. Solve for y: $3(4 + y) = 4y$ 2. Factor: $x^2 - 100$

3. Solve for m: $\dfrac{m}{5} + 5 = 14$

4. What is the greatest common factor of $18x$ and $12x^2$?

5. Find $\sqrt{86}$ to the nearest tenth.

6. Find the solution set of the inequality $2x + 7 > 23$.

7. Express the product $(\sqrt{5})(\sqrt{12})$ in simplest form.

8. A soccer team won 8 games and lost 2. What percent of the games did it lose?

9. Solve for m: $3:2 = m:6$ 10. Solve for y: $0.05y = 2.5$

20

Quadratic Equations

1. MEANING OF A QUADRATIC EQUATION

Recall that the **degree of a polynomial** in one variable is equal to the greatest exponent appearing in the polynomial. Thus, $y^3 + 2y + 7$ is a polynomial in y of the third degree. In an equation involving polynomials in one variable only, the equation is said to be in **standard form** when all terms are collected in one member and the other member is 0. For example, $y^3 + 2y + 7 = 0$ is in standard form. Note that the terms are placed in decreasing order of exponents.

The degree of such an equation is the degree of the polynomial. Thus, the equation $y^3 + 2y + 7 = 0$ is of the third degree.

We referred to an equation such as $5x - 4 = 0$ as a first-degree equation. In this equation, it is understood that x has an exponent of 1.

The equation $x^2 - 8x + 7 = 0$ is of the second degree. Such second-degree equations are also called **quadratic equations.**

ILLUSTRATIVE PROBLEM 1: Write the equation $x^3 - 2x^2 = x^2 - 4$ in standard form and state its degree.

Solution: We must change the given equation into standard form. Since we want all terms in one member and since the other member must be zero, let us add the additive inverse of $x^2 - 4$ to both members of the given equation. The additive inverse of $x^2 - 4$ is $-x^2 + 4$.

$$
\begin{array}{rcl}
x^3 - 2x^2 & = & x^2 - 4 \\
\underline{-x^2 + 4} & & \underline{-x^2 + 4} \\
x^3 - 3x^2 + 4 = & 0 & \quad \text{(standard form)}
\end{array}
$$

Answer: Since the largest exponent in the equation $x^3 - 3x^2 + 4 = 0$ is 3, the equation is of the third degree.

ILLUSTRATIVE PROBLEM 2: Write the equation $r(r + 2) = 5$ in standard form and state its degree.

Solution: Multiply the terms in the left member. Collect all terms in one member.

$$r(r + 2) = 5$$
$$r^2 + 2r = 5$$
$$\underline{-5 = -5} \quad \text{(add } -5 \text{ to both members)}$$
$$r^2 + 2r - 5 = 0 \quad \text{(standard form)}$$

Answer: The equation $r^2 + 2r - 5 = 0$ is a second-degree equation, or a quadratic equation.

ILLUSTRATIVE PROBLEM 3: Write the equation $x(x + 5) = x^2 + 9$ in standard form and state its degree.

Solution:

$$x(x + 5) = x^2 + 9 \quad \text{(multiply terms in left member)}$$
$$x^2 + 5x = x^2 + 9$$
$$\underline{-x^2 \qquad -9 = -x^2 - 9} \quad \text{(add } -x^2 - 9 \text{ to both sides)}$$
$$5x - 9 = 0 \qquad \text{(standard form)}$$

Note that, in standard form, the x^2 term has vanished.

Answer: The equation $x(x + 5) = x^2 + 9$ becomes $5x - 9 = 0$ when written in standard form and is of the first degree.

Exercises ————————————————————————————

In 1 to 6, write each equation in standard form.

1. $x^3 + 7x^2 = 5$ **2.** $3x^2 - 2x = x + 9$ **3.** $y^4 + 3y^3 = y^2 - 1$

4. $n(n - 1) = 2n + 3$ **5.** $2t(t + 3) = 2t^2 + 5$ **6.** $p^2 = 9$

In 7 to 16, write each equation in standard form and state its degree.

7. $a^2 = 6$ **8.** $r^3 = 2r - 7$ **9.** $x(x + 3) = 5$

10. $2y + 4 = 9$ **11.** $x = -2$ **12.** $2n^3 + n^4 = 3n^2 - 7$

13. $s^2 - 5 = s(s + 2)$ **14.** $p^2 - 2(p + 3) = p + 1$

15. $y^2 - 16 = 0$ **16.** $x^2(x - 1) = 3x^2 + 4$

———————————————————————————————————————

2. GRAPH OF A QUADRATIC EQUATION IN TWO VARIABLES

Recall that the graph of a first-degree equation in two variables is a straight line. Thus, the graph of $y = 2x + 3$ is a straight line with a slope of 2 and a y-intercept of 3. For this reason, such equations of first degree are said to be linear equations.

Let us now consider the graph of a second-degree equation (quadratic equation) of the form $y = ax^2 + bx + c$, where a, b, and c are constants $(a \neq 0)$.

ILLUSTRATIVE PROBLEM: Graph the equation $y = x^2 - 3$ from $x = -3$ to $x = 3$. The interval is written $(-3 \leq x \leq 3)$.

Solution: Compute a table of values as follows:

x	$x^2 - 3$	y
-3	$(-3)^2 - 3$	6
-2	$(-2)^2 - 3$	1
-1	$(-1)^2 - 3$	-2
0	$0^2 - 3$	-3
1	$1^2 - 3$	-2
2	$2^2 - 3$	1
3	$3^2 - 3$	6

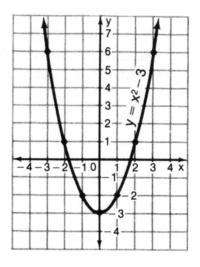

Now plot the points $(-3, 6)$, $(-2, 1)$, $(-1, -2)$, etc. Then draw a smooth curve through the points plotted. This curve is called a **parabola**.

It can be shown that the graph of every quadratic equation of the form $y = ax^2 + bx + c$ $(a \neq 0)$ is a parabola.

Exercises _____

In 1 to 16, graph the quadratic equation in two variables, using the integral values of x in the indicated interval.

1. $y = 2x^2 \ (-3 \leq x \leq 3)$

2. $y = x^2 - 5 \ (-3 \leq x \leq 3)$

3. $y = x^2 + 2x + 1 \ (-4 \leq x \leq 2)$

4. $y = x^2 - 2x + 3 \ (-2 \leq x \leq 4)$

5. $y = 2x^2 - 6 \ (-2 \leq x \leq 2)$ 6. $y = 3x^2 - 5 \ (-2 \leq x \leq 2)$

7. $y = x^2 - 3x \ (-1 \leq x \leq 4)$ 8. $y = -x^2 + 3x \ (-1 \leq x \leq 4)$

9. $y = x^2 - 5x + 4 \ (0 \leq x \leq 5)$ **10.** $y = x^2 + 5x + 4 \ (-5 \leq x \leq 0)$

11. $y = x^2 + 6x - 7 \ (-8 \leq x \leq 2)$

12. $y = -x^2 - 6x + 7 \ (-8 \leq x \leq 2)$

13. $y = 2x^2 - 4x \ (-1 \leq x \leq 3)$ **14.** $y = -2x^2 - 4x \ (-3 \leq x \leq 1)$

15. $y = x^2 + x - 2 \ (-3 \leq x \leq 2)$

16. $y = -x^2 - x + 2 \ (-3 \leq x \leq 2)$

---

3. GRAPHIC SOLUTION OF A QUADRATIC EQUATION

Consider the graph of $y = x^2 - x - 6$ in the interval $-3 \leq x \leq 4$.

x	$x^2 - x - 6$	y
-3	$(-3)^2 - (-3) - 6$	6
-2	$(-2)^2 - (-2) - 6$	0
-1	$(-1)^2 - (-1) - 6$	-4
0	$0^2 - 0 - 6$	-6
1	$1^2 - 1 - 6$	-6
2	$2^2 - 2 - 6$	-4
3	$3^2 - 3 - 6$	0
4	$4^2 - 4 - 6$	6

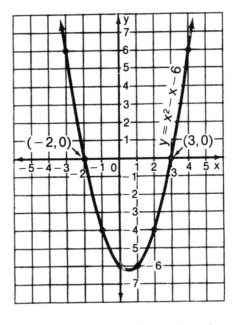

The two x-intercepts of the curve are $(-2, 0)$ and $(3, 0)$. If we let $y = 0$ in the original equation, we obtain the equation $x^2 - x - 6 = 0$. Since $x = 3$ and $x = -2$ satisfy this quadratic equation in x, they are the roots of this equation.

In this case, the roots are integers. But the same method may be used to solve any quadratic equation in one variable. If the roots are fractions or irrational numbers, only an approximation can be obtained from the graph.

   To solve graphically a quadratic equation that has the form $ax^2 + bx + c = 0$ ($a \neq 0$):

1. Set $ax^2 + bx + c = y$.

2. Graph this quadratic equation in two variables (parabola).

3. The abscissas (x-coordinates) of the x-intercepts of the parabola are the roots of the original quadratic equation in x.

Exercises

1. Graph the equation $y = x^2 + x - 6$ in the interval $-4 \leq x \leq 6$. From this graph, determine the roots of $x^2 + x - 6 = 0$.

2. Graph the equation $y = x^2 + 3x - 4$ in the interval $-5 \leq x \leq 2$. From this graph, determine the roots of $x^2 + 3x - 4 = 0$.

3. Graph the equation $y = x^2 - 2$ in the interval $-3 \leq x \leq 3$. From this graph, estimate the roots of $x^2 - 2 = 0$.

   In 4 to 12, solve each quadratic equation graphically. Where necessary, approximate answers are acceptable.

4. $x^2 - 4x - 5 = 0$ 5. $x^2 + 5x - 6 = 0$

6. $x^2 - 3x = 0$ 7. $x^2 + 5x = 0$

8. $x^2 - 5 = 0$ 9. $x^2 + 7 = 6x$

10. $x^2 - x = 0$ 11. $x^2 - x = 5$

12. $2x^2 - x - 6 = 0$

4. SOLVING QUADRATIC EQUATIONS BY FACTORING

   Solving quadratic equations by graphing is time-consuming and frequently inaccurate, particularly when the roots are not integers. Another method often used to solve quadratic equations is *factoring*. This method is demonstrated on the following page.

ILLUSTRATIVE PROBLEM 1: Solve the equation $x^2 - x = 6$.

Solution: First, write the equation in standard form by subtracting 6 from both members.

$$x^2 - x - 6 = 0$$

Now factor the left member.

$$(x - 3)(x + 2) = 0$$

This equation states that the product of two quantities equals zero. If $c \cdot d = 0$, then $c = 0$ or $d = 0$ or both equal zero. Thus, we set each factor equal to zero and solve for x.

$$
\begin{array}{ll}
x - 3 = 0 & x + 2 = 0 \\
\underline{+3 = +3} & \underline{-2 = -2} \\
x = 3 & x = -2
\end{array}
$$

Check by substituting each value of x in the original equation.

Answer: The two roots of the equation are $x = 3$ and $x = -2$.

ILLUSTRATIVE PROBLEM 2: Solve the equation $x^2 - 2x = 0$.

Solution:

$$
\begin{array}{rl}
x^2 - 2x = & 0 \\
x(x - 2) = & 0 \\
x = 0 \,\big|\, x - 2 = & 0 \\
\underline{+2 = } & \underline{+2} \\
x = & 2
\end{array}
$$

Check by substituting each value of x in the original equation.

Answer: The two roots are 0 and 2.

ILLUSTRATIVE PROBLEM 3: Solve the equation $x^2 - 49 = 0$.

Solution:

$$
\begin{array}{rl}
x^2 - 49 = & 0 \\
(x - 7)(x + 7) = & 0 \\
x - 7 = \,0 \,\big|\, x + 7 = & 0 \\
\underline{+7 = +7} \,\big|\, \underline{-7 = -7} \\
x = \,7 \,\big|\, x = -7
\end{array}
$$

Check by substituting each value of x in the original equation.

Answer: The two roots are $+7$ and -7.

Note that, in this case, the two roots are additive inverses of each other.

Every quadratic equation has two roots. In some cases, the two roots may be equal; we call such a root a **double root.**

Remember to put a quadratic equation into standard form before factoring the left member.

Exercises

In 1 to 20, solve and check the quadratic equation.

1. $y^2 + 3y + 2 = 0$ **2.** $n^2 + 9n + 8 = 0$ **3.** $x^2 - 5x - 6 = 0$

4. $p^2 - p - 42 = 0$ **5.** $r^2 - 2r - 3 = 0$ **6.** $t^2 - 4t = 0$

7. $2k^2 + 3k = 0$ **8.** $s^2 - 4s - 12 = 0$ **9.** $2x^2 - 128 = 0$

10. $y^2 - 13y + 12 = 0$ **11.** $c^2 - 36 = 0$ **12.** $5n^2 - 2n = 0$

13. $y^2 = y + 20$ **14.** $r^2 + 5 = 6r$ **15.** $m^2 = 27 - 6m$

16. $2t^2 + t - 6 = 0$ **17.** $p^2 + 4p = 12$ **18.** $2y^2 = 3 - y$

19. $n(n + 5) = 14$ **20.** $c(c - 3) = 28$

In 21 to 28, clear the equation of fractions and solve.

21. $\dfrac{6}{x} + x = 5$ **22.** $\dfrac{5}{x} + 3x = \dfrac{17}{x}$ **23.** $n - 2 = \dfrac{3}{n}$

24. $y - \dfrac{15}{y} = 2$ **25.** $t + \dfrac{3}{t} = 4$ **26.** $\dfrac{8}{d} = \dfrac{d}{2}$

27. $m = \dfrac{15}{m} - 2$ **28.** $g - 2 = \dfrac{35}{g}$

5. SOLVING QUADRATIC EQUATIONS BY FORMULA

Consider the equation $x^2 - x - 1 = 0$. Note that we cannot factor the left member of the equation. The factoring method is suitable only when solving quadratic equations with rational roots. To solve quadratic equations with irrational roots, we must use another method.

Let us consider the standard form of the general quadratic equation.

$$ax^2 + bx + c = 0$$

It can be shown that the roots of this equation can be found by means of the *quadratic formula:*

$$x = \frac{-b \pm \sqrt{b^2 - 4ac}}{2a}$$

ILLUSTRATIVE PROBLEM 1: Solve the following equation by using the quadratic formula: $2x^2 - 3x - 2 = 0$

Solution: Comparing this equation with $ax^2 + bx + c = 0$, we see that $a = 2$, $b = -3$, and $c = -2$. Substitute these values in the quadratic formula.

$$x = \frac{-b \pm \sqrt{b^2 - 4ac}}{2a}$$

$$x = \frac{-(-3) \pm \sqrt{(-3)^2 - 4(2)(-2)}}{2(2)}$$

$$x = \frac{3 \pm \sqrt{9 + 16}}{4} = \frac{3 \pm \sqrt{25}}{4}$$

$$x = \frac{3 \pm 5}{4}$$

$$x = \frac{3 + 5}{4} = \frac{8}{4} = 2; \quad x = \frac{3 - 5}{4} = \frac{-2}{4} = -\frac{1}{2}$$

Answer: The roots are $x = 2$ and $x = -\frac{1}{2}$.

In this case, the roots are rational, and they can be checked by substitution or by solving by the factoring method.

ILLUSTRATIVE PROBLEM 2: Solve the following equation by using the quadratic formula: $x^2 - x - 1 = 0$

Solution:

$$a = 1, b = -1, c = -1$$

$$x = \frac{-b \pm \sqrt{b^2 - 4ac}}{2a}$$

$$x = \frac{-(-1) \pm \sqrt{(-1)^2 - 4(1)(-1)}}{2(1)} = \frac{1 \pm \sqrt{1 + 4}}{2}$$

$$x = \frac{1 \pm \sqrt{5}}{2}$$

Answer: The roots are $\dfrac{1 + \sqrt{5}}{2}$ and $\dfrac{1 - \sqrt{5}}{2}$.

These irrational roots can be approximated by substituting the value of $\sqrt{5}$. Thus,

$$\frac{1 + \sqrt{5}}{2} \approx \frac{1 + 2.24}{2} \approx \frac{3.24}{2} \approx 1.62 \approx 1.6$$

$$\frac{1 - \sqrt{5}}{2} \approx \frac{1 - 2.24}{2} \approx \frac{-1.24}{2} \approx -.62 \approx -.6$$

Answer: To the *nearest tenth*, the roots are 1.6 and $-.6$.

Exercises

In 1 to 14, solve by using the quadratic formula and leave irrational roots in radical form.

1. $x^2 - 3x - 4 = 0$ **2.** $p^2 - 5p + 6 = 0$ **3.** $2n^2 - 3n - 5 = 0$

4. $3t^2 - 5t - 2 = 0$ **5.** $k^2 - 5 = 4k$ **6.** $x^2 - 1 + x = 0$

7. $y^2 - 2y - 2 = 0$ **8.** $r^2 - 4r - 1 = 0$ **9.** $4x^2 - 5 = 0$

10. $2n^2 - 3n = 0$ **11.** $2t^2 - 3t = 9$ **12.** $3s^2 + s = 1$

13. $2q^2 - 5q + 2 = 0$ **14.** $t^2 - 3 = 3t$

In 15 to 20, solve by using the quadratic formula and approximate irrational roots to the nearest tenth.

15. $x^2 + 4x + 2 = 0$ **16.** $y^2 - 5 = 0$ **17.** $r^2 - r - 3 = 0$

18. $2n^2 + 2n - 3 = 0$ **19.** $3p^2 = 21$ **20.** $d^2 + 2d - 1 = 0$

---

6. PROBLEMS INVOLVING QUADRATIC EQUATIONS

In studying mathematics, we find that some problems require the solution of quadratic equations. Some of these problems are illustrated below.

ILLUSTRATIVE PROBLEM 1: If the square of a number is added to 5 times this number, the result is 36. Find the number.

Solution: Let x = the number. Then express the problem as an equation.

$$x^2 + 5x = 36$$

$$x^2 + 5x - 36 = 0 \quad \text{(put into standard form)}$$

$$(x + 9)(x - 4) = 0 \quad \text{(factor)}$$

$x + 9 = 0 \ \bigg| \ x - 4 = 0 \quad$ (set each factor equal to zero)

$x = -9 \ \bigg| \qquad x = 4$

Check both answers in the original problem.

Square of the number: $4^2 = 16$	$(-9^2) = $ 81
Five times the number: $5(4) = 20$	$5(-9) = -45$
$16 + 20 = 36$	$81 + (-45) = $ 36
$36 = 36$ ✔	$36 = $ 36 ✔

Answer: $x = -9$ or $x = 4$

ILLUSTRATIVE PROBLEM 2:   The length of a rectangle exceeds its width by 3 cm. If the area of the rectangle is 40 square centimeters, find the measures of its length and width.

Solution: Let $w = $ width in cm. Then $w + 3 = $ length in cm.

$$\text{area} = \text{length} \times \text{width}$$
$$40 = (w + 3) \times (w), \text{ or}$$
$$w(w + 3) = 40$$
$$w^2 + 3w - 40 = 0 \quad \text{(put into standard form)}$$
$$(w + 8)(w - 5) = 0 \quad \text{(factor)}$$

$w + 8 = $ 0 | $w - 5 = 0$ (set factors equal to zero)

$w = -8$ | $w = 5$

Since the width must be a positive number, we reject $w = -8$ and use only $w = 5$. Thus, w is 5 and $w + 3$ is 8.

The check is obvious since $8 \times 5 = 40$.

Answer: The width is 5 cm and the length is 8 cm.

ILLUSTRATIVE PROBLEM 3:   The formula $s = 16t^2$ gives the distance s, in feet, that an object falls t seconds after it is dropped. How many seconds does it take an object to fall 1600 feet?

Solution: Substitute 1600 for s in the equation $s = 16t^2$ and solve for t.

$$s = 16t^2$$
$$1600 = 16t^2, \text{ or}$$
$$16t^2 = 1600$$
$$t^2 = 100 \quad \text{(divide both members by 16)}$$
$$t = \pm 10$$

Answer: Since only the positive value is meaningful, the solution is $t = 10$ seconds.

The equation $t^2 = 100$ may also be put into the standard form $t^2 - 100 = 0$ and then solved by factoring.

$$t^2 - 100 = 0$$
$$(t + 10)(t - 10) = 0$$
$$t + 10 = 0 \quad | \quad t - 10 = 0$$
$$t = -10 \quad | \quad t = 10$$

Exercises

In 1 to 17, form a quadratic equation. Then solve the problem and check.

1. One of two positive integers is 5 less than the other. If the product of the two integers is 36, find the integers.

2. Find two consecutive positive integers such that the square of the smaller increased by 3 times the larger is equal to 57.

3. Find a positive number that is 42 less than its square.

4. The area of a rectangle is 40. If the length is 6 more than the width, find the dimensions of the rectangle.

5. The length of a rectangle is 5 more than its width. The area of the rectangle equals the area of a square whose side is 6. Find the dimensions of the rectangle.

6. The sum of a positive number and the square of its additive inverse is 30. What is the number?

7. Find three consecutive positive odd integers such that the square of the smallest exceeds twice the largest by 7.

8. If one side of a square is increased by 2 and an adjacent side is decreased by 2, the area of the resulting rectangle is 32. Find a side of the square.

9. The length of a rectangle is two inches more than the width. If the area of the rectangle is 48 square inches, find the number of inches in the width of the rectangle.

10. If the square of a positive number is added to 4 times the number, the result is 12. Find the number.

11. Find the smallest of three consecutive positive *odd* integers such that the product of the second and third is 35.

12. Find three consecutive positive integers such that the square of the smallest integer exceeds the largest by 10.

13. The length of a rectangle is 5 cm more than the width, and the area is 84 square centimeters. Find the width of the rectangle.

14. The difference between a number and its reciprocal is $\frac{8}{3}$. Find the number.

15. The sum of a number and its reciprocal is $\frac{5}{2}$. Find the number.

16. The height of a rectangular solid is 8 cm, and the length of its base is 6 cm more than its width. If the volume of the solid is 216 cubic centimeters, find the length of the base in centimeters. (Recall $V = lwh$.)

17. An object is dropped from a height of 3200 feet. Find, to the *nearest integer*, the number of seconds it takes to reach the ground. (Use $s = 16t^2$, where s is the distance the object falls in feet, and t is the number of seconds after it is dropped.)

---

7. USING THE PYTHAGOREAN RELATIONSHIP FOR THE RIGHT TRIANGLE

Recall that a right triangle is a triangle having one right angle. The side opposite the right angle is the longest side and is called the **hypotenuse** of the triangle (c in the figure). The two sides that form the right angle are called the **legs** of the triangle (a and b in the figure).

About 2500 years ago, a Greek mathematician, Pythagoras, proved the following important relationship among the sides of a right triangle.

In any right triangle, the square of the length of the hypotenuse is equal to the sum of the squares of the lengths of the legs.

$$c^2 = a^2 + b^2$$

By means of this relationship, we can find one side of a right triangle if the other two sides are given.

ILLUSTRATIVE PROBLEM 1: The legs of a right triangle are 6 and 8. Find the length of the hypotenuse.

Solution:
$$c^2 = a^2 + b^2$$
$$c^2 = 8^2 + 6^2 = 64 + 36$$
$$c^2 = 100$$
$$c = \pm 10$$

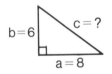

Answer: Since the length of a side of a triangle is always positive, the hypotenuse is 10.

ILLUSTRATIVE PROBLEM 2: The length of the hypotenuse of a right triangle is 15, and the length of one leg is 9. Find the length of the other leg.

Solution:

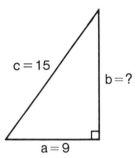

$$c^2 = a^2 + b^2$$
$$15^2 = 9^2 + b^2$$
$$225 = 81 + b^2$$
$$b^2 = 144 \quad \text{(subtract 81 from both members)}$$
$$b = \pm 12 \quad \text{(reject the negative answer)}$$
$$b = 12 \quad (answer)$$

ILLUSTRATIVE PROBLEM 3: Find the length of the diagonal of a rectangle the lengths of whose sides are 7 and 9.

Solution: Let x represent the length of the diagonal.

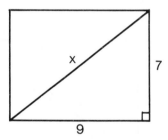

$$x^2 = 7^2 + 9^2$$
$$x^2 = 49 + 81$$
$$x^2 = 130$$
$$x = \pm\sqrt{130}$$
$$x = \sqrt{130} \quad (answer)$$

If we desire the result to the *nearest tenth*, we can show from the table of square roots that $\sqrt{130} \approx 11.4$.

ILLUSTRATIVE PROBLEM 4: Find, correct to the nearest tenth, the side of a square whose diagonal is 12.

Solution: Let x = the length of a side of the square.

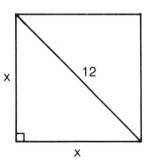

$$x^2 + x^2 = 12^2$$
$$2x^2 = 144$$
$$x^2 = 72$$
$$x = \sqrt{72} \text{ (reject } -\sqrt{72})$$
$$x = \sqrt{72} = \sqrt{36 \cdot 2} = \sqrt{36} \cdot \sqrt{2}$$
$$x \approx 6(1.414) \approx 8.484 \approx 8.5 \quad (answer)$$

Exercises

In 1 to 9, use the data marked in the figure to find the value of x. Leave irrational answers in radical form.

1.

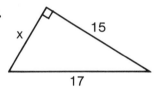

2.

3.

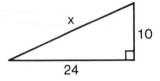

4.

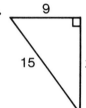

5.

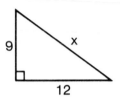

6.

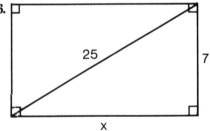

7.

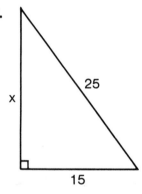

8.

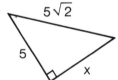

9.

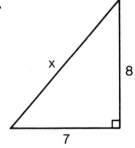

10. A rectangular door is 5 feet wide and 12 feet high. What is the length, in feet, of a diagonal of the door?

11. A ladder 50 feet long leans against a building and reaches a point on the building that is 48 feet above the level ground. How far, in feet, from the bottom of the building is the foot of the ladder?

12. A ship travels 15 miles north and 20 miles west. How many miles is the ship from its starting point?

13. The hypotenuse of a right triangle is 8, and one leg is 4. Find, to the *nearest integer*, the length of the other leg.

14. A 15-foot ladder is leaning against a wall of a building. The bottom of the ladder is 6 feet away from the wall on level ground. Find, to the *nearest foot*, the distance from the top of the ladder to the ground.

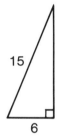

15. The isosceles right triangle shown has two equal sides each of length 20. Find the length of the hypotenuse to the *nearest integer*.

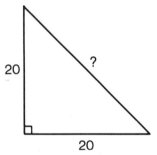

16. The two legs of a right triangle are in the ratio 3:4. If the hypotenuse is 30 inches, find the number of inches in the shorter leg.

17. A wire is used to brace a vertical pole 30 feet high. The wire is attached to a ground stake 16 feet from the base of the pole and run to the top of the pole. How many feet of wire are needed?

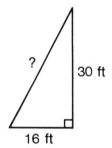

18. A baseball diamond is a square 90 feet on each side. How far is it, to the *nearest foot*, from home plate to second base? (*Hint:* Find the distance measured along a diagonal.)

19. In triangle RST, $RS = ST = 10$ and $RT = 12$. Altitude SM is drawn perpendicular to RT and bisects RT. ($RM = MT$.) Find the length of the altitude SM.

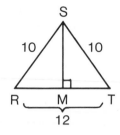

20. In equilateral triangle PQR, $PQ = QR = PR = 20$. Altitude QS is drawn perpendicular to PR and bisects PR. Find, to the *nearest integer*, the length of altitude QS.

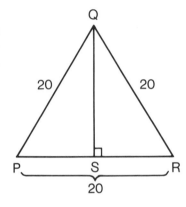

21. A rectangular lot $PQRS$ is 15 yards wide and 36 yards long. How many yards are saved by going from P to R along the diagonal rather than walking along PS and SR?

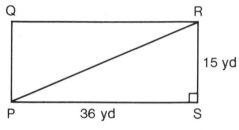

22. The diagonal of a rectangle is 25 and its height is 7.

    a. Find the length of the base of the rectangle.

    b. Find the area of the rectangle.

23. The legs of a right triangle are in the ratio of $2:3$ and the hypotenuse is 26. Find the length of the longer leg to the *nearest tenth*.

24. The hypotenuse of a right triangle is 20 and one of the legs is 4 more than the other. Find the length of the shorter leg.

25. A freighter and a passenger ship started from the same port at the same time. The freighter traveled north at a rate of 15 miles per hour and the passenger ship traveled west at 36 miles per hour. How many miles apart were they at the end of one hour?

Chapter Review Exercises

1. **a.** Graph $y = x^2 - 4x - 5$ in the interval $-2 \le x \le 6$.
   b. From the graph in **a**, find the roots of $x^2 - 4x - 5 = 0$.
   c. Solve the equation in **b** by factoring.

2. **a.** Graph $y = x^2 - 2x - 1$ in the interval $-1 \le x \le 3$.
   b. From the graph in **a**, find the roots of $x^2 - 2x - 1 = 0$ to the *nearest tenth*.
   c. By use of the *quadratic formula*, find the roots of the equation $x^2 - 2x - 1 = 0$ to the *nearest tenth*.

In 3 to 10, solve each equation by factoring.

3. $y^2 - 8y = 20$ **4.** $r^2 = 2r + 8$ **5.** $p^2 + 2 = 3p$

6. $2x^2 = 72$ **7.** $n^2 = 7n$ **8.** $t^2 + 10t + 25 = 36$

9. $2d^2 - d = 1$ **10.** $3k^2 + 2k = 5$

In 11 to 14, solve each equation by use of the quadratic formula. Leave irrational roots in radical form.

11. $s^2 - 2s - 3 = 0$ **12.** $2p^2 + 2p = 1$

13. $n^2 - n - 5 = 0$ **14.** $3k^2 - 2 = -2k$

15. The length of a rectangle is 4 more than its width. The area of the rectangle is 140. Find its width.

16. The ratio of the legs of a right triangle is 1:2. The area of the triangle is 32. Find the length of the shorter leg to the *nearest tenth*.

17. A 15-foot ladder is leaning against the side of a building. If the foot of the ladder is 9 feet from the building on level ground, how many feet above ground is the top of the ladder?

18. The sum of a number and twice its reciprocal is $\frac{11}{3}$. Find the number.

19. The product of two consecutive even numbers is 120. Find the smaller number.

CHECK YOUR SKILLS

1. What value of t will make the expression $\dfrac{t+1}{t-3}$ undefined or meaningless?

2. Find the solution set of $x^2 - 2x = 15$.

3. Find the value of $|-7| - |3|$.

4. Two numbers are in the ratio 3:2 and their sum is 10. Find the smaller number.

5. Find the solution set of $2(m - 3) = -12$.

6. The length of a rectangle is 8 and its width is represented by W. Express the perimeter of the rectangle in terms of W.

7. Find the sum of $2\sqrt{5}$ and $\sqrt{45}$.

8. Carlos buys a television set for \$600. If the sales tax rate is 7%, find the amount of sales tax.

9. Multiply: $3\frac{3}{4} \times 1\frac{1}{5}$ 10. Divide 35 by 0.5.

21

Similar Triangles and the Trigonometry of the Right Triangle

1. SIMILAR TRIANGLES

In the diagram below are two triangles that have the same shape but not the same size. Triangles that have the same shape but do not have the same size are called **similar triangles.**

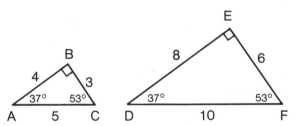

Let us set up a correspondence between triangles ABC and DEF by pairing vertex A and vertex D, vertex B and vertex E, vertex C and vertex F. Every two angles whose vertices we have paired are called a pair of **corresponding angles.** For example, $\angle A$ and $\angle D$, $\angle B$ and $\angle E$, $\angle C$ and $\angle F$ are pairs of corresponding angles. Notice that the corresponding angles are equal.

Now consider the sides that are opposite the two vertices that were paired. We call these sides a pair of **corresponding sides.** For example, AB and DE, BC and EF, AC and DF are pairs of corresponding sides. Notice that the ratios of the lengths of the corresponding sides are equal $\left(\dfrac{AB}{DE} = \dfrac{4}{8} = \dfrac{1}{2}, \dfrac{AC}{DF} = \dfrac{5}{10} = \dfrac{1}{2}, \dfrac{BC}{EF} = \dfrac{3}{6} = \dfrac{1}{2} \right)$. This means that the

lengths of the corresponding sides are *proportional.*

We can use this description as a guide in defining *similar polygons*. The following definition will be useful in our work in this chapter.

Two polygons are *similar* if and only if

(1) their corresponding angles are equal, and

(2) their corresponding sides are proportional; that is, the ratios of the measures of their corresponding sides are equal.

Since triangles are polygons, we can prove that two triangles are similar by showing that they satisfy the two conditions required for similar polygons. In the case of two triangles, however, both conditions do not have to be shown in order to prove that two triangles are similar. We need to show only the *first* condition, namely, that the corresponding angles of the two triangles are equal. It then follows that the corresponding sides will be proportional.

Study the following diagram.

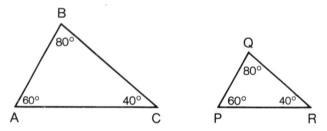

In the diagram, $\angle A = \angle P$, $\angle B = \angle Q$, and $\angle C = \angle R$. Therefore, the two triangles are similar and the sides must be proportional. That is,

$$\frac{AB}{PQ} = \frac{BC}{QR} = \frac{AC}{PR}$$

Note that AB and PQ are corresponding sides because they lie opposite a pair of equal angles, $\angle C$ and $\angle R$. When we pair corresponding angles in two similar triangles, we can then pair the corresponding sides that lie opposite the equal angles.

Recall that, in any triangle, the sum of the measures of the three angles is $180°$. Using this fact, we can easily draw the following conclusion:

If two angles of one triangle are equal to two angles of another triangle, then the third pair of angles must be equal. This idea is demonstrated in the following problem.

ILLUSTRATIVE PROBLEM 1: Use the diagram below to find the measures of ∠ C and ∠ R.

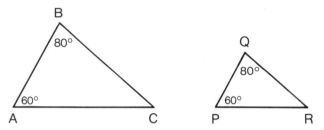

Solution: Since the sum of the measures of three angles of a triangle is 180°, we know that in triangle *ABC*, ∠ *A* + ∠ *B* + ∠ *C* = 180°. Since ∠ *A* + ∠ *B* = 140°, it follows that ∠ *C* = 40°.

Likewise, in triangle *PQR*, we know that ∠ *P* + ∠ *Q* + ∠ *R* = 180°. Since ∠ *P* + ∠ *Q* = 140°, it follows that ∠ *R* = 40°.

Thus, ∠ *C* = ∠ *R* = 40° (*answer*)

We can now state the following principle:

Principle: Two triangles are similar if *two* angles of one triangle are equal to *two* corresponding angles of the other triangle.

To prove that two triangles are *similar*, we only have to show that *two* pairs of corresponding angles are equal.

ILLUSTRATIVE PROBLEM 2: Use the diagram below to find the lengths of the sides *PQ* and *QR*.

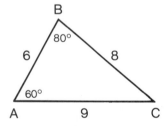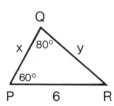

Solution: Since ∠ *A* = ∠ *P* and ∠ *B* = ∠ *Q*, the triangles are similar. Therefore, the corresponding sides are in proportion and we may write

$$\frac{AC}{PR} = \frac{AB}{PQ} = \frac{BC}{QR}$$

This one statement can be written as two separate proportions.

$$\frac{AC}{PR} = \frac{AB}{PQ} \quad \text{and} \quad \frac{AC}{PR} = \frac{BC}{QR}$$

Substitute the values for AC, PR, AB, and BC in each proportion. If we let $PQ = x$ and $QR = y$, we get the following two proportions:

$$\frac{9}{6} = \frac{6}{x} \qquad \frac{9}{6} = \frac{8}{y}$$

Recall that in a proportion, the product of the means equals the product of the extremes. To solve these two proportions, we can cross-multiply.

$$\frac{9}{6} = \frac{6}{x} \qquad \frac{9}{6} = \frac{8}{y}$$

$$9 \cdot x = 6 \cdot 6 \qquad 9 \cdot y = 6 \cdot 8$$

$$9x = 36 \qquad 9y = 48$$

$$x = 4 \qquad y = 5\tfrac{1}{3}$$

Answer: $PQ = 4$ and $QR = 5\tfrac{1}{3}$

ILLUSTRATIVE PROBLEM 3:   A vertical flagpole casts a shadow 15 feet long at the same time that an 8-foot pole casts a shadow 6 feet long. What is the height of the flagpole?

Solution:

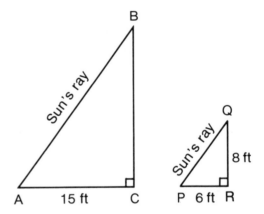

To translate the words of the problem into a workable diagram, we should draw two right triangles as shown. By doing this, we are making the following assumptions:

(1) The ground is a horizontal line.
(2) Each pole is a vertical line.
(3) The two rays from the sun can be thought of as striking the ground at equal angles.

Thus, ∠ *C* = ∠ *R*, since they are both right angles, ∠ *A* = ∠ *P*, and triangle *ABC* is similar to triangle *PQR*. Since the corresponding sides are in proportion, we may write the following proportion:

$$\frac{BC}{QR} = \frac{AC}{PR}$$

If we let side *BC* = *x*, we can now substitute values and solve for *x*.

$$\frac{x}{8} = \frac{15}{6}$$

$$6x = 120$$

$$x = 20$$

Answer: The flagpole is 20 feet high.

Exercises

1. In the two triangles shown below, ∠ *D* = ∠ *A* and ∠ *E* = ∠ *B*. Find the values of *x* and *y*.

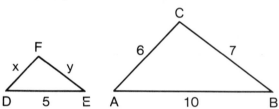

2. The two right triangles below are similar. Find the values of *r* and *s* in the smaller triangle.

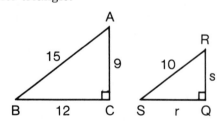

3. A building casts a shadow 120 feet long at the same time that a vertical post 5 feet high casts a shadow 8 feet long. Find the height of the building.

4. A man 6 feet tall casts a shadow 5 feet long at the same time that a nearby building casts a shadow 60 feet long. How high is the building?

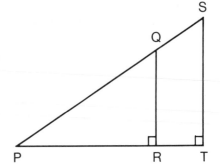

5. a. In the accompanying figure, why is triangle PQR similar to triangle PST?

   b. If $PT = 12$, $ST = 8$, and $PR = 9$, what is the length of QR?

6. A tree 21 feet tall casts a 9-foot shadow. At the same time, a vertical pole casts a 12-foot shadow. What is the height of the pole?

7. In a photograph, the height of a man is 6 centimeters and the height of a boy is 4 centimeters. If the man is 6 feet tall, what is the height of the boy? (*Hint:* The actual heights are proportional to the lengths given in the problem.)

8. A person 5 feet tall is standing near a tree 30 feet high. If the person's shadow is 4 feet long, how long is the shadow of the tree?

9. A student 5 feet tall stands next to a flagpole 45 feet tall. If the shadow cast by the student is 8 feet long, how long is the shadow cast by the flagpole?

---

2. THE TANGENT RATIO

It is sometimes difficult or impossible to measure distances and heights directly. One of the most common techniques of *indirect measurement* is **trigonometry**, which means "measurement of triangles." Distances such as the distance across a pond or the height of a building can be measured indirectly by using trigonometry. Surveyors, engineers, space scientists, and others who use indirect measurement need trigonometry in their daily work.

In this section we will study the trigonometry of a right triangle.

Since we will be discussing the right triangle, it is important to first recall some ways in which the angles and sides of a right triangle are described. Consider right triangle ABC shown here, in which $\angle C$ is the right angle.

Each angle can be referred to by one letter or by three letters. The letter C, for example, is at the *vertex* of the right angle. The *vertex* is the point where the sides of the angle meet. We can call this angle right angle C or $\angle BCA$ or $\angle ACB$. (Note that C, the vertex, is in the middle of the three letters.) Similarly, $\angle A$ can be called $\angle BAC$ or $\angle CAB$. Can $\angle A$ be called $\angle CBA$? In what other ways can $\angle B$ be named?

Each side of triangle ABC can be named by two capital letters or by one lower case letter. The three sides of triangle ABC are side AB (or c), side AC (or b), and side BC (or a). Notice that side a is opposite $\angle A$, side b is opposite $\angle B$, and side c is opposite $\angle C$. Note also that side c, the side opposite the right angle, is called the *hypotenuse* and sides a and b are called the *legs* of right triangle ABC. The hypotenuse, c, is always the longest side of a right triangle.

In naming the parts of a right triangle, we also refer to *adjacent legs*. We say that side b is adjacent to $\angle A$ and side a is adjacent to $\angle B$. When we speak of an adjacent leg, it is only with reference to the acute angles, never the right angle. Side c is never called an adjacent leg since it is always referred to as the hypotenuse.

Starting with $\angle RST$ (shown in the diagram), drop several lines from various points on side SR perpendicular to side ST. Let us call these perpendicular lines AB, CD, EF, etc. Note that triangles ABS, CDS, and EFS are formed and each triangle contains a right angle. Since all these right triangles contain $\angle RST$ (or $\angle S$), it follows that the triangles are all similar to one another. We can now write the following equal ratios:

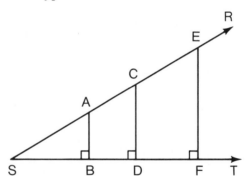

$$\frac{AB}{SB} = \frac{CD}{SD} = \frac{EF}{SF}, \text{ etc.}$$

It appears that, for any given $\angle S$, where $\angle S$ is an acute angle of a right triangle, the *ratio* of the length of the leg *opposite* the angle to the length of the leg *adjacent* to the angle is always the same. This ratio is called the **tangent ratio** and it is defined below:

The *tangent* of an acute angle of a right triangle is the ratio of the length of the leg opposite the acute angle to the length of the leg adjacent to the acute angle.

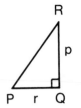

We abbreviate the tangent ratio as "tan." Thus, in the accompanying diagram, the tangent of ∠ P, abbreviated tan P, is

$$\tan \angle P = \frac{\text{length of leg opposite } \angle P}{\text{length of leg adjacent to } \angle P} = \frac{RQ}{PQ} = \frac{p}{r}$$

As the measure of ∠ P changes, the tangent ratio for ∠ P also changes. The tangent ratio for ∠ P depends upon the measure of ∠ P, not upon the size of the right triangle that contains ∠ P.

Mathematicians have prepared a table of values for the tangent ratios of acute angles whose measures are between 0° and 90°. This table, which is called a table of trigonometric functions, appears on page 486.

To find tan 20°, for example, we first look in the column headed "Angle" for the angle 20°. Then, in the column headed "Tangent" on the same horizontal line as 20°, we find the number .3640. Thus, tan 20° is .3640. Similarly, we can find tan 60° to be 1.7321.

We may also use the table to find the measure of an angle when its tangent ratio is known. For example, if tan R = 2.1445, we first look for 2.1445 in the "Tangent" column, then look on the same horizontal line in the "Angle" column to find 65° as the angle. Thus, if tan R = 2.1445, then ∠ R = 65°.

How would we find ∠ P if tan P = ⅔? First, change the fractional value to a decimal value, so that tan P = .6667 (correct to four decimal places). Although this value is not in the "Tangent" column, it is between .6494 (tan 33°) and .6745 (tan 34°). Since .6667 − .6494 = .0173 and .6745 − .6667 = .0078, we see that .6667 is closer to .6745. Thus, we say that if tan P = ⅔, then ∠ P = 34° (to the nearest degree).

Exercises

In 1 to 6, use the table on page 486 to find each value.

1. tan 12° **2.** tan 40° **3.** tan 50°

4. tan 70° **5.** tan 89° **6.** tan 1°

In 7 to 12, use the table on page 486 to find ∠ P.

7. tan P = .4877 **8.** tan P = .9004 **9.** tan P = 1.3270

10. tan P = 2.0503 **11.** tan P = 1.000 **12.** tan P = 7.1154

In 13 to 18, use the table on page 486 to find ∠ R to the nearest degree.

13. $\tan R = .2254$ **14.** $\tan R = .6300$ **15.** $\tan R = .5$

16. $\tan R = \frac{3}{4}$ **17.** $\tan R = \frac{8}{3}$ **18.** $\tan R = \frac{5}{12}$

In 19 to 21, use the definition of the tangent of an acute angle to represent the tangent of each acute angle in the given figures.

19.

20.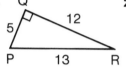

21.

22. Does the tangent of an angle increase or decrease as the measure of the angle varies from 1° to 89°?

23. a. Use the table on page 486 to determine whether tan 50° is twice tan 25°.

b. If the measure of an angle is doubled, is the tangent of the angle also doubled?

24. Using the accompanying figure, find tan ∠x to the nearest ten-thousandth and ∠x to the nearest degree.

3. APPLICATIONS OF THE TANGENT RATIO

Suppose a person is standing at point P and uses a telescope or similar instrument to sight the top of a tree.

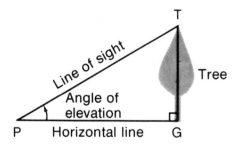

The person must elevate (raise) the instrument from a horizontal position. The line *PT* from the observer, *P*, to the top of the tree, *T*, is called the *line of sight*. The angle formed by the line of sight, *PT*, and the horizontal line, *PG*, is called the **angle of elevation.**

Similarly, suppose a person in a lighthouse tower wishes to sight a boat below.

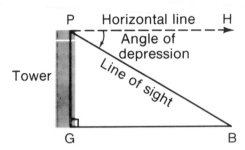

The person must depress (lower) the instrument from a horizontal position. The line *PB* from the observer, *P*, to the boat, *B*, is called the *line of sight*. The angle formed by the horizontal line, *PH*, and the line of sight, *PB*, is called the **angle of depression.**

Note that if we find the angle of elevation of *P* from *B* (∠ *PBG*) and also find the angle of depression of *B* from *P* (∠ *HPB*), we see that both angles are *equal* in measure.

ILLUSTRATIVE PROBLEM 1: From a point on the ground 50 feet from the foot of a flagpole, the angle of elevation of the top of the pole measures 38°. Using the table on page 486, find the height of the pole to the nearest foot.

Solution: Draw a diagram and label all parts of the triangle involved.

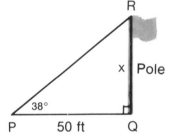

Use the tangent ratio.

$$\tan \angle P = \frac{\text{length of leg opposite } \angle P}{\text{length of leg adjacent to } \angle P} = \frac{RQ}{PQ}$$

$$\tan \angle P = \frac{RQ}{PQ}$$

Let $RQ = x$.

Then $\tan 38° = \dfrac{x}{50}$.

$$.7813 = \frac{x}{50}$$

$x = 50(.7813)$ (multiply both members by 50)

$x = 39.065$

Answer: The height of the pole is 39 feet to the nearest foot.

ILLUSTRATIVE PROBLEM 2: From the top of a lighthouse 180 feet high, the angle of depression of a boat at sea measures 32°. Using the table on page 486, find, to the nearest foot, the distance from the boat to the base of the lighthouse.

Solution: Draw a diagram like the one shown.

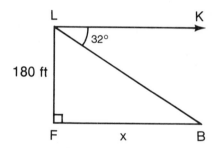

Since $\angle KLB$ is not in right triangle BLF, we must find an angle that is in the right triangle. Since the horizontal line LK and the vertical lighthouse FL form a 90° angle, we subtract 32° from 90°, giving $90° - 32° = 58°$. Thus, $\angle BLF = 58°$.

$$\tan \angle BLF = \frac{\text{length of leg opposite } \angle BLF}{\text{length of leg adjacent to } \angle BLF} = \frac{BF}{LF}$$

If we let $BF = x$, then $\tan 58° = \dfrac{x}{180}$.

$$1.6003 = \frac{x}{180}$$

$x = 180(1.6003)$ (multiply both members by 180)

$x = 288.054$

Answer: The distance from the boat to the base of the lighthouse is 288 feet to the nearest foot.

Alternate Solution: In the preceding problem, we could have used ∠ *B*, whose measure is also 32°. Our equation would then have been

$$\tan 32° = \frac{180}{x}$$

$$.6249 = \frac{180}{x}$$

$$.6249x = 180 \quad \text{(multiply both sides by } x\text{)}$$

$$x = \frac{180}{.6249} \approx 288 \quad (answer)$$

Notice that the final step required long division, which we avoided in the first method. This fact suggests that if the use of the tangent of one of the acute angles in a right triangle requires long division, we can avoid the long division by using the tangent of the other acute angle.

ILLUSTRATIVE PROBLEM 3:   A tower is 140 feet high. What is the angle of elevation to the top of the tower from a point 160 feet from the base of the tower?

Solution: Draw a diagram like the one shown.

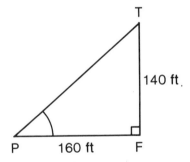

$$\tan \angle P = \frac{\text{length of leg opposite } \angle P}{\text{length of leg adjacent to } \angle P} = \frac{TF}{PF}$$

$$\tan \angle P = \frac{140}{160} = \frac{7}{8}$$

$$\tan \angle P = .8750$$

Notice that .8750 is not in the tangent column of the table on page 486. It is between .8693 (tan 41°) and .9004 (tan 42°). By taking differences, we see that .8750 is nearer to .8693. Thus, ∠ *P* = 41° to the nearest degree.

Answer: The angle of elevation is 41°.

Exercises

In 1 to 6, find, to the nearest integer, the length of the side indicated by x in each of the triangles.

1.

2.

3.

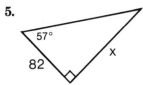

70

41°

x

x

32°

130

x

54°

20

4.

35   x

48°

5.

57°

82   x

6.

x

28°

40

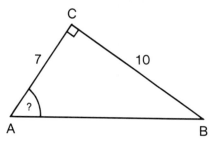

In 7 to 9, find, to the nearest degree, the measure of the angle marked with a "?".

7. R

8

?

Q   5   P

8.   C

7   10

?

A   B

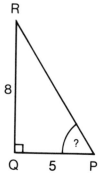

9.   T

R   ?

22   54

S

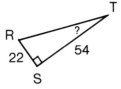

10. The lengths of the legs of a right triangle are 6 and 8. Find, to the nearest degree, the measure of the smaller acute angle.

11. In triangle ABC, $\angle A = 22°$ and $\angle C = 90°$. If $AC = 5$, find BC to the nearest integer.

12. In rectangle $ABCD$, diagonal AC is drawn. If $BC = 9$ and $AB = 15$, find $\angle ACB$ to the nearest degree.

13. Find, to the nearest degree, the measure of the angle of elevation of the sun if a vertical pole 6 feet long casts a horizontal shadow 4 feet long.

14. In triangle ABC, $\angle B = 56°$ and $\angle C = 90°$. If $BC = 16$, find AC to the nearest integer.

15. When the measure of the angle of elevation of the sun is $35°$, the shadow of a tree on level ground is 150 feet long. Find the height of the tree to the nearest foot.

16. An observer in a lookout tower 150 feet high spots a fire on the ground at an angle of depression of $24°$. How many feet is the fire from the foot of the tower if the tower is in a flat plain?

17. What is the angle of elevation of the sun when a vertical pole casts a shadow twice its height?

---

4. THE SINE AND COSINE RATIOS

Consider the accompanying diagram.

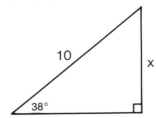

Is the tangent ratio helpful in determining x? This time the hypotenuse is involved. Since the tangent ratio does not involve the hypotenuse, the tangent ratio will not help us to find x in this case.

Thus, we will now define two other ratios involving the lengths of sides of the right triangle that do involve the hypotenuse.

One such ratio is called the **sine ratio.**

The **sine** of an acute angle of a right triangle is the ratio of the length of the leg opposite the angle to the length of the hypotenuse.

We abbreviate the sine ratio as "sin." Thus, in accompanying triangle PQR, the sine of angle P is

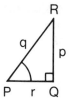

$$\sin \angle P = \frac{\text{length of leg opposite } \angle P}{\text{length of hypotenuse}}$$

$$\sin \angle P = \frac{RQ}{PR} = \frac{p}{q}$$

In like manner, $\sin \angle R = \dfrac{PQ}{PR} = \dfrac{r}{q}$.

The other ratio is called the **cosine ratio.**

The *cosine* of an acute angle of a right triangle is the ratio of the length of the leg adjacent to the angle to the length of the hypotenuse.

We abbreviate the cosine ratio as "cos." In right triangle PQR above,

$$\cos \angle P = \frac{\text{length of leg adjacent to } \angle P}{\text{length of hypotenuse}}$$

$$= \frac{PQ}{PR} = \frac{r}{q}$$

In like manner, $\cos \angle R = \dfrac{RQ}{PR} = \dfrac{p}{q}$.

The table on page 486 lists the values of the sines and cosines. We look in the column labeled "Angle" for the measure of the angle and we look in the columns labeled "Sine" and "Cosine" for the values of these ratios. Thus, $\sin 32° = .5299$ and $\cos 56° = .5592$.

If $\cos B = .7500$, we can find $\angle B$ to the nearest degree in the same way we did when the tangent of an angle was given. Thus, $\angle B$ is between $40°$ and $41°$. By taking differences, we see that $.7500$ is nearer to $.7547$ and $\angle B = 41°$, to the nearest degree.

To summarize the three trigonometric ratios:

$$\tan \angle P = \frac{\text{length of leg opposite } \angle P}{\text{length of leg adjacent to } \angle P} = \frac{p}{r}$$

$$\sin \angle P = \frac{\text{length of leg opposite } \angle P}{\text{length of hypotenuse}} = \frac{p}{q}$$

$$\cos \angle P = \frac{\text{length of leg adjacent to } \angle P}{\text{length of hypotenuse}} = \frac{r}{q}$$

ILLUSTRATIVE PROBLEM 1:   A 20-foot ladder leaning against a building makes an angle of 58° with the ground. Using the table on page 486, find each of the following to the nearest foot.

a. How high up the wall of the building does the ladder reach?

b. How far from the wall is the foot of the ladder?

Solution:

a. In the accompanying diagram, *GH* is the leg opposite ∠ *L*, and *LH* is the hypotenuse. Therefore, we use the sine ratio.

$$\sin \angle L = \frac{\text{leg opposite} \angle L}{\text{hypotenuse}} = \frac{GH}{LH}$$

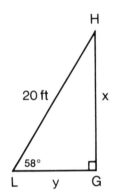

If we let side *GH* = *x*, then

$$\sin 58° = \frac{x}{20}$$

$$.8480 = \frac{x}{20}$$

$$x = 20(.8480) = 16.96$$

$$x = 17 \text{ feet (to nearest foot)} \quad (answer)$$

b. To find *LG*, use the cosine ratio.

$$\cos \angle L = \frac{\text{leg adjacent to} \angle L}{\text{hypotenuse}} = \frac{LG}{LH}$$

If we let side *LG* = *y*, then

$$\cos 58° = \frac{y}{20}$$

$$.5299 = \frac{y}{20}$$

$$y = 20(.5299) = 10.598$$

$$y = 11 \text{ feet (to nearest foot)} \quad (answer)$$

ILLUSTRATIVE PROBLEM 2: A support wire 40 feet long is tied to the top of a telephone pole and to a point on the ground 15 feet from the foot of the pole. What angle does the wire make with the ground? Using the table on page 486, find the angle to the nearest degree.

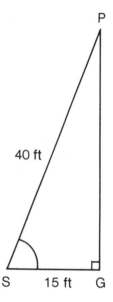

Solution: Since the given information provides the length of a leg adjacent to ∠ S and the *hypotenuse*, we use the cosine ratio.

$$\cos \angle S = \frac{\text{leg adjacent to } \angle S}{\text{hypotenuse}} = \frac{SG}{SP}$$

$$\cos \angle S = \frac{15}{40} = \frac{3}{8}$$

$$\cos \angle S = .3750$$

$$\angle S = 68° \text{ (to nearest degree)} \quad (answer)$$

Exercises

1. Using the table on page 486, find the value of each of the following:
   a. sin 47°   **b.** cos 33°   **c.** sin 86°
   d. cos 89°   **e.** sin 30°   **f.** cos 60°

2. Using the table on page 486, find ∠ P if:
   a. sin P = .4226   **b.** cos P = .9962
   c. cos P = .4540   **d.** sin P = .9744

3. Using the table on page 486, find, to the nearest degree, the measure of ∠ R if:
   a. sin R = .3250   **b.** cos R = .4774
   c. sin R = .8740   **d.** cos R = .9325

4. To the nearest integer, find the length of the side indicated by x in each triangle.

a.

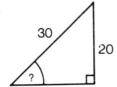

b.

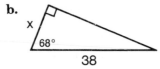

c. 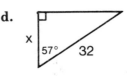 **d.** 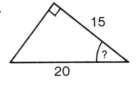 **e.**

5. Find, to the nearest degree, the measure of each angle indicated with a ? in each diagram.

a. **b.** **c.**

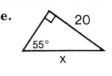

6. The length of the shorter side of a rectangle is 3 and the diagonal is 8. Find, to the nearest degree, the measure of the angle that the diagonal makes with the longer side.

7. In right triangle ABC, with the right angle at C, $AB = 20$ and $\angle A = 53°$. Find the length of AC to the nearest integer.

8. In triangle ABC, $\angle C = 90°$ and $\angle A = 31°$. If $AB = 12$, find the length of BC to the nearest tenth.

9. The length of the hypotenuse of a right triangle is 15 and the length of a leg is 8. Find, to the nearest degree, the measure of the angle included between these two sides.

10. In right triangle ABC, with the right angle at C, AB is 10 and $\angle A = 25°$. Find, to the nearest tenth, the length of BC.

11. In the accompanying diagram, the length of the diagonal of the rectangle is 12 and the width is 8.
    a. Find, to the nearest integer, the length of the rectangle.
    b. Find, to the nearest degree, the measure of the angle formed by a longer side and the diagonal.

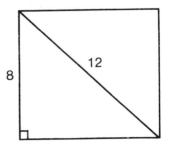

12. As shown in the accompanying diagram, a 15-foot ladder is leaning against a wall of a building. The bottom of the ladder is 6 feet away from the wall on level ground.

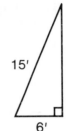

15′

6′

   a. Find, to the nearest degree, the acute angle that the ladder makes with the ground.

   b. Find, to the nearest foot, the distance from the top of the ladder to the ground.

13. A steep mountain road makes an angle of 28° with the horizontal. How many feet above the horizontal does an auto rise when it travels 400 feet up the road? (Round off the answer to the nearest foot.)

14. In right triangle ABC, leg $AC = 28$ and hypotenuse $AB = 42$. Find, to the nearest degree, the measure of angle A.

15. An 18-foot ladder leans against a building and makes an angle of 67° with the ground. Find, to the nearest foot, the distance between the foot of the ladder and the building.

16. In right triangle ABC, angle C is 90°, the measure of AB is 9, and the measure of angle B is 37°.

   a. Find the measure of leg AC to the nearest integer.

   b. Find the measure of side BC to the nearest integer.

17. A guy wire tied to the top of a telephone pole is attached to a stake in the ground 18 feet from the foot of the pole. If the wire makes an angle of 67° with the ground, find, to the nearest foot, the length of the guy wire.

18. In triangle ABC, altitude BD is drawn to side AC, forming two right triangles (ADB and BDC). $\angle A = 62°$, $\angle C = 20°$, and side $BC = 50$ feet.

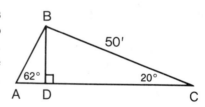

   a. Find BD to the nearest foot.

   b. Find AD to the nearest foot.

Chapter Review Exercises

1. Find the value of:

   a. cos 37°   **b.** tan 81°   **c.** sin 42°

2. Find, to the nearest degree, the measure of angle P for each of the following:

   a. sin $P = .2850$ **b.** tan $P = 1.4570$ **c.** cos $P = .5575$

3. Find, to the nearest degree, the measure of the angle of elevation of the sun when a 27-foot tree casts a 12-foot shadow.

4. In right triangle ABC, $\angle C = 90°$, $\angle A = 32°$, and $AC = 8$. Find the length of hypotenuse AB to the nearest integer.

5. A tree on level ground casts a shadow 12 feet long when the angle of elevation of the sun measures $65°$. Find, correct to the nearest foot, the height of the tree.

6. In right triangle ABC, the right angle is at C. The measure of AC is 5, and the measure of BC is 11.
   a. Find, to the nearest degree, the measure of angle B.
   b. Find, to the nearest integer, the measure of AB.

7. A person 5 feet tall casts a shadow of 12 feet at the same time that a tree casts a shadow of 60 feet. What is the height of the tree in feet?

8. A monument stands on level ground. The angle of elevation of the top of the monument, taken at a point 425 feet from the foot of the monument, is $32°$. Find the height of the monument to the nearest foot.

9. A 25-foot guy wire attached to the top of a pole makes an angle of $62°$ with the ground. Find, to the nearest foot, the distance between the point where the guy wire meets the ground and the foot of the pole.

10. Find, to the nearest degree, the angle of elevation of the sun when a tree 24 feet high casts a shadow of 36 feet.

11. A 40-foot ladder is leaning against a building. The foot of the ladder is 32 feet from the building. Find, to the nearest degree, the angle that the ladder makes with the ground.

12. A television tower is 150 feet high and an observer is 120 feet from the base of the tower. Find, to the nearest degree, the angle of elevation of the top of the tower from the point where the observer is standing.

CHECK YOUR SKILLS

1. Factor $4x^2 - 9$.   **2.** Solve for y: $0.2y = 33$

3. Solve for x in terms of b, m, and y: $y = mx + b$

4. Find the numerical value of $|9 - 3|$.

5. Round 66.45 to the nearest integer.

6. Express $\dfrac{x}{2} + \dfrac{x + 1}{5}$ as a single fraction in simplest form.

7. A ladder 9 feet long leans against a building and reaches a point 6 feet above the ground. Find, to the nearest degree, the angle that the ladder makes with the ground.

8. The hypotenuse of a right triangle is 18 and one leg is 12. Find the length of the other leg. (Answer may be left in radical form.)

9. Solve for x: $4(x - 2) - 3 = 5$

10. Simplify: $\dfrac{x^2 - 9}{4} \cdot \dfrac{8}{x - 3}$

22

Statistics and Probability

1. ORGANIZATION OF DATA

Information, or data, is collected for many reasons. Suppose, for example, that a consumer organization is interested in the gas efficiency of new cars. Tests are conducted and the data is collected and organized. The study of such data is called **statistics.** Statistics refers to the collection, organization, analysis, and interpretation of data. Today much statistical work is done using computers.

For various ways of presenting data in convenient form, see Appendix IV on page 450.

One way of organizing raw data is to use a table.

Suppose a teacher gives a final examination to a class of 30 students and wishes to examine the results in greater detail. When listed by alphabetical order of names, the results appear as follows: 75, 80, 70, 65, 90, 60, 75, 55, 70, 85, 95, 100, 70, 60, 75, 80, 50, 85, 90, 75, 80, 65, 70, 80, 85, 70, 75, 65, 70, 70. The teacher then constructs the following table.

Score	Tally	Frequency
100	|	1
95	|	1
90	||	2
85	|||	3
80	||||	4
75	++++	5
70	++++ ||	7
65	|||	3
60	||	2
55	|	1
50	|	1
		Total: 30

Tally marks are used to indicate the number of times each score occurs. This number is called the **frequency.** Such a table is called a **frequency distribution table (frequency table).**

From such a table, the teacher sees readily that the lowest score is 50 and the highest score is 100. The **range** of scores is from 100 down to 50, or 100 − 50 = 50. The range of scores is determined by subtracting the lowest score (50) from the highest score (100). The score occurring most often is 70.

We can also obtain other information from a table that will help us to analyze data.

Exercises

In 1 to 4, make a frequency table of the scores and determine:
a. the highest score **b.** the lowest score **c.** the range of the scores
d. the score having the greatest frequency

1. The heights, in inches, of 40 students in a class are as follows: 60, 62, 65, 58, 61, 57, 63, 62, 68, 56, 62, 64, 59, 67, 60, 66, 65, 63, 59, 61, 60, 58, 57, 64, 60, 60, 61, 64, 62, 62, 63, 61, 62, 63, 54, 52, 61, 63, 61, 62.

2. The average noon temperatures, in degrees Fahrenheit, for the 30 days of April, in the order of occurrence were as follows: 46, 50, 47, 49, 52, 48, 53, 47, 48, 50, 49, 51, 53, 52, 51, 48, 50, 51, 50, 52, 54, 52, 51, 53, 54, 52, 50, 51, 51, 49.

3. The weights, in pounds, of 30 students in a class were as follows: 125, 121, 127, 127, 124, 124, 119, 125, 122, 125, 123, 125, 126, 124, 125, 122, 124, 120, 124, 126, 128, 119, 124, 123, 123, 122, 120, 120, 126, 121.

4. A science mid-term examination consisting of 50 questions was given to a class of 30 students. The number of questions answered correctly by each student was as follows: 37, 44, 42, 42, 43, 44, 44, 44, 40, 47, 49, 45, 43, 44, 39, 46, 41, 42, 44, 45, 48, 45, 43, 44, 41, 46, 43, 40, 47, 39.

2. THE FREQUENCY HISTOGRAM

A **histogram** is a graph used to show the frequency distribution of grouped data. Frequency numbers are usually plotted on the vertical axis, while data groups are plotted on the horizontal axis.

Since the histogram displays the frequency, or number of scores, we sometimes call this graph a *frequency histogram.*

ILLUSTRATIVE PROBLEM: The data in the following table represents the weekly earnings of 36 factory workers. Construct a frequency histogram based on the data.

Amount in dollars	No. of workers
$200	1
210	2
220	3
230	5
240	8
250	6
260	4
270	3
280	2
290	2
	Total: 36

Solution: Shown below is a frequency histogram for the table.

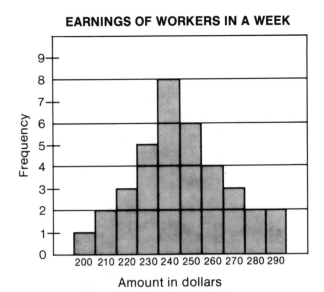

EARNINGS OF WORKERS IN A WEEK

The histogram provides a clear, quick visual presentation of the data. At a glance, the tallest column tells us the most frequent score. If a clustering of columns close in height appears, we say there is a *trend* or *tendency.*

In most collections of data, scores with the greatest frequency tend to cluster around the middle or center. This is called the **central tendency.** Note, for example, how the most frequent scores (the tallest columns) in the histogram appear in the center. In statistics, as we shall soon see, there are several ways to measure the central tendency.

Exercises

1. Construct a histogram for the data in exercise 1 on page 397.
2. Construct a histogram for the data in exercise 2 on page 397.
3. Construct a histogram for the data in exercise 3 on page 397.
4. Construct a histogram for the data in exercise 4 on page 397.
5. The accompanying frequency table shows the scores of the members of a high school bowling team during a practice session. Make a histogram of the data in this table.

Score	*Frequency*
100	1
110	3
120	6
130	9
140	5
150	4
160	2
	Total: 30

3. ARITHMETIC MEAN OR AVERAGE

Ernie has grades in his major subjects of 90, 80, 85, 80, and 85. Maria has grades of 85, 85, 90, 75, and 90 in the same major subjects. Who is doing better? To compare the grades of the two students, we take the average grade for each student. The average is also called the **mean,** or **arithmetic mean.**

Thus, Ernie's average is:

$$\frac{90 + 80 + 85 + 80 + 85}{5} = \frac{420}{5} = 84$$

Maria's average is:

$$\frac{85 + 85 + 90 + 75 + 90}{5} = \frac{425}{5} = 85$$

We see that Maria did slightly better than Ernie.

The **arithmetic mean** of a set of scores is defined as the *sum* of the scores divided by the *number* of scores.

$$\textbf{Mean} = \frac{\textbf{Sum of Scores}}{\textbf{Number of Scores}}$$

The mean is one measure of central tendency.

ILLUSTRATIVE PROBLEM 1:   Carlos gets grades of 75, 84, 85, and 88 in four major subjects. What grade must he get in his fifth major subject to average 85?

Solution: Let x = grade of fifth subject.

$$\text{Average(Mean)} = \frac{\text{Sum of Scores}}{\text{Number of Scores}}$$

$$85 = \frac{75 + 84 + 85 + 88 + x}{5}$$

$$85 = \frac{332 + x}{5}$$

$$85(5) = 332 + x$$

$$425 = 332 + x$$

$$93 = x$$

Answer: Carlos must get 93 in his fifth major subject if he wishes an average of 85.

The preceding problem deals with *individual* scores. We now consider a problem dealing with *groups* of scores.

ILLUSTRATIVE PROBLEM 2:
The final grades in a mathematics
class are shown in the frequency
table at the right. Find the
average (mean) grade.

Grade	Frequency
55	4
60	0
65	3
70	5
75	8
80	6
85	0
90	5
95	4
	Total: 35

Solution: To get the sum of all 35 grades, it is simplest to add the products obtained by multiplying each grade by its frequency. Thus:

$$55 \times 4 = 220$$
$$60 \times 0 = 0$$
$$65 \times 3 = 195$$
$$70 \times 5 = 350$$
$$75 \times 8 = 600$$
$$80 \times 6 = 480$$
$$85 \times 0 = 0$$
$$90 \times 5 = 450$$
$$95 \times 4 = \underline{380}$$
$$\text{Sum} = 2675$$

$$\text{Mean} = \frac{2675}{35} = 76.4 \ (answer)$$

Exercises _____

1. Dolores receives grades of 80, 85, 95, 75, and 65 in her five major subjects. What is her average?

2. The noon temperatures (Celsius) on a certain week in February were: $-4°, -6°, -9°, -5°, 0°, +7°, +3°$. What was the mean noon Celsius temperature for the week?

3. The average of 4, 6, 9, 12, and y is 8. What is the value of y?

4. The weekly payroll of a business firm was as follows:

1 director	$726
2 assistant directors	$620 each
4 foremen	$470 each
30 operators	$340 each
3 office workers	$210 each
2 porters	$180 each

Find the average weekly salary.

5. In exercise 5 on page 399, find the mean score for the bowling team to the nearest integer.

6. Using the frequency table on page 398, find, to the nearest cent, the mean salary of the 36 factory workers for that week.

7. On a science test, 20 students in the class averaged 82% and 10 students in the class averaged 76%. What was the average for all 30 students?

8. Using the frequency table on page 396, find the mean grade for the class of 30 students.

---

4. THE MEDIAN

The *mean*, or *average*, is not always the best representation of a set of data. The mean is useful when the scores in a set of data are fairly evenly distributed, as in the grades used to obtain a report card average. The mean is less useful, however, when scores are very *unevenly* distributed.

Suppose that a manager in a company earns $1000 a week in salary and five workers each receive $200 a week in salary. What is the most representative salary? In such a situation, the mean would give us a distorted measure of the most representative salary.

If the distribution is such that a few very large or very small scores are piled up at one end or the other of the scale, then another measure of central tendency may be used to supplement the mean. This measure of central tendency is called the *median*.

The *median* is the middle number of a set of data arranged in order.

For example, let us find the median noon temperature for a week in which the noon temperatures from lowest to highest were 41°, 52°, 55°, 57°, 58°, 70°, 71°. Since there are seven temperature readings, the fourth would be the middle one (three on each side). Thus, the median would be 57°.

$$41, 52, 55, \mathbf{57}, 58, 70, 71$$

This procedure is easily applied when there is an odd number of scores. How do we find the median where there is an even number of scores?

Eight test scores arranged in order of size are: 62, 65, 77, 78, 79, 81, 82, 98. Since there is no actual middle score, we average the two middle scores, 78 and 79, to find the median.

$$62, 65, 77, \mathbf{78}, \mathbf{79}, 81, 82, 98$$

The median is therefore $\dfrac{78 + 79}{2}$ or $78\frac{1}{2}$. Note that the first step in finding the median is always to arrange the data in order of size.

Exercises _____

1. Find the median height of a group of students whose heights in inches are: 62, 50, 61, 64, 63, 68, 66, 62, 77.

2. Find the median weight of a group of girls whose weights in pounds are: 121, 98, 125, 127, 123, 132, 131, 124.

3. A group of students received the following grades in a final examination in science: 78, 75, 80, 62, 83, 76, 95, 74, 79. Find the median grade.

4. Find the median weekly salary of a group of workers whose weekly salaries in dollars are as follows: 112, 218, 214, 330, 224, 215, 218, 226, 429, 213.

5. A basketball team made the following number of points in each of the last 12 games it played: 82, 74, 84, 78, 52, 85, 70, 86, 68, 76, 78, 86. Find the median number of points.

_____

5. THE MODE

Another measure of the central tendency that is occasionally used is the *mode*.

The *mode* is the item or score in a set of data that occurs most frequently.

The mode can be readily obtained from a frequency table by reading the largest frequency or from a histogram by noting the tallest column. (The adjective form of mode is *modal*.)

Suppose we determine the cost in cents of a dozen eggs in several supermarkets and find them to be

87, (89, 89) , 91, 93, 94, 96

We see that 89 is the mode since it occurs twice, while all other costs occur only once. (Note that some sets of data have no mode. This is the case when no score or item is repeated.)

Some sets of data may have more than one mode. For example, the heights in inches of boys in a class are

57, 58, 59, 59, 61, 62, 64, 64, 65, 67

Thus, we see that both 59 and 64 are modes for this set of data.

Exercises

1. One day in a clothing store, suits of the following sizes are sold: 37, 38, 40, 40, 40, 41, 41, 43, 43, 45. What is the mode?

2. The noon temperatures for a certain month, in degrees Fahrenheit, are listed in the following frequency table. What is the mode?

Temperature	Frequency
56°	3
58°	5
59°	6
60°	8
62°	7
64°	2

3. Twelve students in a ninth-grade class worked after school one week and earned the following amounts in dollars: 54, 56, 58, 55, 53, 54, 59, 58, 60, 58, 57, 53. What is the mode?

4. Using the frequency table on page 396, find the modal test grade for the class.

5. Using the frequency table on page 398, find the modal amount earned by the factory workers.

6. Using the frequency table in exercise 5 on page 399, find the modal bowling score.

7. In illustrative problem 2 in section 3, page 401, what is the modal final grade for the class?

8. A boy makes the following grades in mathematics tests: 72, 74, 75, 75, 77, 78, 81, 81, 83. What are the modes?

---

6. THE MEANING OF PROBABILITY

In ordinary conversation, we use such terms as "impossibility," "certainty," "an even chance," etc. In mathematics, we define this concept of chance in a very specific way.

Consider the scale shown below. We refer to the numerical measure of chance as **probability,** and we assign values of probability from 0 to 1.

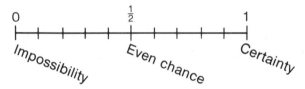

Thus, the probability of snow falling in Florida in July would be 0, since there is no record of such an event occurring. A probability of 0 means an event is impossible, while a probability of 1 means an event is certain.

If a heavy weight is thrown out of an aircraft in flight, what is the probability that the weight will fall? All of our previous observations indicate that it will certainly fall. Thus, the probability is 1, which signifies certainty.

These examples of statistical probability are based on our previous observations or data.

Now, if we toss an ordinary coin, what is the probability that the coin will fall with the head side up? There are only two possible outcomes when we toss a coin: it will fall either head side up or tail side up. These outcomes are equally likely. Thus, we say that the chance or probability of getting a head is 1 out of 2 or $\frac{1}{2}$. A probability of $\frac{1}{2}$ refers to an even chance.

Suppose a jar contains 2 red chips and 3 white chips. Without looking, a girl picks 1 chip from the jar. What is the probability that she picks a *red* chip? There are five possible outcomes. Two of these are *favorable* to the result we are seeking. Thus, we say the probability is 2 out of 5, or 2:5, or $\frac{2}{5}$.

These examples lead us to the following principle:

Principle: The probability (p) of an event occurring is the ratio of the number of favorable possible outcomes (f) to the total number of possible outcomes (t).

$$p = \frac{f}{t}$$

ILLUSTRATIVE PROBLEM 1: Without looking, Carlos picked a marble out of a bag containing 2 red, 2 blue, and 2 green marbles. What is the probability that he picked a green marble?

Solution: The total number of possible outcomes (t) is 6, two of which are favorable (f). Thus:

$$p = \frac{f}{t} = \frac{2}{6} = \frac{1}{3} \quad (answer)$$

ILLUSTRATIVE PROBLEM 2: In a game using dice, one die (singular of dice) is thrown. What is the probability of getting a 6 on one throw of the die?

Solution: Since a die has six faces, there is a total of 6 possible outcomes (t), only 1 of which is favorable (f). Thus:

$$p = \frac{f}{t} = \frac{1}{6} \quad (answer)$$

ILLUSTRATIVE PROBLEM 3: In a standard deck of playing cards, there are four suits—hearts and diamonds, which are red, and spades and clubs, which are black. Each suit consists of 13 cards—numbered cards 2 through 10, and "face" cards Jack, Queen, King, and Ace.

Without looking, a person draws a card from a shuffled deck of 52 playing cards. What is the probability that the card drawn "blind" (that is, without the person's looking) is an ace?

Solution: There is a total of 52 possible outcomes (t). Since there are 4 aces in a deck, there are 4 favorable possibilities (f). Thus:

$$p = \frac{f}{t} = \frac{4}{52} = \frac{1}{13} \quad (answer)$$

ILLUSTRATIVE PROBLEM 4: A spinner has 3 equal areas labeled R, S, and T, as shown in the accompanying figure. What is the probability that the arrow on the spinner will stop in area S on the next spin?

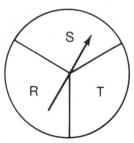

Solution: There is a total of 3 possible outcomes, of which only 1 is favorable. Thus:

$$p = \frac{f}{t} = \frac{1}{3} \quad (answer)$$

Exercises

1. Without looking, a person draws a card from a shuffled deck of 52 playing cards. What is the probability that the drawn card is a spade?

2. What is the probability of drawing a winning ticket in a lottery if there are 3 winners among 600 tickets?

3. Two dimes are tossed at the same time. What is the probability that they both turn up tails?

4. Two students in a class of 30 are to be chosen at random to go to see a play. If all 30 names are thrown into a hat and 2 are drawn, what is the probability that Dolores, a student in the class, is chosen?

5. A bag contains 3 white balls and 4 green balls. What is the probability of drawing **a.** a white ball? **b.** a green ball? **c.** What is the sum of the probabilities in **a** and **b**?

6. A card is drawn "blind" from a shuffled deck of 52 playing cards. What is the probability that the card drawn is the ace of spades?

7. The probability of snow falling in Miami, Florida during the month of August is **a.** close to 0. **b.** close to $\frac{1}{2}$. **c.** close to 1.

8. The probability that a newly born baby will be a boy is **a.** close to 0. **b.** close to $\frac{1}{2}$. **c.** close to 1.

9. Two coins are tossed. Find the probability of getting one head and one tail.

10. A die is tossed. Find the probability of an odd number showing.

11. A card is drawn "blind" from a shuffled standard deck of 52 playing cards. Find the probability that the drawn card is a red queen.

12. Louisa has 3 pennies, 5 nickels, and 7 dimes in a drawer. She picks one of the coins without looking. Find the probability of its **a.** being a nickel. **b.** not being a nickel. **c.** being a penny. **d.** not being a penny.

13. An urn, or large jar, contains 10 slips of paper numbered consecutively from 1 to 10. Find the probability of **a.** drawing 5. **b.** drawing an odd number. **c.** drawing a number divisible by 4. **d.** drawing a number greater than 6.

14. On a clock face, straight lines are drawn from the center to each of the 12 numbers on the clock, dividing the face into 12 equal areas. A spinner is placed at the center. On a given spin, what is the probability that the spinner's arrow will stop in the area between 3 and 4?

15. In a lottery, 50 tickets (numbered 1 to 50) are sold. To select the winner, one ticket is drawn. Find the probability that **a.** it bears a number greater than 25. **b.** it bears a number divisible by 5. **c.** it bears a number divisible by 7.

Chapter Review Exercises

1. In the last 10 years, a baseball team won the following number of games each season: 82, 84, 89, 101, 92, 89, 97, 89, 84, 90. Find: **a.** the mean **b.** the median and **c.** the mode.

2. The heights of boys (in inches) in a class are given by the accompanying frequency table:
   a. Draw a histogram of the data.
   b. What is the modal height?
   c. What is the mean height?
   d. What is the median height?

Height	Frequency
64	2
65	3
66	5
67	6
68	8
69	4
70	2
71	1
	Total: 31

3. A card is drawn "blind" from a shuffled standard deck of 52 playing cards. Find the probability that the card is **a.** a heart. **b.** a queen. **c.** a black card. **d.** the king of diamonds. **e.** a black jack.

4. On 6 science tests, Nina received grades of 82, 88, 90, 92, 91, and 89. What grade must she receive on her next test to have an average of 90?

5. Ten students in a class average 84% on a particular test. Twenty students average 78% on the same test. What is the average grade for all 30 students?

6. A bag contains 20 jelly beans, of which 5 are black, 8 are red, and 7 are white. A student picks a jelly bean without looking. Find the probability that

   a. a black one is picked.

   b. a white one is picked.

   c. a blue one is picked.

   d. a black or white one is picked.

   e. a black or white or red one is picked.

CUMULATIVE REVIEW

1. Find, in simplest radical form, the side of a square whose area is 18.

2. Find the negative root of $x^2 + 6x - 16 = 0$.

3. What is the average of the three numbers represented by $n + 3$, $2n - 1$, and $3n + 4$?

4. The value of $\sqrt{130}$ lies between two consecutive integers x and y ($x < \sqrt{130} < y$). What is the value of x?

5. Express the sum of $4\sqrt{3}$ and $3\sqrt{27}$ in simplest form.

6. Find, in radical form, the length of the hypotenuse of a right triangle whose legs have lengths of 3 cm and 5 cm.

7. The width of a rectangle is 2 less than a side of a square and the length is 1 more than the side of the square. If the area of the rectangle is 28, find the length of a side of the square.

8. A card is drawn "blind" from an ordinary shuffled deck of 52 playing cards. What is the probability that the card is a queen?

9. Belle receives test grades of 82, 87, 78, 74, 90, 68, 81, 85, 79. What is her median grade?

10. If x is 60% of y, what percent is y of x?

Answers to Cumulative Review

If you get an incorrect answer, refer to the chapter and section shown in brackets for review. For example, [8-3] means chapter 8, section 3.

1. $3\sqrt{2}$ [19-7]

2. -8 [20-4]

3. $2n + 2$ [22-3]

4. 11 [19-3]

5. $13\sqrt{3}$ [19-10]

6. $\sqrt{34}$ cm [20-7]

7. 6 [20-6]

8. $\frac{1}{13}$ [22-6]

9. 81 [22-4]

10. $166\frac{2}{3}\%$ [7-6]

Glossary

The explanations in this glossary are intended to be brief descriptions of the terms listed.

abscissa The x-coordinate of an ordered pair of numbers (always the first number), or the distance from the y-axis. (See *ordinate*.)

absolute value The absolute value of a number is the number or its opposite, whichever is positive. The absolute value of -5, written $|-5|$, is 5.

acute angle An angle whose measure is less than $90°$.

additive identity Zero is the additive identity because adding zero to a number leaves the original number unchanged.

additive inverse The additive inverse of a number is the opposite of the number. For example, -2 is the additive inverse of 2, and 4 is the additive inverse of -4.

algebra A branch of mathematics in which variables are often used to express numerical relationships. (See *variable*.)

altitude of a triangle A line segment drawn from any vertex and perpendicular to the opposite side.

area The area of a geometric figure is the number of units of square measure contained in the figure.

associative principle A rule stating that the way in which three numbers are grouped for adding (or multiplying) does not affect the result.

average The sum of a set of quantities divided by the number of quantities. (See *mean*.)

bar graph A statistical graph in which vertical or horizontal bars are used as a method of comparing quantities.

base In 5^2, the 5 is called the *base*.

binomial A polynomial with two terms, such as $2x + y$.

broken-line graph A statistical graph in which points are joined by straight line segments.

central angle   An angle whose vertex is at the center of a circle and whose sides are radii.

circle   A closed curve in a plane such that every point is the same distance from the center.

circumference of a circle   The measure of the distance around a circle.

coefficient   A numerical factor of a term. In the term $8x$, 8 is the coefficient of x.

commutative principle   A rule stating that the order in which two numbers are added (or multiplied) does not affect the result.

complementary angles   Two angles the sum of whose measures is 90°.

composite number   An integer that has factors other than itself and one. (See *prime number*.)

coordinates of a point   The ordered pair of numbers that describes the position of a point on a graph in a coordinate plane.

cosine (cos)   The cosine of an acute angle in a right triangle is the ratio of the length of the leg adjacent to the angle to the length of the hypotenuse of the triangle.

$$\text{Cos } A = \frac{\text{adjacent leg}}{\text{hypotenuse}}$$

cube   A three-dimensional solid figure with six square faces.

decimal   Decimal fraction, in which the denominator is a power of 10, written in shortened form using a decimal point. The decimal form of $\frac{25}{100}$ is .25.

degree   The degree of a polynomial is the highest exponent of any of its terms. The degree of polynomial $7x^3 - 4x + 1$ is 3.

denominator   The term of a fraction that is below the fraction line and that indicates the number of equal parts into which a unit is divided. In the fraction $\frac{3}{4}$, 4 is the denominator.

diameter   A line segment whose midpoint is at the center of a circle and whose endpoints are on the circle.

discount   The amount by which the list price or original price is reduced.

distributive property of multiplication over addition   A rule stating that when a number is multiplying a sum, you can start with either the multiplication or the addition.

empty set   A set that contains no elements; also called the *null set* and written $\varnothing$ or { }.

equation A mathematical sentence using an equals sign to indicate that two expressions are equal.

equilateral triangle A triangle in which all three sides are equal in length.

equivalent fraction The result of multiplying (or dividing) both the numerator and the denominator of a fraction by the same number.

exponent A number in an expression that indicates the number of times the base occurs as a factor. In 5^3, the 3 is the exponent and $5^3 = 5 \cdot 5 \cdot 5$.

extremes The first and fourth terms of a proportion. In the proportion $\dfrac{a}{b} = \dfrac{c}{d}$, a and d are the extremes. (See *means.*)

factor If two or more numbers or expressions are multiplied, each one is a factor of the product.

formula An expression in which letters representing words are related by mathematical symbols.

frequency For a collection of data, the number of items in a given category.

gram The basic unit of weight in the metric system.

greatest common factor (GCF) The largest number by which each of a given set of two or more numbers is divisible. For example, 4 is the GCF of 8, 16, and 20.

histogram A bar graph of a frequency distribution.

hypotenuse The longest side of a right triangle; the side of a right triangle opposite the right angle.

inequality A mathematical sentence using one or more of the symbols >, <, ≠, ≥, ≤.

integers The positive and negative whole numbers and zero.

interest The amount paid for the use of borrowed money.

isosceles triangle A triangle with exactly two sides equal in length.

least common denominator The smallest counting number into which two or more denominators will divide exactly.

least common multiple (LCM) The least common multiple of two or more numbers is the smallest number that has each of the original numbers as a factor. The LCM of 2, 3, and 4 is 12.

leg The legs of a right triangle are the two sides that form the right angle. The legs of an isosceles triangle are the two equal sides.

like fractions Fractions having the same denominator. For example, $\dfrac{3}{7x}$ and $\dfrac{5}{7x}$ are like fractions.

like terms Terms that have identical variable factors. For example, $9ab$ and $3ab$ are like terms.

linear equation An equation that contains no terms higher than the first degree and no terms with variables in the denominator. For example, $3x + 2y = 8$ is a linear equation. The graph of a linear equation is a straight line.

liter The basic unit of liquid measure of capacity in the metric system. A liter is slightly more than a quart.

lowest terms of a fraction A fraction is in lowest terms when the numerator and denominator have no common factor other than 1.

mean (arithmetic mean) Same as average; the sum of a set of quantities divided by the number of quantities.

means The second and third terms of a proportion. In the proportion $\dfrac{a}{b} = \dfrac{c}{d}$, b and c are the means. (See *extremes*.)

median The middle score of a set of scores when arranged according to size; for an even number of scores, the median is the average of the middle two scores.

meter The basic unit of length in the metric system. A meter is about 1.1 yards.

metric system Measurements based on the decimal system. Each unit is $\frac{1}{10}$ of the next larger unit.

mode The number that occurs with greatest frequency in a set of scores. There may be two or more modes or no mode.

monomial A polynomial of only one term. Examples of monomials are 4, x, $3x$, and $9ab^2$.

multiplicative identity The number 1 is the multiplicative identity since the product of 1 and any number leaves the original number unchanged.

multiplicative inverse (reciprocal) When the product of two numbers is 1, each number is called the multiplicative inverse of the other. For example, $\frac{3}{5}$ and $\frac{5}{3}$ are reciprocals.

negative number   Any number less than zero.

net price   The price paid after a discount or reduction is made.

numerator   The term of a fraction that is above the fraction line. In the fraction $\frac{3}{4}$, 3 is the numerator.

obtuse angle   An angle whose measure is greater than 90° and less than 180°.

opposite of a number   Same as the additive inverse of a number.

order of operations   The sequence in which the arithmetic is done in finding the value of a numerical expression. First evaluate expressions in parentheses, then do exponents, then do multiplication and division (from left to right), and finally addition and subtraction (from left to right).

ordered pair   A pair of numbers (a, b) in which the order is considered, so that (a, b) is different from (b, a). Each ordered pair of numbers corresponds to a point in a coordinate plane and vice versa.

ordinate   The y-coordinate of an ordered pair of numbers (always the second number), or the distance from the x-axis. (See *abscissa*.)

origin   The point $(0, 0)$, or the intersection of the x-axis and y-axis in a coordinate plane.

parallel lines   Two or more straight lines in the same plane that do not intersect.

parallelogram   A quadrilateral (four-sided plane figure) in which both pairs of opposite sides are parallel.

percent   A fraction whose denominator is 100, or a decimal in hundredths. For example, 42% means 42 out of a hundred and may be written $\frac{42}{100}$ or .42.

perfect square   A number with two equal factors. 9, 64, and 100 are examples of perfect squares.

perimeter   The measure of the distance around a polygon.

perpendicular lines   Two intersecting lines that form right angles.

polygon   A closed plane figure whose sides are line segments.

polynomial   A polynomial contains one or more terms. For example, $3x$ (a monomial), $x + 7$ (a binomial), and $x^3 + 7x + 4$ (a trinomial) are polynomials.

positive number   Any number greater than zero.

prime number   A number having no factors other than itself and one; the first prime number is 2. (See *composite number*.)

principal   The amount of money, borrowed or invested, on which interest is paid.

principal square root The positive square root of a number.

probability The probability of an event occurring is the ratio of the number of favorable outcomes to the total number of possible outcomes.

proportion A proportion is an equation that states that two ratios are equal. $\dfrac{a}{b} = \dfrac{c}{d}$ or $a:b = c:d$; read: a is to b as c is to d.

Pythagorean relationship In any right triangle, the square of the hypotenuse equals the sum of the squares of the other two sides. If a and b represent the legs of a right triangle and c represents the hypotenuse, then $a^2 + b^2 = c^2$.

quadrilateral A polygon with four sides.

radical An expression using the symbol $\sqrt{}$ (radical sign). Examples are $\sqrt{4}$ (the square root of 4) and $\sqrt[3]{8}$ (the cube root of 8).

radicand The number under a radical sign.

radius A line segment from the center of a circle to any point on the circle.

range The range of a set of scores is the difference between the highest score and lowest score in the set.

ratio The ratio of two numbers is a comparison of the two numbers obtained by forming their quotient.

rational number A number that can be expressed as the ratio of two integers, $\dfrac{p}{q}$ where $q \neq 0$.

reciprocal Same as multiplicative inverse.

rectangle A quadrilateral with four right angles.

regular polygon A polygon that is both equilateral (equal sides) and equiangular (equal angles).

relatively prime Two numbers are relatively prime if their only common factor is 1. For example, 2 and 5 are relatively prime, while 2 and 6 are not relatively prime.

right angle An angle whose measure is $90°$.

right triangle A triangle having one right angle.

root A solution of an equation or an inequality.

scalene triangle A triangle in which no two sides are equal in length.

scientific notation A convenient way of writing very large or very small numbers. For example, 20,000,000 can be written as 2×10^7.

similar polygons Two or more polygons whose corresponding sides are proportional and whose corresponding angles are equal.

simultaneous equations Two or more equations with more than one variable that have a common solution. The equations $x + y = -2$ and $2x - y = 8$ have the solution $(2, -4)$, or $x = 2$, $y = -4$.

sine (sin) The sine of an acute angle in a right triangle is the ratio of the length of the side opposite the angle to the length of the hypotenuse of the triangle.

$$\text{Sin } A = \frac{\text{opposite leg}}{\text{hypotenuse}}$$

slope of a line The slope of a line in a coordinate plane refers to the steepness of the line. The slope of a horizontal line is zero. The slope of a vertical line is undefined. The slope of a line is determined from two points on the line:

$$\frac{\text{difference of } y\text{-coordinates}}{\text{difference of } x\text{-coordinates}}$$

square of a number The square of a number is the product of the number and itself.

square root x is a square root of a number, n, if $x \cdot x = n$. For example, 6 is a square root of 36 since $6 \cdot 6 = 36$. Also, -6 is a square root of 36 since $(-6)(-6) = 36$.

statistics The mathematics of collecting, organizing, and analyzing numerical information.

supplementary angles Two angles the sum of whose measures is $180°$.

tangent (tan) The tangent of an acute angle in a right triangle is the ratio of the length of the side opposite the angle to the length of the side adjacent to the angle.

$$\text{Tan } A = \frac{\text{opposite leg}}{\text{adjacent leg}}$$

term One or more numerals and variables connected by multiplication or division only. For example, 5, x, $5x$, x^2, $2x^2y^3$, and $\dfrac{3xy^2}{2}$ are terms.

triangle A polygon with three sides.

trinomial A polynomial with three terms, such as $12x^2 + 8x - 4$.

variable A variable takes the place of a number. In algebra, an alphabet letter is generally used as a variable. In $5x - 3 = 7$, x is the variable.

vertex The point where the two sides of an angle meet.

volume The measure of the amount of space occupied by a figure in three dimensions.

x-axis The horizontal number line in a coordinate plane.

x-coordinate The x-coordinate, or abscissa, of an ordered pair of numbers is the first number in the pair. For $(8, -3)$, 8 is the x-coordinate.

y-axis The vertical number line in a coordinate plane.

y-coordinate The y-coordinate, or ordinate, of an ordered pair of numbers is the second number in the pair. For $(5, -2)$, -2 is the y-coordinate.

y-intercept The y-intercept of a line in a coordinate plane is the y-coordinate of the point of intersection of the line with the y-axis. In the equation of a line, $y = mx + b$, b is the y-intercept.

Appendix I:
Arithmetic of
Whole Numbers

PRELIMINARY NOTE

If you are having difficulty with some of the computational skills or concepts of the arithmetic of whole numbers, you should review this appendix. Read carefully the descriptive material in each section and work out the exercises at the end. In each section, an attempt is made to give an understanding of the rules of arithmetic and why these rules are true.

1. SYSTEMS OF NUMERATION

Early in history, humans developed and used various simple systems of numeration. When the need arose for more complex computation, however, most of these systems proved inadequate.

The ancient Romans developed a system that used seven capital letters to stand for numbers: **I** for 1, **V** for 5, **X** for 10, **L** for 50, **C** for 100, **D** for 500, and **M** for 1000.

The system is essentially *repetitive*. For example, **II** = 2, **III** = 3, **XXX** = 30, **CC** = 200, and **MM** = 2000. The letters **V**, **L**, and **D** are not repeated. The letters **I**, **X**, and **C** may each be repeated as many as three times in succession, but **M** may be repeated as many times as needed. Thus, **XXIII** = 23, and **CCCXII** = 312.

Note that the system is *additive*. For example, **VI** = 5 + 1 = 6, and **MC** = 1000 + 100 = 1100. It is also *subtractive*. For example, **IV** = 5 − 1 = 4, and **CM** = 1000 − 100 = 900. In other words, the letter of *lesser* value written *before* a letter of *greater* value *reduces* the value of the larger number by that amount.

The Roman numeral system is still used for certain purposes: chapter headings, dates on cornerstones of buildings, numbering of various volumes of a series of books, numbering systems for outlines, etc. For purposes of computation, however, it is cumbersome and inadequate.

The system of numeration we use today is the Hindu-Arabic system. We call it a *decimal system* (or base 10 system) because it is based on ten digits. By using a system of "place value" or positional notation, this system permits the writing of any number, no matter how large or small, using only ten symbols (digits): **0, 1, 2, 3, 4, 5, 6, 7, 8, 9.** It also simplifies the more complex problems of computation that humans have encountered in recent history.

Exercises

In 1 to 10, write the Roman numerals as Hindu-Arabic numerals.

1. XXI   **2.** XXXV   **3.** XLIV   **4.** XCVI   **5.** MDCX

6. CXVI   **7.** MMDXC   **8.** LXXV   **9.** MCMXC   **10.** CMXVII

In 11 to 20, write the Hindu-Arabic numerals as Roman numerals.

11. 17   **12.** 74   **13.** 92   **14.** 1982   **15.** 150

16. 755   **17.** 2001   **18.** 350   **19.** 580   **20.** 182

2. THE SYSTEM OF WHOLE NUMBERS

Whole numbers are formed from various combinations of the digits from 0 to 9. When we write the number 734, we mean

7 hundreds plus 3 tens plus 4 units

or

$7 \times 100 + 3 \times 10 + 4 \times 1$

The value of each digit in a whole number is determined by its placement, so that we call our number system a "place value" system. The **units** or **ones** are shown in the first place on the right, the **tens** in the second place from the right, and so on.

The following table shows place values in whole numbers:

10th	9th	8th	7th	6th	5th	4th	3rd	2nd	1st
BILLIONS	HUNDRED MILLIONS	TEN MILLIONS	MILLIONS	HUNDRED THOUSANDS	TEN THOUSANDS	THOUSANDS	HUNDREDS	TENS	ONES

To read a number, separate its digits into groups of three, beginning at the right side. For 5-digit numbers and larger, commas are used to separate these groups. The comma is optional when dealing with 4-digit numbers. A number is read from the left side after the proper grouping is recognized. Thus, 2,375,208 is read two million, three hundred seventy-five thousand, two hundred eight. The word "and" is not used in reading whole numbers.

Here are some examples of reading and writing whole numbers:

1. Read the number 85,042 and write it as a word statement.
   Answer: eighty-five thousand, forty-two
      Note: The "0" serves as a place-holder to indicate that there are no hundreds.

2. Write the word statement as a number: eight thousand, four hundred ninety-two.
   Answer: 8492

3. What is the value of the 7 in the number 4703?
   Answer: Since the 7 is the third digit from the right, it is in the hundreds place and represents a value of 7 × 100, or 700.

Exercises

In 1 to 10, read the number and write it as a word statement.

1. 582   2. 304   3. 9254
4. 18,723   5. 7008   6. 423,601
7. 6,000,000   8. 12,472,000
9. 6,425,000,000   10. 22,500,000,000

In 11 to 20, write the word statement as a number.

11. seven hundred twenty-one
12. two thousand, eight hundred forty-two
13. nine thousand, fifty-seven   14. ninety-three million
15. two hundred twenty-three million, eight hundred fifty thousand
16. fourteen billion
17. six thousand, two hundred forty-five
18. fifteen thousand, eight
19. three billion, two hundred sixty-eight million
20. eight million, eight thousand, eight hundred

21. Write the number that means 7 hundreds, 5 tens, and 8 ones.

22. Write the number that means 2 thousands, 6 hundreds, and 3 ones.

In 23 to 25, give the value of the 7 in each number.

23. 372   **24.** 27,245   **25.** 58,702

---

3. ADDITION OF WHOLE NUMBERS

The operation of addition is indicated by a plus sign (+), as in 7 + 4 = 11. The numbers 7 and 4 are called the **addends** and the result, 11, is called the **sum** or **total.**

Most of us are now well aware that 4 + 7 also equals 11; that is, it does not matter in what *order* we add the two numbers. Likewise, 8 + 5 = 5 + 8 = 13. The sum is the same regardless of the order of the addends. We say that addition is **commutative,** and we refer to this order principle as the **commutative principle.**

When we add three or more numbers, we may use the principle of *grouping.* We use parentheses ( ) as symbols of grouping; they indicate that the operations within the parentheses are to be done first.

Thus, (4 + 5) + 6 = 9 + 6 = 15.

And 4 + (5 + 6) = 4 + 11 = 15.

Note that the sums are the same no matter how we group them. We say that addition is **associative**; that is, it does not matter whether we *associate* the 4 and 5 first or the 5 and 6 first when adding 4 + 5 + 6.

Using the commutative and associative principles of addition, we conclude that

$$(4 + 5) + 6 = 4 + (5 + 6)$$
$$= 4 + (6 + 5)$$
$$= (5 + 4) + 6$$
$$= 5 + (4 + 6), \text{ etc.}$$

Since the sum is 15 in any case, we generally omit the parentheses and write 4 + 5 + 6 = 15.

The commutative principle and the associative principle justify our adding three or more numbers in *any order* and *grouping* them any way we please.

If several large numbers are to be added, the addends are placed beneath one another so that the digits are lined up correctly: ones digits in the right-hand column, tens digits in the next column to the left, etc. For example, 548 + 85 + 1324 + 459 is arranged like this:

$$
\begin{array}{r}
548 \\
85 \\
1324 \\
459 \\
\text{carry} \quad \underline{1\,2\,2} \\
2416 \quad (sum)
\end{array}
$$

When we add the digits in the ones column, we get 26. We write the 6 in the ones column and carry the "2" to the tens column, since the "2" really means 2 tens or 20. We then add the tens column and get 21, which now means 21 tens or 210. Again, we place the 1 in the tens column and carry the "2" into the hundreds column. We then add the hundreds column and get 14, which now means 14 hundreds or 1400. We place the 4 in the hundreds column and carry the "1" into the thousands column. We then add the thousands column and get 2, which now means 2000.

This method of column addition, with its use of the carrying of digits, shows us one of the great advantages of place value in our number system.

We may check the addition by using the commutative and associative principles. One way is merely to reverse the addition by adding upward:

$$
\begin{array}{r}
\text{carry} \quad 1\,2\,2 \\
548 \\
85 \\
1324 \\
\underline{459} \\
2416 \ \checkmark
\end{array}
$$

Exercises

1. Add the following columns and check:

a.	b.	c.	d.
43	36	148	347
27	29	321	85
52	84	707	1140
60	98	423	139

2. Arrange the following numbers in columns and add:
   a. 235 3423 57 109   **b.** 1147 532 84 903
   c. 4283 20,147 709 3283   **d.** 304 29 1612 75 237

3. Bill spent $1.96 for a notebook, $1.24 for paper, 89 cents for a ruler, 69 cents for pencils, and 98 cents for graph paper. How much did he spend altogether?

4. In a junior high school, there are 420 seventh graders, 503 eighth graders, and 487 ninth graders. Find the total number of students in the school.

5. How much does a set of bedroom furniture cost if the bed costs $189, the dresser $240, the night table $68, the vanity $312, and a book rack $45?

6. In taking a trip, a husband and wife drove 420 miles the first day, 388 miles the second day, 175 miles the third day, 412 miles the fourth day, and 78 miles the fifth day. How many miles did they travel altogether?

7. Mr. Burns makes the following purchases in a department store: a suit for $140, a topcoat for $128, a shirt for $12, a tie for $7, and a pair of shoes for $35. How much did he spend?

8. Henry bought a new car for $8250. He also bought air-conditioning for $415, power steering for $184, power brakes for $97, and a stereo radio for $132. Find his total cost.

---

4. SUBTRACTION OF WHOLE NUMBERS

The operation of subtraction is indicated by a minus sign ($-$), as in $9 - 3 = 6$. The number 9 is called the **minuend**, 3 is called the **subtrahend**, and 6 is called the **difference** or **remainder**. The number zero is the difference whenever the minuend and subtrahend are equal. Thus, $7 - 7 = 0$.

Subtraction is the opposite (or inverse) of addition. When we write $9 - 3 = ?$, we are really asking for the number that must be added to 3 to give us 9. Thus, if we know our addition combinations, we can readily determine our subtraction combinations.

In simple arithmetic, the subtrahend must be less than the minuend. Hence, we cannot speak of a *commutative principle* for subtraction. For example, $3 - 9$ would have no meaning at this point. A simple example would readily convince us that the *associative principle* does not apply to subtraction either. Is $10 - (6 - 3)$ equal to $(10 - 6) - 3$?

The following illustrative problems will serve to explain the actual process of subtraction.

ILLUSTRATIVE PROBLEM 1: Subtract 43 from 58.

Solution:

$$\frac{\begin{array}{r}58\\-43\end{array}}{} = \frac{\begin{array}{l}5 \text{ tens} + 8 \text{ ones}\\4 \text{ tens} + 3 \text{ ones}\end{array}}{1 \text{ ten}\ + 5 \text{ ones} = 15\ (answer)}$$

This shows that we are subtracting ones from ones and tens from tens.

ILLUSTRATIVE PROBLEM 2: Find the difference between 53 and 28.

Solution:

$$\frac{\begin{array}{r}53\\-28\end{array}}{} = \frac{5 \text{ tens} + 3 \text{ ones}}{2 \text{ tens} + 8 \text{ ones}}$$

Since we cannot take 8 ones from 3 ones, we therefore change 53 to $4 \text{ tens} + 13 \text{ ones}$. We are then able to subtract 8 ones from 13 ones.

$$\frac{\begin{array}{r}53\\-28\end{array}}{} = \frac{\begin{array}{l}4 \text{ tens} + 13 \text{ ones}\\2 \text{ tens} +\ \ 8 \text{ ones}\end{array}}{2 \text{ tens} +\ \ 5 \text{ ones} = 25\ (answer)}$$

Note: We usually indicate subtraction like this:

$$\begin{array}{r}{}^{4\ 13}\\ \cancel{5}\,\cancel{3}\\ -2\,8\\ \hline 2\,5\end{array}$$

ILLUSTRATIVE PROBLEM 3: How much greater is 830 than 397?

Solution:

$$\frac{\begin{array}{r}830\\-397\end{array}}{} = \frac{8 \text{ hundreds } 3 \text{ tens } 0 \text{ ones}}{3 \text{ hundreds } 9 \text{ tens } 7 \text{ ones}}$$

$$= \frac{8 \text{ hundreds } 2 \text{ tens } 10 \text{ ones}}{3 \text{ hundreds } 9 \text{ tens }\ \ \ 7 \text{ ones}}$$

$$= \frac{\begin{array}{l}7 \text{ hundreds } 12 \text{ tens } 10 \text{ ones}\\3 \text{ hundreds }\ \ 9 \text{ tens }\ \ \ 7 \text{ ones}\end{array}}{4 \text{ hundreds }\ \ \ 3 \text{ tens }\ \ \ 3 \text{ ones} = 433\ (answer)}$$

Note: This subtraction example would usually be done like this:

$$
\begin{array}{r}
{\scriptstyle 7\;12\;10} \\
8\,3\,0 \\
-3\,9\,7 \\
\hline
4\,3\,3
\end{array}
$$

Here we first changed 830 to 8 hundreds, 2 tens, and 10 ones. When we could not subtract 9 tens from 2 tens, we then further changed 830 to 7 hundreds, 12 tens, and 10 ones.

ILLUSTRATIVE PROBLEM 4: Subtract 372 from 700.

Solution:

$$
\begin{array}{r}
{\scriptstyle 6\;9\;10} \\
7\,0\,0 \\
-3\,7\,2 \\
\hline
3\,2\,8
\end{array}
$$

Here, when we increased the first zero on the right by 10, we could not "borrow" 10 from the tens column. Instead, we changed the two left digits in the minuend from 70 to 69. In effect, we are changing 700 to 690 + 10. The column-by-column digit subtraction is then possible.

All of these subtraction problems may be checked by adding the subtrahend to the remainder. The sum should then be the minuend.

For instance, in illustrative problem 4:

$$
\begin{array}{rl}
Check: & 372 \quad \text{(subtrahend)} \\
& +328 \quad \text{(remainder)} \\
\hline
& 700 \;\checkmark\; \text{(minuend)}
\end{array}
$$

Exercises

1. Subtract and check:

   a. 58   b. 428   c. 781   d. 902
       −24   −113   −256   −427

2. Subtract and check:

   a. 892   b. 5374   c. 7620   d. 5400
       −374   −249   −359   −728

3. Subtract and check:

   a. 297 − 58   b. 5275 − 1492   c. 64,000 − 9700

4. Subtract 329 from 1100.

5. Take 847 from 1252.

6. From 1848 take 1088.

7. Find the difference between 685 and 278.

8. A boy weighed 128 pounds one year and 144 pounds the following year. How many pounds did he gain?

9. Mr. Tobin bought an apartment for $38,500 and sold it five years later for $46,250. How much more did he sell it for?

10. Mrs. Andrews bought a color television set for $528. She made a down payment of $150. How much more does she still have to pay?

11. Roberto went to the supermarket with a $20 bill. He spent $6.20 for meats, $2.40 for fruits and vegetables, $3.37 for detergents, and $2.76 for baked goods. How much did he have left?

---

5. MULTIPLICATION OF WHOLE NUMBERS

If a classroom has 4 rows of seats with 5 seats in each row, then we can say it has a total of $5 + 5 + 5 + 5 = 20$ seats. We can also say that 4 is **multiplied** by 5, or $4 \times 5 = 20$. Written vertically, the problem appears as

$$\begin{array}{rl} 5 & \text{(multiplicand)} \\ \underline{\times 4} & \text{(multiplier)} \\ 20 & \text{(product)} \end{array}$$

The **multiplier** indicates the number of times the **multiplicand** is to be written down and added. The result of multiplication is the **product.** The multiplication symbol ($\times$) is read *times*; we sometimes also use a dot placed in the middle of the line of writing—thus, $5 \cdot 4 = 20$. We may also indicate multiplication by putting one or both of the numbers in parentheses; thus, $5(4) = 20$. Numbers that are multiplied together to form a product are also called **factors** of the product. Thus, 5 and 4 are factors of 20.

In the example above, note that 4×5 also yields 20, the total number of seats in the room; that is, $5 \times 4 = 4 \times 5$. This indicates to us that the **commutative principle** also applies to multiplication. This principle permits us to check a multiplication problem by interchanging the multiplicand and the multiplier.

A simple illustration shows that the **associative principle** of multiplication also holds true. Note that $(3 \times 4) \times 5 = 12 \times 5 = 60$.

Note also that $3 \times (4 \times 5) = 3 \times 20 = 60$.

Thus, $(3 \times 4) \times 5 = 3 \times (4 \times 5)$.

Since the manner in which we *group* the factors does not matter, we usually write the product above as 3 × 4 × 5.

If we wish to find the product 2 × (3 + 4), we may add 3 and 4 and multiply the sum by 2. Thus,

$$2 \times (3 + 4) = 2 \times 7 = 14$$

Note that, if we multiply each addend by 2 and then add, we get the same result. Thus,

$$2 \times (3 + 4) = 2 \times 3 + 2 \times 4 = 6 + 8 = 14$$

This example illustrates that **multiplication is distributive over addition.**

In the example above, the multiplier, 2, is "distributed"—given out—to each of the numbers in the parentheses. This means that both the numbers 3 and 4 (connected here by addition) must have the number 2 act on them as a multiplier.

The distributive principle is used when we multiply numbers with two or more digits. Thus,

$$3 \times 23 = 3 \times (20 + 3) = \quad 3 \times 20 + 3 \times 3$$
$$= 60 + 9 = 69$$

Written vertically, this appears as

$$\begin{array}{r} 23 \\ \times 3 \\ \hline 69 \end{array}$$

Note that the ones digits are aligned and that the multiplier (usually the smaller number) is the bottom number. Then multiply from right to left, starting with the ones digit in the multiplicand. "Carrying" will usually be necessary, as shown in the following illustrative problems:

ILLUSTRATIVE PROBLEM 1: Multiply 237 by 5.

Solution: First, 5 × 7 = 35. Write the 5 under the ones column and carry the 3 to the tens column.

$$\begin{array}{r} \overset{3}{} \\ 237 \\ \times 5 \\ \hline 5 \end{array}$$

Then 5 × 3 = 15, plus the 3 that was carried, equals 18. Write the 8 in the tens column and carry the 1 to the hundreds column.

$$\begin{array}{r} \overset{1\,3}{} \\ 237 \\ \times 5 \\ \hline 85 \end{array}$$

Then 5 × 2 = 10, plus the 1 that was carried, equals 11. Write 1 in the hundreds column and 1 in the thousands column.

$$\begin{array}{r} \overset{1\,3}{} \\ 237 \\ \times 5 \\ \hline 1185 \end{array}$$

Answer: 237 × 5 = 1185

The following illustrates how we multiply two numbers with two or more digits:

ILLUSTRATIVE PROBLEM 2: Multiply 345 × 32.

Solution: First multiply the 2 in the multiplier by 345.

This product is called a **partial product.**

$$\begin{array}{r} 345 \\ \times 32 \\ \hline 690 \end{array}$$

Now multiply the 3 in the multiplier by 345 to get another partial product. Start by placing the 5 in the partial product under the 3 in the multiplier (tens column) and the other digits of the partial product to the left. Then draw a line and add the two partial products.

$$\begin{array}{r} 345 \\ \times 32 \\ \hline 690 \\ 10\ 35 \\ \hline 11,040 \end{array}$$

Answer: 345 × 32 = 11,040

The same procedure is followed when the multiplier has 3 or more digits. Multiply with every digit in the multiplier, beginning with the ones digit. Place the first right-hand digit of each partial product directly under the digit you are multiplying by.

You may check your result by reversing the multiplicand and multiplier.

Remember that any number multiplied by zero yields zero. This fact can be used to simplify problems in multiplication where zero digits appear. When there are one or more zeros in the multiplier, there is no need to write down horizontal lines of zeros. If you come to a zero in the multiplier, just bring down the zero *in a straight line* as a place holder and then continue the multiplication with the next digit to the left. See examples 1 and 2 below.

EXAMPLE 1: Multiply 74 × 30.

Solution:

$$\begin{array}{r} {}^{1} \\ 74 \\ \times 30 \\ \hline 2220 \end{array}$$

EXAMPLE 2: Multiply 345 by 201.

Solution:

$$\begin{array}{r} {}^{1} \\ 345 \\ \times 201 \\ \hline 34\,5 \\ 69\,00 \\ \hline 69,345 \end{array}$$

A simple rule for multiplying a number by 10, 100, 1000, etc., is merely to write 1, 2, 3, etc., zeros, respectively, after the extreme right-hand digit of the number. See example 3 below.

EXAMPLE 3: Multiply 47 by 10, by 100, and by 1000.

Solution:

$$47 \times 10 = 470$$
$$47 \times 100 = 4700$$
$$47 \times 1000 = 47,000$$

Exercises _____

1. Multiply and check:
   a. 72
      ×6
   b. 42
      ×27
   c. 325
      ×58
   d. 719
      ×32

2. Find the product and check:
   a. 409
      ×78
   b. 732
      ×30
   c. 547
      ×203
   d. 76,008
          ×45

3. Multiply:
   a. 763 × 200
   b. 4752 × 101
   c. 854 × 1000
   d. 8720 × 230

4. An auto mechanic earns $410 per week. What does he earn in a year (52 weeks)?

5. An appliance dealer bought 18 toasters, each costing $27. How much did he pay for all of them?

6. A used-car dealer sells 134 cars at $4250 each. How much did he receive for all of them?

7. How many envelopes are there in 145 boxes, packed 500 to the box?

8. A hotel manager is carpeting 125 rooms and needs 37 square yards of carpet per room. How many square yards of carpet must she order?

9. A housing development is having air-conditioning units installed in 137 houses. If each installation costs $1450, what is the total cost?

10. A clothing store sold the following items during the course of one month:

     178 shirts at \$12 each   235 suits at \$142 each

     256 ties at \$8 each   43 raincoats at \$52 each

     How much money did the store take in that month?

11. A construction firm sells 245 mobile homes for \$48,500 each. Find the total amount it receives for the sale.

12. A trucking company purchases 352 small new trucks costing \$9270 each. How much does it pay for all of them?

---

6. DIVISION OF WHOLE NUMBERS

In a class of 18 students, a gym teacher wants to form squads of 6 each; we say that he is **dividing** the 18 students into groups of 6. In this case, he divides 18 by 6 and sees that he gets exactly 3 squads. This division may be indicated in any of three different ways:

$$(1) \quad 18 \div 6 = 3$$

$$(2) \quad \frac{18}{6} = 3$$

$$(3) \quad 6\overline{)18}^{\,3}$$

Division is the *opposite*, or *inverse*, of multiplication. When we say 18 divided by 6 (18 ÷ 6), we mean: What number multiplied by 6 will give us 18? The number 18 is called the **dividend,** 6 is called the **divisor,** and the result 3 is called the **quotient.** The quotient tells us how many times the divisor is contained in the dividend.

As in multiplication, division is also *distributive over addition.*

For example, $\dfrac{18 + 12}{6} = \dfrac{30}{6} = 5.$

Also, $\dfrac{18 + 12}{6} = \dfrac{18}{6} + \dfrac{12}{6} = 3 + 2 = 5.$

Note that distributing the divisor among the addends in the dividend gives the same result. The *distributive principle* provides the basis for the process of dividing whole numbers.

For example, if we wish to divide 36 by 3, we may write

$$\frac{36}{3} = \frac{30 + 6}{3} = \frac{30}{3} + \frac{6}{3} = 10 + 2 = 12$$

Using the division box, we usually write this division example as

$$\frac{12}{3)\,36}$$

This is an abbreviated form for

$$\frac{1 \text{ ten } + 2 \text{ ones} = 12}{3)\,3 \text{ tens} + 6 \text{ ones}}$$

The following illustrative problems show how to proceed in more difficult cases of division:

ILLUSTRATIVE PROBLEM 1:   Divide 52 by 4.

Solution: We say 4 is contained in 5 once and write 1 in the tens place over the 5. Since $1 \times 4 = 4$, write 4 in the tens place under 5, take the difference, and bring down 2. Now 4 is contained in 12 three times. Write 3 in the units place over the 2 in the dividend.

$$\begin{array}{r} 13 \\ 4)\overline{52} \\ \underline{4} \\ 12 \\ \underline{12} \end{array}$$

Thus, $52 \div 4 = 13$.

We may check by multiplying:

$$4 \times 13 = 52 \ \checkmark$$

ILLUSTRATIVE PROBLEM 2:   Divide 352 by 11.

Solution: Remember that 35 here is really 35 tens; therefore, 35 tens $\div$ 11 is about 3 tens. We place the 3 in the tens column above the 5. Since $3 \times 11 = 33$, place 33 under 35 and subtract, giving 2. Carry down the 2 in the dividend. Since $22 \div 11 = 2$ exactly, place 2 over the 2 in the dividend. Then multiply 2 by 11, giving 22.

$$\begin{array}{r} 32 \\ 11)\overline{352} \\ \underline{33} \\ 22 \\ \underline{22} \end{array}$$

Thus, $352 \div 11 = 32$.

Check:
$$\begin{array}{r} 32 \\ \times 11 \\ \hline 32 \\ 32 \\ \hline 352 \ \checkmark \end{array}$$

ILLUSTRATIVE PROBLEM 3: Divide 15,750 by 25.

Solution: Since 25 is greater than 15, the first 3 digits in the dividend are required to obtain the first digit in the quotient. A first rough estimate can be made by considering $15 \div 2 \approx 7$; but a trial multiplication shows this to be too large. $157 \div 25$ is about 6. Proceed as in problem 2 above. When there is no remainder after the second digit in the quotient, we place a zero in the quotient over the final 0 in the dividend.

```
        630
25) 15,750
    150
     75
     75
     00
```

$$15{,}750 \div 25 = 630$$

ILLUSTRATIVE PROBLEM 4: Divide 42 by 5.

Solution: Here we see that the quotient is 8, and we are left with a *remainder* of 2. We may write the result as 8 R 2, or we may write the result as $8\frac{2}{5}$, where the remainder is above the fraction line and the divisor below it. We check that this latter result is correct by showing that divisor $\times$ quotient = dividend. Hence,

```
     8
5) 42
   40
    2
```

$$5 \times 8\tfrac{2}{5} = 42$$

Thus, $42 \div 5 = 8$ R 2 or $8\frac{2}{5}$.

ILLUSTRATIVE PROBLEM 5: Divide 6592 by 74.

Solution:

1. Divide. How many 74's are in 659? Consider how many 7's are in 65. Try 9; then multiply 9 by 74. The product is larger than 659. Try 8. Write the 8 over the 9.

2. Multiply 74 by 8. Place the product 592 under 659.

3. Subtract 592 from 659. The remainder must be smaller than 74, the divisor.

```
       89
74) 6592
    592
    672
    666
      6
```

4. Bring down the next digit, 2, and place it to the right of the remainder 67.

5. Now repeat the 4 steps above. Divide 672 by 74. Write the 9 of the quotient over the 2 of the dividend, multiply 9 by 74, and write the product under 672. The remainder is then 6. The quotient is 89 R 6 or $89\frac{6}{74}$.

To check, multiply the quotient 89 by the divisor 74, obtaining the product 6586. Then add the remainder 6 to the product. The sum should be the dividend, 6592.

$$\begin{array}{r} 89 \\ \times 74 \\ \hline 356 \\ 623 \\ \hline 6586 \\ +6 \\ \hline 6592 \end{array}$$ ✔

ILLUSTRATIVE PROBLEM 6:   Divide 64,528 by 128.

Solution: $645 \div 128$ is about 5. (Think of $600 \div 100 = 6$. Six is too large since $128 \times 6 = 768$, so try $64 \div 12 \approx 5$.) When we multiply 5 by 128, the product is 640. When we subtract and bring down the 2, we are left with 52, which is smaller than the divisor, 128. In this case, we place a zero over the 2 in the dividend to indicate that, having considered this division, 0 is the partial answer and must be entered. Then we bring down the next digit, 8. Then we repeat the process.

$$\begin{array}{r} 504 \\ 128{\overline{\smash{\big)}\,64{,}528}} \\ \underline{640} \\ 528 \\ \underline{512} \\ 16 \end{array}$$

The quotient is 504 R 16 or $504\frac{16}{128}$.

Exercises _____

1. Divide and check:
   a. $4{\overline{\smash{\big)}\,484}}$ **b.** $3{\overline{\smash{\big)}\,936}}$ **c.** $7{\overline{\smash{\big)}\,518}}$
   d. $6{\overline{\smash{\big)}\,4596}}$ **e.** $34{\overline{\smash{\big)}\,7582}}$ **f.** $26{\overline{\smash{\big)}\,910}}$
   g. $48{\overline{\smash{\big)}\,3648}}$ **h.** $37{\overline{\smash{\big)}\,2072}}$ **i.** $79{\overline{\smash{\big)}\,6557}}$

2. Divide and check:
   a. $127{\overline{\smash{\big)}\,896}}$ **b.** $620{\overline{\smash{\big)}\,4340}}$
   c. $183{\overline{\smash{\big)}\,4758}}$ **d.** $4590 \div 135$
   e. $24{,}909 \div 437$ **f.** $49{,}296 \div 632$
   g. $87{,}710 \div 358$ **h.** $249{,}888 \div 274$
   i. $24{,}800 \div 800$ **j.** $325{,}000 \div 5000$

3. Divide and express each remainder as a fraction:
   a. $5{\overline{\smash{\big)}\,658}}$ **b.** $3{\overline{\smash{\big)}\,245}}$ **c.** $8{\overline{\smash{\big)}\,247}}$
   d. $7{\overline{\smash{\big)}\,503}}$ **e.** $9{\overline{\smash{\big)}\,435}}$ **f.** $100 \div 37$
   g. $54{\overline{\smash{\big)}\,2130}}$ **h.** $78{\overline{\smash{\big)}\,5796}}$ **i.** $4508 \div 98$
   j. $528{\overline{\smash{\big)}\,38{,}676}}$ **k.** $64{\overline{\smash{\big)}\,74{,}816}}$ **l.** $218{,}541 \div 42$

4. A bushel basket contains 936 eggs, which are then placed in boxes holding one dozen eggs each. How many boxes are filled?

5. Mrs. Petrillo earns $24,440 per year. What are her average weekly earnings? (Use 52 weeks per year.)

6. A tourist drives 544 miles in one day and uses 32 gallons of gasoline. How many miles does he travel on one gallon?

7. There are 16 ounces in 1 pound. How many pounds are there in 2960 ounces?

8. On a field day, a school orders buses, each of which carries 38 students. How many buses should it order for 722 students?

9. There are 12 inches in a foot. How many feet are there in 157 inches?

10. A car is traveling at constant speed on a highway. The car goes 287 miles in 5 hours. How many miles does it travel each hour?

11. The *average* of a group of numbers is obtained by *dividing* the sum of the numbers by the number of items in the group. Find the average of the following test marks: 82, 75, 89, 68, 78.

12. During four weeks in February, a sales representative earned the following commissions: $285, $340, $315, $296. What was her average weekly commission?

13. In a certain junior high school, the ninth-grade classes have the following class registers: 32, 29, 27, 31, 26, 30. What is the average class size?

14. At the end of one term, Jennifer got the following grades in her five major subjects: 85, 92, 84, 75, 93. What was her grade average for the term?

Appendix II: Problem Solving in Arithmetic

One of the most important applications of arithmetic skills is to the solution of *word problems*. Word problems arise in daily life situations and in science and industry.

1. STEPS IN SOLVING PROBLEMS

To solve *word problems*, follow these six steps:

(1) *Read* the problem *carefully*. If necessary, read it two or three times to make sure you understand it.
(2) Determine the facts *given* in the statement of the problem. Decide how these known facts are related to each other.
(3) Make sure you clearly understand what you are to *find*.
(4) Decide what arithmetic *operation* (or operations) you must use to solve the problem. Your calculations may involve addition, subtraction, multiplication, or division, or a combination of these operations.
(5) Do the calculation.
(6) *Check* your answer. *Estimation* will frequently tell you if your answer seems reasonable.

See how to apply the six steps above to the solution of a simple word problem.

A mechanic earns $9 per hour. If he works 38 hours one week, what are his entire wages for that week?

Solution:

(1) The problem states the mechanic's hourly rate of pay and the number of hours he worked in a week.

(2) The given *facts* are:
    a. The mechanic earns $9 per hour.
    b. He worked 38 hours in the week.

(3) You must *find* the mechanic's total wages for the week.

(4) *Total wages* = rate per hour × number of hours
                  = $9   ×   38

(5) *Calculate*   38
                 × 9
                 $ 342

(6) *Check* the answer by *estimation*. Round off the figures.

38 is approximately 40

$9 is approximately $10

40 × $10 is $400

You expect the answer to be somewhat *less* than $400, since you over-estimated the original two factors.

Answer: $342

2. DECIDING WHICH OPERATIONS TO USE

The most important and difficult step in the process of problem solving is trying to determine which arithmetic operation or operations to use. Some problems contain *key words* that help you determine which operation to use. Some examples follow:

Use *addition* if the problem states:

a. Find the *sum*. . .   c. How many in *all*. . .
b. What is the *total*. . .

Use *subtraction* if the problem states:

a. How many *more*. . .   d. How much *less*. . .
b. How much *more*. . .   e. How much *greater*. . .
c. Find the *difference*. . .

Use *multiplication* if the problem states:

a. Find the *product* of. . .
b. Find the cost of ____ *items* if the cost of *one item* is. . .
c. Find a *fraction of* a given number.
d. Find a *percent of* a given number.

Use *division* if the problem states:

a. Find the *quotient* of. . .
b. Find the cost of *one*. . .
c. Find the *average* of. . .
d. Find the amount for *each* item.
e. Find the percent equivalent to the fraction.

ILLUSTRATIVE PROBLEMS: In the following problems, tell which operation you would use to find the answer and explain your choice of operation.

1. A man buys 8 shirts selling for $15 per shirt. Find the total cost.

Answer: The operation is *multiplication* because you are finding the cost of 8 *items* when the cost of *one item* is given (8 × $15).

2. Terry is 16 years old and his father is 41 years old. What is the difference in their ages?

   Answer: The operation is *subtraction* because you are finding a *difference* (41 − 16).

3. Mrs. Comito earned $28,000 one year. How much did she earn per month?

   Answer: The operation is *division* because you are asked to find the amount of pay for *each* month. The year consists of 12 months with the same pay each month ($28,000 ÷ 12).

4. The weights of three packages are 7 pounds, 4 pounds, and 6 pounds. What is the total weight of the packages?

   Answer: The operation is *addition* because you are finding a *total* (7 + 4 + 6).

3. PROBLEMS REQUIRING TWO OR MORE OPERATIONS

Many word problems require more than one arithmetic operation to obtain the answer. Use the same six steps for solving problems mentioned above. In step 4, however, you must also decide on the sequence of operations.

ILLUSTRATIVE PROBLEM 1: Mr. Delia bought a TV set for $425. He made a cash payment of $100 and agreed to pay the balance in 10 equal monthly payments. How much was each payment?

Solution: This problem involves two operations. First, you must find the balance left after the cash payment. Then you must find the amount of each monthly payment.

To determine the remaining balance, *subtract* $100 from $425. Thus, $425 − $100 = $325.

To obtain the amount of each payment, *divide* $325 by 10. Thus, $325 ÷ 10 = $32.50.

He makes 10 payments of $32.50 each. (*answer*)

ILLUSTRATIVE PROBLEM 2: Kenny takes four math tests and obtains grades of 74, 82, 86, and 90. What is his average grade?

Solution:

Sum of grades	÷	*Number of grades*	=	*Average*
74 + 82 + 86 + 90				
332	÷	4		= 83 (*answer*)

4. EXTRA INFORMATION IN A PROBLEM

In some verbal problems, there may be more given information than is needed to solve the problem. You must recognize this extra information and discard it from your solution.

ILLUSTRATIVE PROBLEM 1: Bernie drove 250 miles at an average rate of 50 miles per hour. Gasoline cost him $1.10 per gallon. How many hours did the trip take him?

Solution: Note that the cost of a gallon of gasoline is not needed to determine the time for the trip. You need only to divide 250 by 50, giving an answer of 5 hours for the trip.

ILLUSTRATIVE PROBLEM 2: Find the perimeter of the rectangle below.

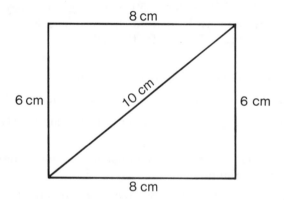

Solution: The perimeter is the sum of the length of the four sides of the rectangle. The length of the diagonal is not needed.

Perimeter = 6 + 8 + 6 + 8 = 28 cm   (*answer*)

Review Exercises

In 1–10: (**a**) Tell what operation or operations you would use to solve the problem. (**b**) Solve the problem.

1. Mr. Lawson can rent a house for a total annual cost of $6,780. How much rent must he pay per month?

2. What is the cost of 220 items at $1.50 each?

3. Mrs. Levin bought a rug 3 yards by 4 yards at a cost of $12.50 per square yard. What was the cost of the rug?

4. Buddy drove 175 miles at an average speed of 50 mph. How many hours did the trip take?

5. Ramona can type 12 pages an hour. How many hours will it take her to type 54 pages at this rate?

6. A piece of paper is .012 cm thick. If a package contains 200 of these sheets, how many centimeters thick is the package?

7. A store is selling four bars of candy for 80 cents. The regular price of these bars is 22 cents each. How much is saved by buying eight bars at the sale price?

8. A suit selling for $124 was marked $\frac{1}{4}$ off. How much would Carlos pay for the suit if he bought it at the reduced price?

9. Mrs. Bernard deposited $800 in a bank that paid annual simple interest of 8%. What amount did she have in the bank after $1\frac{1}{2}$ years?

10. Bert may buy a TV set for $450 cash, or he may pay $100 down and then pay 12 monthly installments of $35 each. How much does he save by paying $450 cash for the set?

In 11 to 15, do *not* solve the problem. Write:

   (A) if there is *too little* given information to solve the problem.

   (B) if there is *just enough* given information to solve the problem.

   (C) if there is *too much* given information (more than is needed to solve the problem).

11. Rita buys an $80 dress for $60. What is the rate of discount she receives?

12. How many quarts of paint should you buy to cover a rectangular area that measures 20 feet by 18 feet?

13. The perimeter of a square is 32 inches, and its area is 64 square inches. What is the length of each side of the square?

14. On a road map, one inch represents 20 miles. The distance on the map from Bayville to Smithtown is $6\frac{1}{4}$ inches. Find the distance in miles from Smithtown to Bayville.

15. Lester earned $1800 one summer. He spent $600 and put the remainder in a savings account that paid interest semiannually. At the end of six months, how much was in his account?

Appendix III: Sets

1. MEANING OF TERMS AND SYMBOLS

An understanding of **sets** is helpful in the study of mathematics. Since the use of sets frequently helps to clarify ideas in elementary mathematics, we will review here some of the basic concepts and terminology.

In ordinary speech, we talk about a set of dishes, a set of books, a set of opinions, and so on. A **set** is a collection of objects or ideas. In mathematics, we speak of a set of numbers or a set of triangles. The set of digits, for example, consists of the whole numbers from 0 to 9 inclusive. Each of these numbers is called an **element** or **member** of the set of digits.

One way of indicating a set is to list the elements between braces. Thus, we may represent the set of digits as {0, 1, 2, 3, 4, 5, 6, 7, 8, 9}. The set of vowels is {a, e, i, o, u}. This way of representing the elements of a set is called the **list method** or the **roster method**.

We frequently use a capital letter to refer to a particular set. For example, $P = \{1, 3, 5\}$ may be read "P is the set whose elements are 1, 3, and 5." To indicate that 3 is an element of P, we write $3 \in P$ (read also as "3 belongs to P"). To indicate that 4 is not an element of P, we write $4 \notin P$.

It does not matter in what order the elements are listed. For example, the set of vowels may be written {e, i, a, u, o} or {u, o, i, e, a}.

If there are many elements in a set, we do not use the roster method. Instead, we *describe* the set. We may write {counting numbers from 1 to 50}, which is read "the set of counting numbers from 1 to 50." We may also write {1, 2, 3, ... 48, 49, 50}. The dots indicate that we have not bothered to write all the counting numbers from 4 through 47. This set is also read "the set of counting numbers from 1 to 50."

Sometimes a set has no element. For example, the set of Presidents of the United States under 25 years of age has no elements. We refer to a set with no elements as the **empty set** or **null set**. We represent the empty set by empty braces { } or by the symbol $\varnothing$. If M is the set of months beginning with the letter Z, we write $M = \varnothing$ or $M = \{\ \}$. Note that {0} is not the empty set since it has the number 0 as a single element.

We can readily tell by counting that the set of digits has 10 elements and that the set of vowels has 5 elements. If the number of elements in a set can be counted, we call the set a **finite set**.

Let us consider the set E of positive even numbers. We may write

$$E = \{2, 4, 6, 8, 10, \ldots\}$$

Here the dots indicate that we are to go on indefinitely writing consecutive even numbers. Since the elements of the set go on forever, we cannot count them. Such a set is called an **infinite set.** The set of counting numbers (or natural numbers) is also an infinite set:

$$\{1, 2, 3, 4, 5, 6, \ldots\}$$

To write the month "August," we use six letters. But the letter "u" is repeated. We write the set of letters of this word: {A, u, g, s, t}. We do not repeat the "u." When we describe a set by the roster method, we list the same element only once.

If two sets have the same number of elements, they are said to be **equivalent sets**. The set $A = \{1, 3, 5, 7\}$ and the set $B = \{p, q, r, s\}$ are equivalent, since they both contain 4 elements.

If two sets have *exactly* the same elements, they are said to be **identical sets** or **equal sets**. Thus, $P = \{a, e, i, o, u\}$ and $Q = \{u, e, i, a, o\}$ are equal sets. Note that they have the same number of members as well as the same members.

Exercises

In 1 to 10, represent each set by the roster method (list the elements of each set between braces).

1. The set of all months that begin with the letter J.
2. The set of even numbers between 1 and 11.
3. The set of months that have fewer than 30 days.
4. The set of days of the week spelled with 6 letters.
5. The set of odd numbers between 1 and 33 that are divisible by 5.
6. The set of all female U.S. Presidents.
7. The set of all days of the week beginning with the letter T.
8. The set of letters in the word "null."
9. The set of whole numbers between 7 and 8.
10. The set of even numbers between 9 and 19.

In 11 to 15, describe in words the set whose elements are given by the roster method.

EXAMPLE: {1, 3, 5, 7, 9} is the set of odd numbers from 1 to 9.

11. {2, 4, 6, 8}   **12.** {Alaska, Oregon, Washington, California}

13. {1, 2, 3, 4, 5, 6}   **14.** {March, May}   **15.** {3, 6, 9, 12, 15}

In 16 to 25, state whether the set is an *infinite set*, a *finite set*, or an *empty set*.

16. The set of odd numbers.

17. The set of people living in Arizona.

18. The set of states of the U.S. smaller in size than Rhode Island.

19. The set of counting numbers greater than 1000.

20. The set of months of the year beginning with K.

21. The set of rectangles.

22. The set of 3-digit numbers beginning with 5.

23. The set of whole numbers that are multiples of 7.

24. {months of the year having 25 days}   **25.** {5, 10, 15, 20, 25, ...}

In 26 to 34, state whether the two sets in each exercise are *equivalent*, *equal*, or *neither*.

26. {p, q, r, s} and {r, s, q, p}

27. {Bill, Mary, Ed, Sue} and {a, b, c, d}

28. {5, 7, 9, 11, 13} and {p, q, r, s}

29. {a, b, l, e} and {e, l, b, a}

30. {2, 3, 4, 5} and $\{\frac{1}{2}, \frac{1}{3}, \frac{1}{4}, \frac{1}{5}\}$

31. $\varnothing$ and {0}   **32.** {M, I, T} and {T, I, M}

33. {counting numbers from 1 to 10} and {digits from 0 to 9}

34. {days in the week} and {months in the year}

---

2. SUBSETS

In a particular discussion, the set of all objects being considered is called the **universal set** or the **universe**. If we are talking about numbers, the set of counting numbers could be the universal set. If we are talking about American citizens, the set of residents of the U.S. could be the universal set. We designate such a set by the letter U.

Now let us consider the sets $U = \{$citizens of U.S.$\}$ and $T = \{$citizens of Texas$\}$.

All the elements of T are also elements of U. In such a case, we say that T is a **subset** of U, and we write $T \subset U$.

If we consider U as $\{a, e, i, o, u\}$, we can see that there are many possible subsets of U; for example, $\{a, e, i\}$ or $\{i, o\}$. Also, we can write $\{e\} \subset U$ or $\{i\} \subset U$. We may also consider $\{a, e, i, o, u\}$ as a subset of U since *all* of its elements are also elements of U. Such a subset is called an **improper subset** of U, whereas subsets with fewer elements than U are called **proper subsets** of U.

The example above illustrates the fact that *every set is a subset of itself.* Mathematicians have also agreed to consider the empty set to be a subset of every set.

Let us list all the subsets of the set $\{a, b, c\}$. These would be: $\{a\}$, $\{b\}$, $\{c\}$, $\{a, b\}$, $\{a, c\}$, $\{b, c\}$, $\{a, b, c\}$, and $\{\ \}$.

Let us consider the sets $V = \{a, e, i, o, u\}$ and $L = \{a, b, c, d, e\}$. Clearly, V is not a subset of L and L is not a subset of V. However, V and L have 2 elements in common, a and e. V and L are called **overlapping sets**. Two sets are overlapping if they have elements in common but neither set is a subset of the other.

Two sets are called **disjoint sets** if no element of one is an element of the other. Thus, $\{a, b, c\}$ and $\{1, 2, 3\}$ are disjoint sets.

Exercises

In 1 to 10, two sets are given in each exercise. State whether the two sets are *overlapping sets*, *disjoint sets*, or one is a *subset* of the other. In the latter case, indicate which set is the subset.

EXAMPLE: $P = \{a, b, c\}$ and $Q = \{a, b, c, d\}$

Answer: subset; $P \subset Q$

1. $A = \{5, 6, 7\}$ and $B = \{5, 6, 7, 8\}$
2. $\{$dogs$\}$ and $\{$birds$\}$
3. $C = \{$cows$\}$ and $A = \{$animals$\}$
4. $A = \{2, 4, 6, 8\}$ and $B = \{1, 2, 3, 4\}$
5. $S = \{$Seniors in the school$\}$ and $J = \{$Juniors in the school$\}$

6. $\varnothing$ and A = {residents of Alabama}

7. S = {ships} and B = {battleships}

8. A = {p, q, r} and B = {q, r, p}

9. C = {Mary, Harry, Ed} and D = {Jane, Tony, Harry}

10. E = {all even numbers} and F = {4, 8, 12, 16}

In 11 to 14, list all subsets of the given set.

11. {p}   **12.** {3, 4}   **13.** {r, s, t}   **14.** {Sue, Tom}

In 15 to 18, state whether the first set is a *proper* subset or an *improper* subset of the second set.

15. {mackerel} $\subset$ {fish}   **16.** {b, c, d} $\subset$ {d, b, c}

17. {w, x, y, z} $\subset$ {z, w, y, x}

18. {all multiples of 6} $\subset$ {all even numbers}

---

3. OPERATIONS ON SETS

Let us consider two overlapping sets, such as P = {a, e, i, o, u} and Q = {a, b, c, d, e}. Note that a and e are elements of both sets. We refer to the set {a, e,} as the **intersection** of sets P and Q.

The intersection of two sets P and Q is the set containing only those elements of P that also belong to Q.

The intersection of set P and set Q is represented by the symbol $P \cap Q$, which is read "P intersection Q" or "P cap Q." Thus, if P is the set of counting numbers from 1 to 10 and Q = {8, 10, 12, 14}, then $P \cap Q$ = {8, 10}.

It is convenient to represent the relationships of sets by means of **Venn diagrams**. We represent the universal set by a rectangle and its interior. We usually label the rectangle with a U in one corner, as shown in Fig. 1.

Fig. 1

If we want to picture a subset P of U, we place a circle within the rectangle and mark its interior with a P, as shown in Fig. 2.

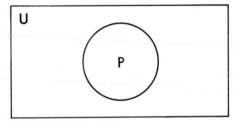

Fig. 2

Now let us consider the possible diagrams for $P \cap Q$, when P and Q are subsets of U.

When P and Q are overlapping sets, we draw overlapping circles, as shown in Fig. 3. The shaded part of the two circles represents $P \cap Q$.

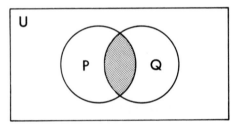

Fig. 3. P ∩ Q when P and Q
are overlapping sets

Fig. 4 pictures two disjoint sets. Since the sets do not overlap, there is no shaded area. Thus, $P \cap Q = \varnothing$.

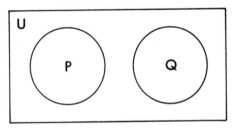

Fig. 4. P ∩ Q when P and Q
are disjoint sets

Fig. 5 shows the situation when $Q \subset P$. The circle representing Q is completely within the circle representing P. The shaded area represents $P \cap Q$. Note that $P \cap Q = Q$ in this case.

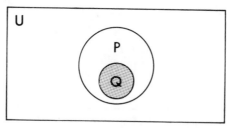

Fig. 5. P ∩ Q when Q ⊂ P

The **union** of two sets P and Q is the set of all elements that belong to P or Q or both. The union of P and Q is represented by $P \cup Q$, read "P union Q" or "P cup Q."

The Venn diagrams for $P \cup Q$ are shown in Figs. 6, 7, and 8. The shaded region in each diagram shows $P \cup Q$.

In Fig. 6, if $P = \{6, 7, 8, 9\}$ and $Q = \{8, 9, 10, 11\}$, then $P \cup Q = \{6, 7, 8, 9, 10, 11\}$. Note that P and Q are both subsets of $P \cup Q$.

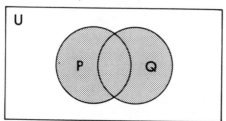

Fig. 6. P ∪ Q when P and Q
are overlapping sets

In Fig. 7, if $P = \{6, 7, 8, 9\}$ and $Q = \{10, 11, 12, 13\}$, then $P \cup Q = \{6, 7, 8, 9, 10, 11, 12, 13\}$.

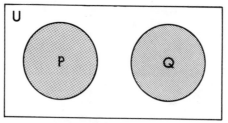

Fig. 7. P ∪ Q when P and Q
are disjoint sets

In Fig. 8, if $P = \{6, 7, 8, 9\}$ and $Q = \{7, 8\}$, $P \cup Q = \{6, 7, 8, 9\}$. When $Q \subset P$, note that $P \cup Q = P$.

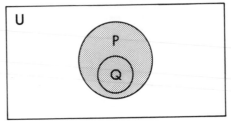

Fig. 8. P ∪ Q when Q ⊂ P

ILLUSTRATIVE PROBLEM: If $U = \{$all the letters of the alphabet$\}$, $V = \{a, e, i, o, u\}$, and $L = \{a, b, c, d, e, f\}$, find:

a. $V \cap L$

b. $V \cup L$

c. Draw Venn diagrams to show **a** and **b**.

Solution:

a. $V \cap L = \{a, e\}$

b. $V \cup L = \{a, e, i, o, u, b, c, d, f\}$

c.

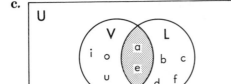

V ∩ L (shaded area)

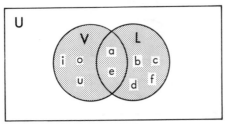

V ∪ L (shaded area)

Exercises

1. If $P = \{2, 4, 6\}$ and $Q = \{1, 2, 3, 4\}$, find:
   a. $P \cap Q$ **b.** $P \cup Q$

2. If $A = \{p, q, r, s\}$ and $B = \{r, s, t\}$, find:
   a. $A \cap B$ **b.** $A \cup B$

3. If $C = \{10, 11, 12\}$ and $D = \{13, 14\}$, find:
   a. $C \cap D$ **b.** $C \cup D$

4. If $U = \{\text{counting numbers}\}$, $A = \{3, 4, 5\}$, and $B = \{3, 4, 5, 6, 7\}$, find:
   a. $A \cap B$ **b.** $A \cup B$

5. If $T = \{\text{triangles}\}$, $R = \{\text{rectangles}\}$, and $U = \{\text{geometric figures}\}$:
   a. Find $T \cap R$.
   b. On separate paper, draw a Venn diagram to show $T \cap R$.

6. **a.** Find $\{2, 4, 6, 8\} \cap \{4, 6\}$.   **b.** Find $\{2, 4, 6, 8\} \cup \{4, 6\}$.

7. Find: **a.** $P \cap \varnothing$ **b.** $P \cup \varnothing$ **c.** $P \cap P$ **d.** $P \cup P$

8. If $U = \{\text{people}\}$, $L = \{\text{blue-eyed people}\}$, and $R = \{\text{brown-eyed people}\}$, draw a Venn diagram to show:
   a. $L \cap R$ **b.** $L \cup R$

9. If P and Q are disjoint sets, what is $P \cap Q$?

10. If $A \subset B$, find:
    a. $A \cap B$.   **b.** $A \cup B$.

Appendix IV:
Interpretation of
Statistical Graphs

PRELIMINARY NOTE

Organized data, or *statistics*, are frequently presented in the form of tables and charts. However, in order to save time and effort in analyzing statistics and tables, it is often desirable to present numerical data in visual form by means of graphs. The types of graphs that are commonly used are bar graphs, line graphs, and circle graphs. These graphs are discussed in this appendix.

1. BAR GRAPHS

A bar graph may be made up of all vertical bars or all horizontal bars. This type of graph is used mainly for purposes of comparison.

ILLUSTRATIVE PROBLEM: A vertical bar graph indicating the daily sales of Mr. Stein for one week is shown on the next page.

a. By what amount did Friday's sales exceed Wednesday's sales?

b. What was the amount of Mr. Stein's average daily sales for the week?

c. If Mr. Stein earns an 8% commission, what was the commission on Tuesday's sales?

d. What was the percentage of increase in Mr. Stein's sales on Saturday over Monday?

Solution:
a. Friday's sales $840
    Wednesday's sales   <u>−600</u>
                $240   (*answer*)

Daily Sales of Mr. Stein for One Week

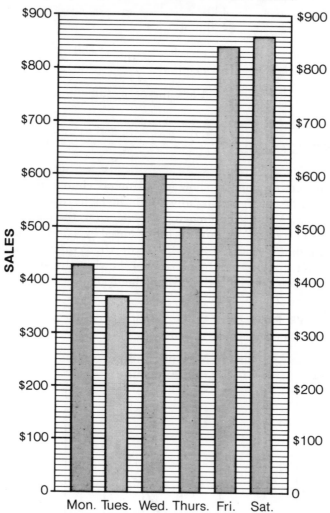

b.

Mon.	$430
Tues.	370
Wed.	600
Thur.	500
Fri.	840
Sat.	860
Total 6 days	$3600

$600 Average (*answer*)
6)$3600

c. Tuesday's sales = $370
                          ×.08
                         $29.60   Commission (*answer*)

d. Saturday's sales = $860
    Monday's sales   −430
               $430 increase

$$\% \text{ increase} = \frac{\text{increase}}{\text{original}} \times 100 = \frac{430}{430} \times 100$$

$$= 100\% \text{ increase} \quad (answer)$$

2. LINE GRAPHS

Line graphs frequently show a continuous change in a particular variable. A line graph moving *upward* to the right represents a continuous *increase*. A line graph moving *downward* to the right indicates a continuous *decrease*. *No* change is represented by a *horizontal* line. The steeper the line segment *rises* to the right, the *greater* the *slope* of the segment and the greater the *increase*. Likewise, the steeper the line segment *falls* to the right, the *greater* the *decrease*.

ILLUSTRATIVE PROBLEM: A line graph indicating average temperatures for six months is shown below.

Average Monthly Temperatures
(January–June)

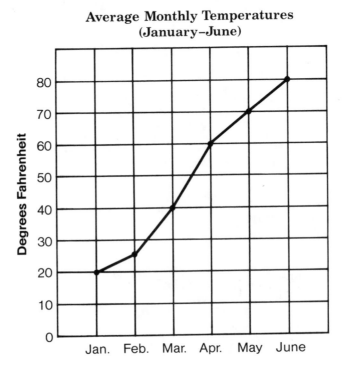

a. Between what two months does the average monthly temperature increase the *most?*

b. Between what two months does the average monthly temperature increase the *least?*

c. What is the percentage of increase from the lowest temperature to the highest temperature?

Solution:

a. Look for the *steepest* part of the graph (greatest slope). This occurs between March and April. *(answer)*

b. Look for the *flattest* part of the graph (least slope). This occurs between January and February. *(answer)*

c. The highest temperature was in June: 80°.
The lowest temperature was in January: 20°.

$$\text{The percent increase} = \frac{\text{increase}}{\text{original}} \times 100$$

$$= \frac{80 - 20}{20} \times 100 = \frac{60}{20} \times 100$$

$$= 300\% \quad (answer)$$

3. CIRCLE GRAPHS

When parts of a whole are to be represented, a *circle graph* or *pie chart* is used. These graphs are commonly used to show some type of distribution.

In such a graph, a circle is used to represent the whole quantity. The circle is then divided into *sectors*, where each sector represents a proportional part of the whole. A sector of a circle is a portion of the circle bounded by two radii and the arc that they cut off on the circumference. The angle formed by the two radii, ∠ *POQ*, is called the central angle.

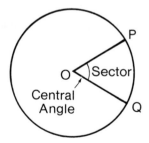

To find the number of degrees, *n*, in the central angle of a particular sector, we form the proportion:

$$\frac{n°}{360°} = \frac{\text{part}}{\text{whole}}$$

ILLUSTRATIVE PROBLEM: The annual sales of the Globe Manufacturing Company amounted to $240,000 last year. The circle graph shown below indicates how each sales dollar was used.

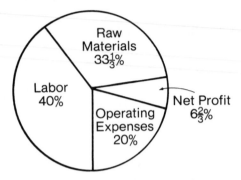

a. What amount of money was spent for labor?

b. What was the amount of money spent for raw materials?

c. How many degrees are there in the central angle of the sector representing operating expenses?

Solution:

a. Find 40% of $240,000.

$$240,000$$
$$\underline{\times .40}$$
$$\$96,000.00 \quad (answer)$$

b. Find $33\frac{1}{3}$% of $240,000.

$$\frac{1}{3} \times \$240,000 = \$80,000 \quad (answer)$$

c. Let n = the number of degrees in the central angle.

$$\frac{n}{360} = \frac{20}{100} \quad (20\%)$$

$$\frac{n}{360} = \frac{1}{5} \quad (cross\text{-}multiply)$$

$$5n = 360$$

$$n = 72° \quad (answer)$$

Review Exercises

1. The circle graph
   shows how each tax
   dollar is spent in
   one community.

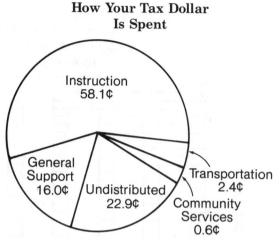

**How Your Tax Dollar
Is Spent**

   a. On which item is about $\frac{1}{6}$ of the tax dollar spent?
   b. What is the ratio of the amount spent on "community services"
   to that spent on "transportation"?
   c. About how many degrees are in the central angle of the sector
   for "undistributed"?
   d. What percent of the tax dollar is spent on "transportation" and
   "community services" combined?

2. Base your answers to questions **a**, **b**, and **c** on the graph below, which
   shows the results of seven typing tests taken by a student.

   a. What was the increase in
   words per minute between
   test one and test three?
   b. What average number of
   words per minute did the
   student type on these
   seven tests?
   [Express your answer to
   the *nearest whole number.*]
   c. Between which two consecutive test numbers was
   there no improvement?

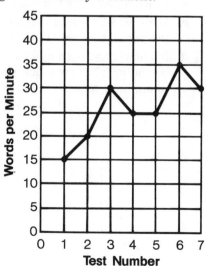

3. Base your answers to questions **a** through **d** on the graph below:

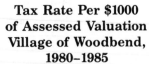

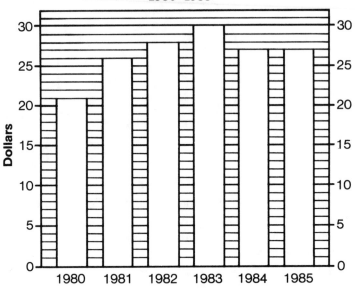

a. What was the tax per $1000 of assessed valuation in 1982?

b. In what year was the tax rate the highest?

c. What was the lowest tax per $1000 of assessed valuation during the period covered by the graph?

d. If a house had been assessed at $70,000 during the period covered by the graph, how much more money would a taxpayer in this community have paid in 1985 than in 1980 for village taxes?

4. The projected total world population in the year 2000 is 6.35 billion people. Use this projection and the graph below to answer questions **a** through **d**.

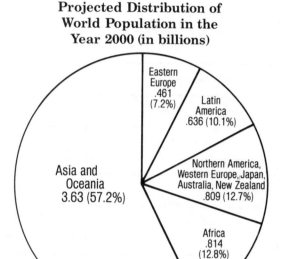

Projected Distribution of World Population in the Year 2000 (in billions)

a. If world population today is about 5 billion, then the projected percent increase in population from now to the year 2000 is approximately (1) 54%   (2) 38%   (3) 27%   (4) 20%.

b. What fraction of the projected world population in the year 2000 is expected to be living in Africa, Northern America, Western Europe, Japan, Australia, and New Zealand?

c. Africa, Asia, and Oceania make up the "less-developed" regions of the world. What percent of world population in the year 2000 will be living in "less-developed" regions?

d. How many degrees are in the central angle of the sector for Latin America?

5. Base your answers to questions **a**, **b**, and **c** on the graph below.

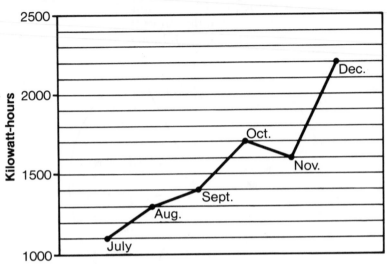

**Kilowatt-Hours of Electricity
Used for a Six-Month Period by T. J. Watson Co.**

a. What was the average monthly number of kilowatt-hours used by the Watson Company for the six-month period?

b. If the price per kilowatt-hour is $.0375, what was the Watson Company's electric bill for the month of December?

c. Between which two months did the number of kilowatt-hours used by the company decrease?

6. Base your answers to questions **a** through **d** on the following graph:

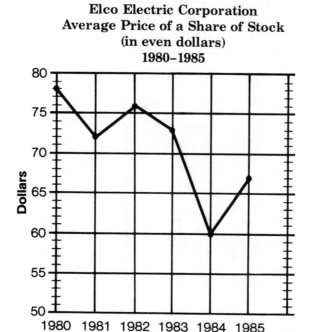

Elco Electric Corporation
Average Price of a Share of Stock
(in even dollars)
1980–1985

a. What was the average price of a share of Elco Electric Corporation stock in 1984?

b. If 15 shares of this stock had been bought at the average price in 1981, what amount of money would have been paid? (Disregard brokerage charges.)

c. What was the average selling price of this stock for the six years shown on the graph?

d. What was the amount of the decline between the average price in 1980 and the average price in 1985?

7. Base your answers to questions **a** through **d** on the following graph:

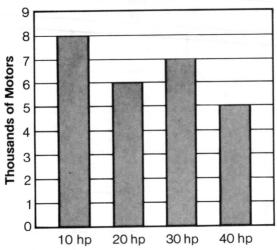

**Turbo Outboard Motor Company
Production of Motors by Horsepower
(Annual Report for Last Year)**

a. What was the total number of motors produced last year by the Turbo Outboard Motor Company?

b. By what percent did the production of 30-horsepower motors exceed the production of 40-horsepower motors?

c. If the selling price of 10-horsepower motors is $115 each, what was the total amount of sales for 10-horsepower motors?

d. It is estimated that 40-horsepower motors will show a 20% increase in sales this year compared to last year. How many 40-horsepower motors should be planned for production this year?

Appendix V:
Introduction to
Geometry

PRELIMINARY NOTE

In our work with algebra, we often use words that belong to the branch of mathematics called *Geometry*. In this appendix, explanations, definitions, and properties of many geometric terms and figures are grouped for reference.

1. POINT, LINE, AND PLANE

The basic words of geometry are *point*, *line*, and *plane*. These are the only words of geometry that have no formal definitions.

A *point* has no length or width; it only marks position. In everyday speech, we often use the word "point" to mean "place" or "position." A point is usually represented by a dot. We place a capital letter near the dot so that we may refer to it. The point shown is called point *P*.

• P

A *line* is an infinite set of points. A line has only the *one dimension* of length; it has no width. Usually, when we speak of a line in geometry, we mean a *straight* line, that is, one that may be drawn with a ruler or straightedge.

We can name a straight line by referring to two points on the line. In the figure below, we can refer to the infinite line as line *RS* or line *SR*.

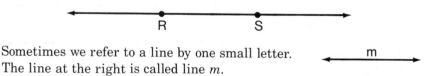

Sometimes we refer to a line by one small letter. The line at the right is called line *m*.

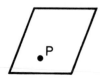

A *plane* is a flat surface that extends in all directions without end. It can be named by one capital letter, representing one point in the plane. The figure at the left represents plane *P*. Actually, we have pictured only part of the plane because a plane has no boundaries.

A plane can also be named by three points that are not on the same straight line. The plane at the right is called plane *ABC*.

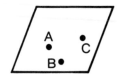

2. MORE ABOUT LINES

If you examine the two figures pictured below, you will see that it is reasonable to accept the following facts about straight lines:

1. In a plane, an infinite number of straight lines can be drawn through a given point.

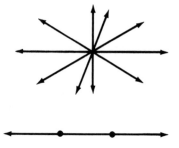

2. One and only one straight line can be drawn through two given points. (We say that two points determine a straight line.)

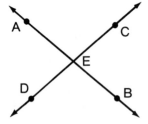

Lines that meet or cross each other are called *intersecting lines.* Two straight lines can intersect in only one point.

In the figure at the left, line *AB* and line *CD* intersect at point *E*.

Two lines in the same plane that never intersect are called *parallel lines.* In the figure at the right, *AB* is parallel to *CD*. Using the symbol "∥" for "is parallel to," we write *AB* ∥ *CD*.

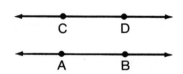

A *line segment* is a part of a line that consists of two points on the line and the set of points between them. In the figure at the right, *AB* is a line segment on the infinite line *m*. *A* and *B* are called *endpoints*.

A *ray* is also a part of a line. It has only one endpoint and continues indefinitely in one direction. In the figure at the right, *AB* is a ray.

The *length of a line segment* can be measured with the use of a ruler. The length is the distance between the two endpoints of the line segment.

Note that neither a ray nor a line has a definite length, since each continues without end.

If two line segments have the same length, we say they are *equal in measure* (=). In the figure at the right, *AB* = *CD*.

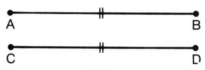

The *midpoint* of a line segment is the point on the line segment that divides the line segment into two equal segments. In the figure at the left, *M* is the midpoint of *AB* if *AM* = *MB*.

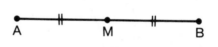

Also, each of the equal segments is one-half of the entire line segment ($AM = \frac{1}{2}AB$ and $MB = \frac{1}{2}AB$) or *AB* is *bisected* by *M*.

3. ANGLES

An *angle* is the union of two rays that extend from a common endpoint. The common endpoint is called the *vertex* (plural: *vertices*) of the angle, and the two rays are called the *sides* of the angle. The symbol for angle is " ∠."

There are four ways to name the angle shown at the right.

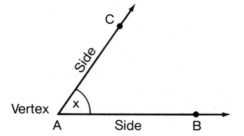

1. ∠ *CAB*   2. ∠ *BAC*
3. ∠ *A*   4. ∠ *x*

Notice that when three letters are used to name an angle, the vertex is the middle letter.

We usually refer to an angle by mentioning just the letter at the vertex, but only if there is no doubt about which angle we are referring to. Thus, in the figure at the right, when we refer to ∠ P, there is no doubt which angle we mean. However, there is doubt if we refer to ∠ Q. If we mean the small angle at point Q, we would refer to ∠ SQR or ∠ y. There are also ∠ RQP and ∠ SQP at point Q.

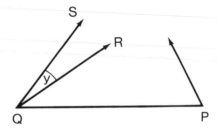

A common unit of measure for an angle is a *degree* (°). The instrument used to measure an angle is called a *protractor*, shown below.

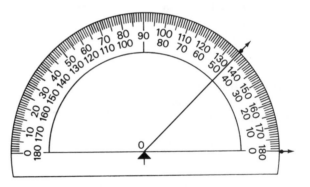

Angles are named according to the number of degrees they contain.

An angle that measures exactly 90° is called a *right angle*. ∠ RST is a right angle. All right angles are equal.

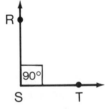

An angle that measures more than 0° but less than 90° is called an *acute angle*. ∠ UVW is an acute angle. ∠ UVW = 60°. (When we say that an angle equals n°, we mean that its *measure* equals n°.)

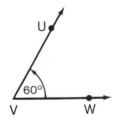

An angle that measures more than 90° but less than 180° is called an *obtuse angle.* ∠ *XYZ* is an obtuse angle and ∠ *XYZ* = 137°.

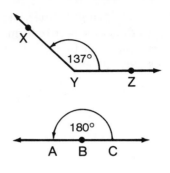

An angle that measures exactly 180° is called a *straight angle.* ∠ *ABC* is a straight angle. All straight angles are equal.

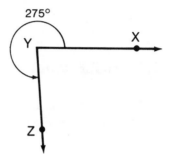

An angle that measures more than 180° but less than 360° is called a *reflex angle.* ∠ *XYZ* is a reflex angle and ∠ *XYZ* = 275°.

Note that the lengths of the rays that form the sides of an angle are not what tell us the size of the angle. The rays are infinitely long. The size of the angle is the number of degrees contained in the angle.

Two angles that have the same measure are called *equal angles.* In the figure below ∠ *BCD* = ∠ *EFG.* (*Note:* The double marks in the angles show that the angles are equal.)

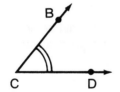

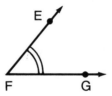

The *bisector of an angle* is a ray through the vertex of the angle that divides the angle into two equal angles.

In the figure, *BQ* is the bisector of ∠ *ABC.* Thus, ∠ *ABQ* = ∠ *QBC.* Also,

$$\angle ABQ = \tfrac{1}{2} \angle ABC$$

$$\angle QBC = \tfrac{1}{2} \angle ABC$$

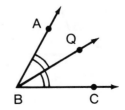

Two straight lines, line segments, or rays that intersect to form right angles are *perpendicular* to each other. The symbol for "is perpendicular to" is "⊥." If *AB* ⊥ *CD* at point *E*, then the four angles at *E* (∠ *AED*, ∠ *AEC*, ∠ *CEB*, and ∠ *BED*) are right angles.

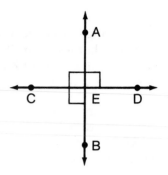

4. ANGLE RELATIONSHIPS

If two angles have a common vertex and a common side, but no interior points in common, they are called *adjacent angles*.

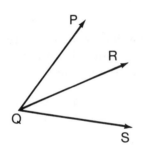

∠ *PQR* and ∠ *RQS* are adjacent angles.

∠ *PQR* and ∠ *PQS* are not adjacent angles because they have an infinite number of interior points in common.

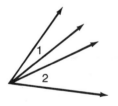

∠ 1 and ∠ 2 are not adjacent angles because they do not have a common side.

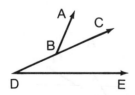

∠ *ABC* and ∠ *CDE* are not adjacent angles because they do not have a common vertex.

If two straight lines intersect, the pairs of opposite angles formed are called *vertical angles*.

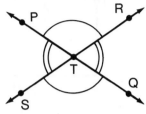

In the figure at the right, ∠ *PTR* and ∠ *QTS* are a pair of vertical angles. Also, ∠ *RTQ* and ∠ *PTS* are vertical angles. Note that vertical angles are *not* adjacent. Vertical angles are equal. Thus, ∠ *PTR* = ∠ *QTS* and ∠ *RTQ* = ∠ *PTS*.

If the sum of two angles is 180°, the two angles are called *supplementary angles*. Each angle is the *supplement* of the other.

Supplementary angles may be adjacent or non-adjacent, as shown in the figures below.

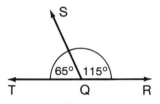

Adjacent Supplementary
Angles

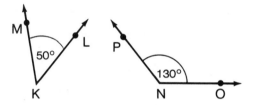

Non-Adjacent Supplementary
Angles

ILLUSTRATIVE PROBLEM 1: An angle is five times its supplement. Find each angle.

Solution: Let x = the smaller angle.

Then $5x$ = the larger angle.

$5x + x = 180°$ (The sum of two supplementary angles is 180°.)

$6x = 180°$

$x = 30°$ (smaller angle)

$5x = 150°$ (larger angle)

Answer: The two angles are 30° and 150°.

If the sum of two angles is 90°, the two angles are called *complementary angles*. Each angle is called the *complement* of the other. Complementary angles may be adjacent or non-adjacent, as shown in the figures on the following page.

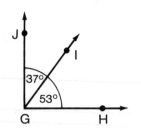 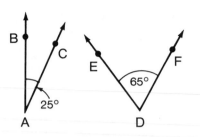

Adjacent Complementary   Non-Adjacent Complementary
Angles   Angles

ILLUSTRATIVE PROBLEM 2: An angle is 40° more than its complement. Find the larger angle.

Solution: Let x = the smaller angle.

Then $40 + x$ = the larger angle.

$x + (40 + x) = 90°$ (The sum of two complementary
                               angles is 90°.)

$x + 40 + x = 90°$

$2x + 40 = 90°$

$2x = 50°$

$x = 25°$ (smaller angle)

$40 + x = 65°$ (larger angle)

Answer: The larger angle is 65°.

5. POLYGONS

A *polygon* is a closed figure formed by three or more straight lines,
all of which intersect in a plane.

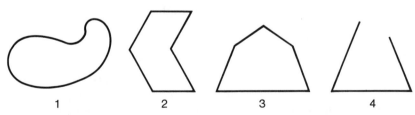

Of the figures above, only the second and third are polygons. The
first figure is not a polygon because it is composed of a curved line

rather than straight lines. The fourth figure is not a polygon because the straight lines do not all intersect and, thus, the figure is not closed.

The straight line segments forming the polygon are called the *sides*, the intersection points of the sides are called the *vertices*, and the angles formed by the sides are called the *angles* of the polygon.

As shown in the figures below, polygons are classified according to the number of sides.

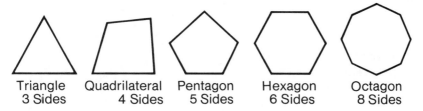

| Triangle | Quadrilateral | Pentagon | Hexagon | Octagon |
| 3 Sides | 4 Sides | 5 Sides | 6 Sides | 8 Sides |

(*Note:* In a *regular polygon*, all of the sides are equal and all of the angles are equal.)

The *perimeter* of a polygon is the distance around the polygon. Thus, the perimeter of a polygon is equal to the sum of all its sides.

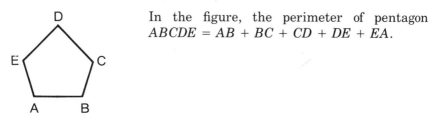

In the figure, the perimeter of pentagon $ABCDE = AB + BC + CD + DE + EA$.

The *area* of a polygon is the amount of the plane enclosed by the polygon. Area is measured in square units, such as square inches (in.²) or square centimeters (sq cm). There is a formula for computing the area of each type of polygon. These formulas will be given in the following sections.

6. TRIANGLES

A polygon with three sides and three angles is called a *triangle*. The symbol for a triangle is "△." The figure at the right shows △ *ABC*. Note that each vertex is named by a capital letter. A triangle is usually named by its vertices.

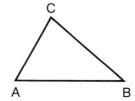

Triangles may be classified according to the relative lengths of their sides.

A triangle with three sides of equal length is called an *equilateral triangle*. An equilateral triangle also has three angles of equal measure. $\triangle LMN$ is an equilateral triangle. (*Note*: The double marks on the sides show that the sides are equal.)

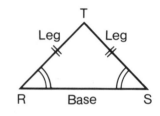

A triangle with two sides of equal length is called an *isosceles triangle*. The two sides of equal length are called *legs* of the triangle. The third side is called the *base* of the triangle. The angles opposite the two equal sides are called the *base angles* and have the same measure. The angle opposite the base is called the *vertex angle*. $\triangle RST$ is an isosceles triangle.

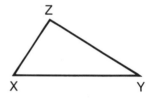

A triangle with no two sides of equal length is called a *scalene triangle*. All three angles of a scalene triangle are unequal. $\triangle XYZ$ is a scalene triangle.

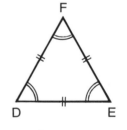

Triangles may also be classified according to their angles.

A triangle whose angles are equal is called an *equiangular triangle*. An equiangular triangle is also an equilateral triangle. $\triangle DEF$ is an equiangular triangle.

A triangle with a right angle is called a *right triangle*. The other two angles are acute angles. The longest side of a right triangle is the side opposite the right angle. It is called the *hypotenuse* of the triangle. The other two sides are called *legs* of the triangle. $\triangle GHI$ is a right triangle.

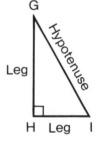

A triangle with an obtuse angle is called an *obtuse triangle*. $\triangle KLM$ is an obtuse triangle.

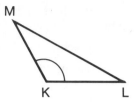

A triangle with three acute angles is called an *acute triangle*. $\triangle ABC$ is an acute triangle.

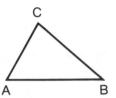

It can be shown that the sum of the three angles of any triangle is equal to 180°.

ILLUSTRATIVE PROBLEM 1: The vertex angle of an isosceles triangle is 50°. Find the number of degrees in each base angle.

Solution: Let x = the number of degrees in each base angle. (The base angles of an isosceles triangle are equal.)

$$x + x + 50 = 180° \quad \text{(The sum of the three angles}$$
$$\text{of any triangle is 180°.)}$$

$$2x + 50 = 180°$$

$$2x = 130°$$

$$x = 65°$$

Answer: Each base angle is 65°.

ILLUSTRATIVE PROBLEM 2: The three angles of a triangle are in the ratio 1:2:3. Find the number of degrees in each angle.

Solution: Let x, $2x$, and $3x$ represent the three angles in the ratio 1:2:3.

$$x + 2x + 3x = 180° \quad \text{(The sum of the three angles}$$
$$\text{of any triangle is 180°.)}$$

$$6x = 180°$$

$$x = 30°$$

$$2x = 60°$$

$$3x = 90°$$

Answer: The three angles are 30°, 60°, and 90°.

Since the sum of the three angles of any triangle is 180°, it follows that:

(1) The acute angles of a right triangle are complementary and

(2) Each angle of an equilateral (equiangular) triangle is 60°.

There are several line segments that are associated with triangles.

An *angle bisector* of a triangle is a line segment that bisects an angle of the triangle and terminates in the opposite side. In the figure at the right, *BD* is the angle bisector of ∠ *ABC*.

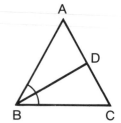

There are three angle bisectors in every triangle, and it can be shown that the three are *concurrent* (meet in a common point).

An *altitude* of a triangle is a line segment that is drawn from a vertex and is perpendicular to (and terminating in) the opposite side or in an extension of the opposite side.

The figures below show that an altitude may be inside or outside a triangle or may even be one of its sides.

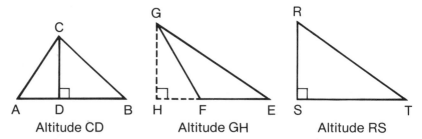

Altitude CD   Altitude GH   Altitude RS

There are three altitudes in every triangle, and it can be shown that the three altitudes are concurrent.

A *median* of a triangle is a line segment that is drawn from a vertex of the triangle to the midpoint of the opposite side. In the figure, *CD* is the median to side *AB*.

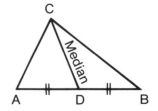

There are three medians in every triangle, and it can be shown that the three are concurrent.

ILLUSTRATIVE PROBLEM 3: In $\triangle PQR$, $PQ = PR$. If $PQ = 5x + 14$, $PR = 30 - 3x$, and $QR = 5x$, find the perimeter (the sum of the three sides) of $\triangle PQR$.

Solution: Since the triangle is isosceles with $PQ = PR$, it follows that $5x + 14 = 30 - 3x$.

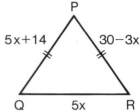

$$\text{Thus, } 8x = 16$$
$$x = 2$$
$$PQ = 5x + 14 = 5(2) + 14 = 24$$
$$PR = 30 - 3x = 30 - 3(2) = 24$$
$$QR = 5x = 5(2) = 10$$

Answer: Perimeter $= 24 + 24 + 10 = 58$.

ILLUSTRATIVE PROBLEM 4: In $\triangle PQR$, PM is the bisector of $\angle P$. If $\angle MPR = (3y - 20)°$ and $\angle MPQ = (2y + 5)°$, find $\angle QPR$.

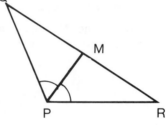

Solution: Since PM is the angle bisector,

$\angle MPR = \angle MPQ$

$3y - 20 = 2y + 5$

$\quad\quad y = 25$

$\angle MPQ = 2y + 5$

$\quad\quad\quad = 2(25) + 5 = 55°$

$\angle MPR = 3y - 20$

$\quad\quad\quad = 3(25) - 20 = 55°$

$\angle QPR = 55 + 55 = 110°$

Answer: $\angle QPR = 110°$

The area of a triangle is computed by the formula:

$$\text{Area of triangle} = \tfrac{1}{2} \text{ base} \cdot \text{height or } A = \tfrac{1}{2}b \cdot h$$

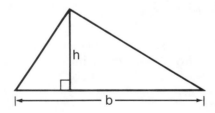

(*Note:* Any side of a triangle may be considered the base. In an area problem, be sure to use the altitude drawn to the particular side mentioned. That side will be the "base," and that altitude will be the "height" in the area formula.)

ILLUSTRATIVE PROBLEM 5: The sides of a triangle are 6 cm, 8 cm, and 9 cm, respectively. If the altitude to the 8-cm side is 4 cm, find the area of the triangle.

Solution:

$$\text{Area of triangle} = \frac{1}{2}b \cdot h$$

$$= \frac{1}{2}(8)(4)$$

$$= 16$$

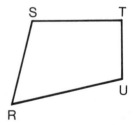

Answer: The area of the triangle is 16 square centimeters.

ILLUSTRATIVE PROBLEM 6: Find the area of a right triangle whose legs are 5 cm and 12 cm.

Solution: Since the legs of a right triangle are perpendicular, either leg can be considered as the altitude and the other leg as the base.

$$\text{Area of triangle} = \frac{1}{2}b \cdot h$$

$$= \frac{1}{2}(12)(5)$$

$$= 30$$

Answer: The area of the triangle is 30 square centimeters.

7. QUADRILATERALS

A polygon with four sides is called a *quadrilateral.* Thus, *RSTU* is a quadrilateral whose sides are *RS*, *ST*, *TU*, and *UR*, and whose angles are ∠ *RST*, ∠ *STU*, ∠ *TUR* and ∠ *URS*.

When we vary the shape of the quadrilateral by making some of its sides parallel, some of its sides equal in length, or its angles right angles, we get different members of the family of quadrilaterals, as shown below:

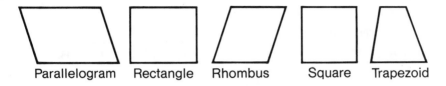

Parallelogram   Rectangle   Rhombus   Square   Trapezoid

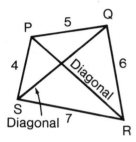

A quadrilateral also has two diagonals. A *diagonal* is a straight line segment joining two opposite vertices of the quadrilateral. In *PQRS*, *PR* and *QS* are diagonals. The *perimeter* of a quadrilateral is the sum of the lengths of the four sides. In the figure to the left, the perimeter is $4 + 5 + 6 + 7 = 22$.

A *parallelogram* is a quadrilateral in which both pairs of opposite sides are *parallel*. In the figure at the right, *CD* ∥ *FE* and *CF* ∥ *DE*. (*Note:* The arrowheads on the opposite sides show that the opposite sides are parallel.) Line segments *CH* and *DJ* are

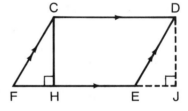

perpendicular to base *FE* of the parallelogram (extended to *J*). Thus, *CH* and *DJ* are altitudes to base *FE* of the parallelogram.

Some properties of a parallelogram are the following:

1. Opposite sides are equal.
2. Opposite angles are equal.
3. Altitudes to the same base are equal.
4. The diagonals bisect each other.

The formula for the area of a parallelogram is

$$\text{Area of parallelogram} = \text{base} \cdot \text{height or } A = bh$$

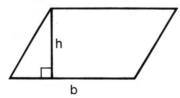

(*Note:* In an area problem, be sure to use the altitude drawn to the particular base mentioned in the problem.)

ILLUSTRATIVE PROBLEM 1: Using the lengths given in the figure at the right, find the area of parallelogram *ABCD*.

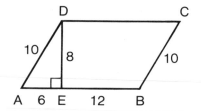

Solution: The altitude to base *AB* is the one given.

$$\text{Area of } ABCD = b \cdot h$$
$$= AB \cdot DE$$
$$= (6 + 12)8$$
$$= (18)8$$
$$= 144$$

Answer: The area of the parallelogram is 144 square units.

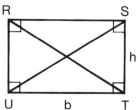

When all the angles of a parallelogram are right angles, the figure is a *rectangle*. The diagonals of a rectangle are equal. In rectangle *RSTU* at the left, *RT* = *SU*.

We frequently refer to the base and height of the rectangle. In *RSTU*, *UT* = base = *b* and *ST* = height = *h*; or we may call them the length = *l* and width = *w*, respectively. Since the opposite sides are equal, the perimeter $p = 2b + 2h = 2l + 2w$.

ILLUSTRATIVE PROBLEM 2: Find the perimeter of a rectangle whose dimensions are 15 meters and 10 meters.

Solution:
$$p = 2l + 2w$$
$$= 2(15) + 2(10)$$
$$= 30 + 20$$
$$= 50$$

Answer: The perimeter of the rectangle is 50 meters.

The formula for the area of a rectangle is

$$\text{Area of rectangle} = \text{base} \cdot \text{height or } A = bh$$

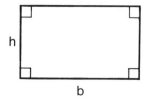

ILLUSTRATIVE PROBLEM 3: Find the height of a rectangle whose area is 60 sq cm and whose base is 10 cm.

Solution: Area of rectangle $= b \cdot h$

$$60 = 10h$$

$$6 = h$$

Answer: The height of the rectangle is 6 cm.

A rectangle with all four sides equal is a *square*. The perimeter, p, of the square is given by $p = 4s$.

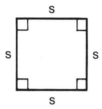

ILLUSTRATIVE PROBLEM 4: Find the perimeter of a square of side 14 inches.

Solution: $p = 4s$

$$= 4(14)$$

$$= 56$$

Answer: The perimeter of the square is 56 inches.

The formula for the area of a square is

$$\text{Area of square} = \text{side} \cdot \text{side or } A = s^2$$

ILLUSTRATIVE PROBLEM 5: Find the area of a square whose perimeter is 20 inches.

Solution: First find a side of the square.

$$p = 4s$$

$$20 = 4s$$

$$5 = s$$

Now, use a side to determine the area of the square.

$$A = s^2$$
$$= 5^2$$
$$= 5 \cdot 5$$
$$= 25$$

Answer: The area of the square is 25 square inches.

A quadrilateral in which only one pair of opposite sides is parallel is called a *trapezoid.* If the non-parallel sides (the legs) are equal, the trapezoid is isosceles.

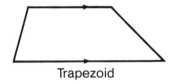

Trapezoid

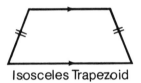

Isosceles Trapezoid

The formula for the area of a trapezoid is

Area of trapezoid = $\frac{1}{2}$ height (sum of the bases) or $A = \frac{1}{2}h(b_1 + b_2)$

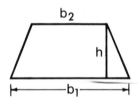

ILLUSTRATIVE PROBLEM 6: Find the area of the trapezoid whose altitude is 6 cm and whose bases are 10 cm and 12 cm.

Solution:

$$A = \frac{1}{2}h(b_1 + b_2)$$
$$= \frac{1}{2} \cdot 6(10 + 12)$$
$$= 3(22)$$
$$= 66$$

Answer: The area of the trapezoid is 66 square centimeters.

8.   CIRCLES

A *circle* is a plane figure bounded by a curved line every point of which is the same distance from the center of the figure.

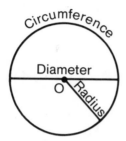

The *circumference* of a circle is the line that forms its outer boundary. It may be considered the "perimeter" of the circle. A *radius* of a circle is a line segment joining the center to any point on the circumference. A *diameter* is a line segment joining two points on the circumference and passing through the center. (*Note:* diameter = 2 × radius or $d = 2r$.)

The circumference, C, of a circle is given by the formula $C = 2\pi r$, where r is the radius and $\pi \approx \frac{22}{7} \approx 3.14$.

The area, A, of a circle is given by the formula $A = \pi r^2$.

ILLUSTRATIVE PROBLEM 1:   Find the circumference of a circle whose diameter is 20 cm. (Use $\pi = 3.14$.)

Solution: Since the diameter, d, of the circle is 20, the radius is $\frac{1}{2}d = \frac{1}{2}(20) = 10$.

$$C = 2\pi r$$
$$= 2(3.14)10$$
$$= 62.8$$

Answer: The circumference of the circle is 62.8 centimeters.

ILLUSTRATIVE PROBLEM 2:   Find the area of a circle whose radius is 7 cm. (Use $\pi = 3.14$.)

Solution:
$$A = \pi r^2$$
$$A = (3.14)(7^2)$$
$$A = (3.14)(49)$$
$$= 153.86$$

Answer: The area of the circle is 153.86 square centimeters.

9. IDENTIFYING SOLID FIGURES

Below are pictured some of the more common solid figures that appear about us in nature and in objects and structures made by people. Note how dashed lines are used to show hidden edges and surfaces in the illustrations of solid figures.

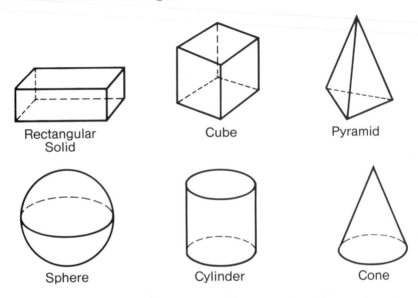

Rectangular Solid

Cube

Pyramid

Sphere

Cylinder

Cone

The *rectangular solid* has the shape of a book or a brick. Its shape appears all about us in many sizes and forms. The *cube* is a rectangular solid with all of its edges equal and all of its faces identical. Note that its six faces are all squares. A lump of sugar usually has the shape of a cube. A die (singular for dice) is also *cubical* in shape.

The *pyramid* has the shape of the famous buildings in ancient Egypt. The figure shown above represents a *triangular pyramid*, since its base is a triangle. If the base were a square, the pyramid would be called a *square pyramid*. The base can be any polygon.

The *sphere* is the general shape of the earth and other planets. It is also the shape of a ball. Every point on the surface of the sphere is the same distance from a point inside the sphere called the *center*. The distance from the center to any point on the sphere is the *radius* of the sphere.

The *cylinder* is generally the shape of a tin can or of most water glasses. When the bases are circles, we call the solid a *circular cylinder*. If a line joining the two bases is perpendicular to both bases, we call the solid a *right circular cylinder*.

The *cone* is in the shape of a tepee. If the base of the cone is a circle, the cone is called a *circular cone*. The top point of the cone is called the *apex*. If the line joining the apex to the center of the circular base is perpendicular to the base, the solid is called a *right circular cone*.

Review Exercises

1. In the figure at the right, name three angles.

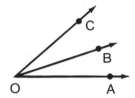

2. What type of angle (acute, right, obtuse, reflex) is formed when:
   a. a straight angle is bisected?
   b. a right angle is bisected?
   c. a right angle is added to a straight angle?
   d. an acute angle is subtracted from a straight angle?

3. Find the measure of each of the following angles:
   a. $\frac{3}{4}$ of a straight angle
   b. $\frac{1}{3}$ of a right angle
   c. $\frac{2}{5}$ of a straight angle

4. What kind of angle is equal to its supplement?

5. Find the complement of an angle that is $18°$.

6. Find an angle whose supplement is eight times the angle.

7. If two angles are both complementary and equal, what is the measure of each angle?

8. If an altitude of a triangle falls outside the triangle, is the triangle acute, right, or obtuse?

9. Find the supplement of an angle that is $42°$.

10. Express, in terms of y, the complement of an angle that is $2y°$.

11. The supplement of an angle equals three times the complement of the angle. Find the angle.

12. An angle exceeds its supplement by $48°$. Find the angle.

13. The sum of the supplement and complement of an angle is $120°$. Find the angle.

14. Four times an angle equals the supplement of the angle. Find the angle.

15. Find an angle that is five times its supplement.

16. Represent, in terms of y, the supplement of an angle that is $3y°$.

17. How many degrees are there in each angle of an equilateral (equiangular) triangle?

18. The vertex angle of an isosceles triangle is $80°$. How many degrees are there in each of the base angles?

19. How many degrees are there in each acute angle of an isosceles right triangle?

20. In triangle PQR, angle Q is twice angle P, and angle R is three times angle P. Find angle P.

In 21 to 33, write *always, sometimes,* or *never* in the blank space to complete each statement correctly.

21. The supplement of an angle is _____ greater than the angle.

22. The complement of an acute angle is _____ greater than the supplement of the angle.

23. A plane is _____ determined by three points that are on the same straight line.

24. A right triangle is _____ isosceles.

25. If two angles have a common side, they are _____ adjacent angles.

26. The supplement of an obtuse angle is _____ a right angle.

27. An equilateral triangle is _____ equiangular.

28. In an isosceles triangle, the base angles are _____ equal.

29. An altitude of a triangle _____ falls inside the triangle.

30. An obtuse triangle is _____ isosceles.

31. Parallel lines _____ intersect in the same plane.

32. The points on the circumference of a circle are _____ the same distance from the center.

33. If two straight lines intersect, the vertical angles formed are _____ equal.

34. In isosceles triangle PQR at the right, $PQ = PR$. Find the base of the triangle.

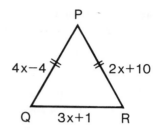

35. The two legs of an isosceles triangle are $(3r - 10)$ and $(r + 24)$. If the base is $(4r - 50)$, find the perimeter of the triangle.

36. Which figure does *not always* have its opposite sides equal?
(*a*) parallelogram (*b*) rectangle (*c*) quadrilateral (*d*) square

37. Which figure does *not always* have its diagonals bisecting each other?
(*a*) parallelogram (*b*) rectangle (*c*) square (*d*) quadrilateral

38. Find the perimeter of a rectangle whose dimensions are 18 cm by 14 cm.

39. If the perimeter of a square is 50 meters, what is each side of the square?

40. The perimeter of a rectangle is 72 cm and its length is twice its width. What is the width?

41. The area of a triangle is 60. If one side of the triangle measures 24, find the altitude drawn to that side.

42. Find the area of a right triangle whose legs are 6 cm and 9 cm.

43. The area of a triangle is 40. If a side of the triangle is represented by $2x + 2$ and the altitude drawn to that side is 8, find the value of x.

44. Find the area of a parallelogram whose base is 8 in. and whose altitude is 2.5 in.

45. The area of a parallelogram is 36, and a base is 4. Find the altitude drawn to the base.

46. In a parallelogram whose area is 60, one of the sides is represented by $4x - 4$ and the altitude drawn to that side is 5. Find the value of x and the side represented by $4x - 4$.

47. Find the area of a rectangle whose base is $3\frac{1}{2}$ *feet* and whose altitude is 6 *inches*.

48. The dimensions of a rectangular living room are 13 feet by 21 feet. How many square feet of carpet are needed to cover the whole floor?

49. The perimeter of a rectangle is 30 cm. Find the area of the rectangle if one of its sides is 4 cm.

50. Find a side of a square whose area is 49 square centimeters.

51. Find the area of a square whose perimeter is 32 feet.

52. The altitude of a trapezoid is 6 and its bases are 8 and 12. Find the area of the trapezoid.

53. The area of a trapezoid is 36 and the sum of its bases is 18. Find the altitude.

54. The radius of a circle is 3. Using $\pi = 3.14$, find the area.

55. A side of a regular hexagon is 5. What is the perimeter?

Tables and Formulas

SQUARES AND SQUARE ROOTS

No.	Square	Square Root	No.	Square	Square Root	No.	Square	Square Root
1	1	1.000	51	2,601	7.141	101	10,201	10.050
2	4	1.414	52	2,704	7.211	102	10,404	10.100
3	9	1.732	53	2,809	7.280	103	10,609	10.149
4	16	2.000	54	2,916	7.348	104	10,816	10.198
5	25	2.236	55	3,025	7.416	105	11,025	10.247
6	36	2.449	56	3,136	7.483	106	11,236	10.296
7	49	2.646	57	3,249	7.550	107	11,449	10.344
8	64	2.828	58	3,364	7.616	108	11,664	10.392
9	81	3.000	59	3,481	7.681	109	11,881	10.440
10	100	3.162	60	3,600	7.746	110	12,100	10.488
11	121	3.317	61	3,721	7.810	111	12,321	10.536
12	144	3.464	62	3,844	7.874	112	12,544	10.583
13	169	3.606	63	3,969	7.937	113	12,769	10.630
14	196	3.742	64	4,096	8.000	114	12,996	10.677
15	225	3.873	65	4,225	8.062	115	13,225	10.724
16	256	4.000	66	4,356	8.124	116	13,456	10.770
17	289	4.123	67	4,489	8.185	117	13,689	10.817
18	324	4.243	68	4,624	8.246	118	13,924	10.863
19	361	4.359	69	4,761	8.307	119	14,161	10.909
20	400	4.472	70	4,900	8.367	120	14,400	10.954
21	441	4.583	71	5,041	8.426	121	14,641	11.000
22	484	4.690	72	5,184	8.485	122	14,884	11.045
23	529	4.796	73	5,329	8.544	123	15,129	11.091
24	576	4.899	74	5,476	8.602	124	15,376	11.136
25	625	5.000	75	5,625	8.660	125	15,625	11.180
26	676	5.099	76	5,776	8.718	126	15,876	11.225
27	729	5.196	77	5,929	8.775	127	16,129	11.269
28	784	5.292	78	6,084	8.832	128	16,384	11.314
29	841	5.385	79	6,241	8.888	129	16,641	11.358
30	900	5.477	80	6,400	8.944	130	16,900	11.402
31	961	5.568	81	6,561	9.000	131	17,161	11.446
32	1,024	5.657	82	6,724	9.055	132	17,424	11.489
33	1,089	5.745	83	6,889	9.110	133	17,689	11.533
34	1,156	5.831	84	7,056	9.165	134	17,956	11.576
35	1,225	5.916	85	7,225	9.220	135	18,225	11.619
36	1,296	6.000	86	7,396	9.274	136	18,496	11.662
37	1,369	6.083	87	7,569	9.327	137	18,769	11.705
38	1,444	6.164	88	7,744	9.381	138	19,044	11.747
39	1,521	6.245	89	7,921	9.434	139	19,321	11.790
40	1,600	6.325	90	8,100	9.487	140	19,600	11.832
41	1,681	6.403	91	8,281	9.539	141	19,881	11.874
42	1,764	6.481	92	8,464	9.592	142	20,164	11.916
43	1,849	6.557	93	8,649	9.644	143	20,449	11.958
44	1,936	6.633	94	8,836	9.695	144	20,736	12.000
45	2,025	6.708	95	9,025	9.747	145	21,025	12.042
46	2,116	6.782	96	9,216	9.798	146	21,316	12.083
47	2,209	6.856	97	9,409	9.849	147	21,609	12.124
48	2,304	6.928	98	9,604	9.899	148	21,904	12.166
49	2,401	7.000	99	9,801	9.950	149	22,201	12.207
50	2,500	7.071	100	10,000	10.000	150	22,500	12.247

VALUES OF THE TRIGONOMETRIC RATIOS

Angle	Sine	Cosine	Tangent	Angle	Sine	Cosine	Tangent
1°	.0175	.9998	.0175	46°	.7193	.6947	1.0355
2°	.0349	.9994	.0349	47°	.7314	.6820	1.0724
3°	.0523	.9986	.0524	48°	.7431	.6691	1.1106
4°	.0698	.9976	.0699	49°	.7547	.6561	1.1504
5°	.0872	.9962	.0875	50°	.7660	.6428	1.1918
6°	.1045	.9945	.1051	51°	.7771	.6293	1.2349
7°	.1219	.9925	.1228	52°	.7880	.6157	1.2799
8°	.1392	.9903	.1405	53°	.7986	.6018	1.3270
9°	.1564	.9877	.1584	54°	.8090	.5878	1.3764
10°	.1736	.9848	.1763	55°	.8192	.5736	1.4281
11°	.1908	.9816	.1944	56°	.8290	.5592	1.4826
12°	.2079	.9781	.2126	57°	.8387	.5446	1.5399
13°	.2250	.9744	.2309	58°	.8480	.5299	1.6003
14°	.2419	.9703	.2493	59°	.8572	.5150	1.6643
15°	.2588	.9659	.2679	60°	.8660	.5000	1.7321
16°	.2756	.9613	.2867	61°	.8746	.4848	1.8040
17°	.2924	.9563	.3057	62°	.8829	.4695	1.8807
18°	.3090	.9511	.3249	63°	.8910	.4540	1.9626
19°	.3256	.9455	.3443	64°	.8988	.4384	2.0503
20°	.3420	.9397	.3640	65°	.9063	.4226	2.1445
21°	.3584	.9336	.3839	66°	.9135	.4067	2.2460
22°	.3746	.9272	.4040	67°	.9205	.3907	2.3559
23°	.3907	.9205	.4245	68°	.9272	.3746	2.4751
24°	.4067	.9135	.4452	69°	.9336	.3584	2.6051
25°	.4226	.9063	.4663	70°	.9397	.3420	2.7475
26°	.4384	.8988	.4877	71°	.9455	.3256	2.9042
27°	.4540	.8910	.5095	72°	.9511	.3090	3.0777
28°	.4695	.8829	.5317	73°	.9563	.2924	3.2709
29°	.4848	.8746	.5543	74°	.9613	.2756	3.4874
30°	.5000	.8660	.5774	75°	.9659	.2588	3.7321
31°	.5150	.8572	.6009	76°	.9703	.2419	4.0108
32°	.5299	.8480	.6249	77°	.9744	.2250	4.3315
33°	.5446	.8387	.6494	78°	.9781	.2079	4.7046
34°	.5592	.8290	.6745	79°	.9816	.1908	5.1446
35°	.5736	.8192	.7002	80°	.9848	.1736	5.6713
36°	.5878	.8090	.7265	81°	.9877	.1564	6.3138
37°	.6018	.7986	.7536	82°	.9903	.1392	7.1154
38°	.6157	.7880	.7813	83°	.9925	.1219	8.1443
39°	.6293	.7771	.8098	84°	.9945	.1045	9.5144
40°	.6428	.7660	.8391	85°	.9962	.0872	11.4301
41°	.6561	.7547	.8693	86°	.9976	.0698	14.3007
42°	.6691	.7431	.9004	87°	.9986	.0523	19.0811
43°	.6820	.7314	.9325	88°	.9994	.0349	28.6363
44°	.6947	.7193	.9657	89°	.9998	.0175	57.2900
45°	.7071	.7071	1.0000	90°	1.0000	.0000	

MEASURES

Length	Capacity
1 kilometer (km) = 1000 meters 1 hectometer (hm) = 100 meters 1 dekameter (dam) = 10 meters 1 decimeter (dm) = 0.1 meter 1 meter (m) 1 centimeter (cm) = 0.01 meter 1 millimeter (mm) = 0.001 meter	1 kiloliter (kL) = 1000 liters 1 hectoliter (hL) = 100 liters 1 dekaliter (daL) = 10 liters 1 liter (L) 1 deciliter (dL) = 0.1 liter 1 centiliter (cL) = 0.01 liter 1 milliliter (mL) = 0.001 liter

Length continued:

1 foot (ft) = 12 inches (in.)

$1 \text{ yard (yd)} = \begin{cases} 3 \text{ feet} \\ 36 \text{ inches} \end{cases}$

$1 \text{ mile (mi)} = \begin{cases} 5280 \text{ feet} \\ 1760 \text{ yards} \end{cases}$

Capacity continued:

5 milliliters = 1 teaspoon

12.5 milliliters = 1 tablespoon

1 liter = 1000 cubic centimeters (cu cm)

1 milliliter = 1 cubic centimeter

1 cup (c) = 8 fluid ounces (fl oz)
1 pint (pt) = 2 cups
1 quart (qt) = 2 pints
1 gallon (gal) = 4 quarts

Mass/Weight

1 kilogram (kg) = 1000 grams
1 hectogram (hg) = 100 grams
1 dekagram (dag) = 10 grams
1 gram (g)
1 decigram (dg) = 0.1 gram
1 centigram (cg) = 0.01 gram
1 milligram (mg) = 0.001 gram
1 metric ton (t) = 1000 kilograms

1 pound (lb) = 16 ounces (oz)
1 ton (T) = 2000 pounds

Time

1 minute = 60 seconds
1 hour = 60 minutes
1 day = 24 hours
1 week = 7 days

$1 \text{ year} = \begin{cases} 12 \text{ months} \\ 365 \text{ days} \end{cases}$

1 decade = 10 years
1 century = 100 years

Area

1 sq ft = 144 sq in.
1 sq yd = 9 sq ft
1 acre = 4840 sq yd

Volume

1 cu ft = 1728 cu in.
1 cu yd = 27 cu ft

IMPORTANT FORMULAS AND RELATIONSHIPS

Perimeter	Interest
Rectangle $P = 2l + 2w$ Square $\quad P = 4s$ Triangle $\quad P = a + b + c$	Interest equals principal $\times$ rate $\times$ time in years ($i = prt$).
	Slope
Circumference	The slope (m) of a line passing through the points (x_1, y_1) and (x_2, y_2) is $m = \dfrac{y_2 - y_1}{x_2 - x_1}$.
Circle $\quad C = 2\pi r$ or $C = \pi d$	
	Pythagorean Relationship
Area	If a and b represent the legs of a right triangle and c represents the hypotenuse, then $c^2 = a^2 + b^2$.
Rectangle $\quad A = lw$ or bh Square $\qquad A = s^2$ Triangle $\qquad A = \frac{1}{2}bh$	
	Trigonometric Ratios
Parallelogram $\;A = bh$ Trapezoid $\quad A = \frac{1}{2}(B + b)h$ Circle $\qquad A = \pi r^2$	The following relationships can be used for either acute angle of a right triangle. $\dfrac{\text{sine (sin)}}{\text{of the angle}} = \dfrac{\text{length of leg opposite the angle}}{\text{length of hypotenuse}}$
Volume	$\dfrac{\text{cosine (cos)}}{\text{of the angle}} = \dfrac{\text{length of leg adjacent to the angle}}{\text{length of hypotenuse}}$
Rectangular Prism $V = lwh$ Cube $\qquad\qquad V = s^3$ or $V = e^3$	
Distance	$\dfrac{\text{tangent (tan)}}{\text{of the angle}} = \dfrac{\text{length of leg opposite the angle}}{\text{length of leg adjacent to the angle}}$
Distance equals rate $\times$ time ($d = rt$).	
General Quadratic Equation $ax^2 + bx + c = 0$	**Quadratic Formula** $x = \dfrac{-b \pm \sqrt{b^2 - 4ac}}{2a}$

Index